Optical Materials and Applications - Volume **1**

Novel Optical Materials

Optical Materials and Applications

Print ISSN: 3029-1089
Online ISSN: 3029-1038

Series Editor: Francesco Simoni (*Università Politecnica Delle Marche, Italy*)

Published:

Vol. 1   *Novel Optical Materials*
          edited by Iam Choon Khoo, Francesco Simoni and Cesare Umeton

Optical Materials and Applications - Volume **1**

Novel Optical Materials

editors

Iam Choon Khoo
The Pennsylvania State University, USA

Francesco Simoni
Università Politecnica delle Marche, Italy

Cesare Umeton
Università della Calabria, Italy

W⊖ **World Scientific**

NEW JERSEY · LONDON · SINGAPORE · BEIJING · SHANGHAI · HONG KONG · TAIPEI · CHENNAI · TOKYO

Published by

World Scientific Publishing Co. Pte. Ltd.
5 Toh Tuck Link, Singapore 596224
USA office: 27 Warren Street, Suite 401-402, Hackensack, NJ 07601
UK office: 57 Shelton Street, Covent Garden, London WC2H 9HE

Library of Congress Cataloging-in-Publication Data
Names: Khoo, Iam-Choon, editor. | Simoni, Francesco, editor. | Umeton, Cesare, editor.
Title: Novel optical materials / editors Iam Choon Khoo, the Pennsylvania State University, USA,
    Francesco Simoni, Università Politecnica delle Marche, Italy,
    Cesare Umeton, Università della Calabria, Italy.
Description: New Jersey : World Scientific, [2024] | Series: Optical materials and applications ;
    volume 1 | Includes bibliographical references and index.
Identifiers: LCCN 2023051524 (print) | LCCN 2023051525 (ebook) |
    ISBN 9789811280597 (hardcover) | ISBN 9789811280603 (ebook for institutions) |
    ISBN 9789811280610 (ebook for individuals)
Subjects: LCSH: Optical materials. | Optical materials--Technological innovations.
Classification: LCC QC374 .N688 2024 (print) | LCC QC374 (ebook) |
    DDC 620.1/1295--dc23/eng/20231120
LC record available at https://lccn.loc.gov/2023051524
LC ebook record available at https://lccn.loc.gov/2023051525

British Library Cataloguing-in-Publication Data
A catalogue record for this book is available from the British Library.

For any available supplementary material, please visit
https://www.worldscientific.com/worldscibooks/10.1142/13523#t=suppl

Typeset by Stallion Press
Email: enquiries@stallionpress.com

Preface

This book is the first of a new series titled *Optical Materials and Applications* that aims at covering hot topics in the wide landscape of research related to this field. In this series, a broad spectrum of materials will be considered: from semiconductors to polymers and liquid crystals as well as plasmonic and optical metamaterials. Applications may span from micro- and nano-optics to optical information technology, optofluidics, biophotonics, imaging, holographic technologies, and more. The aim of each book should be to present in a comprehensive way the state-of-the-art of the covered topic in order to be a useful read for researchers in the field and for students and scientists approaching it for the first time.

The series starts with this collection of chapters that is a kind of miscellanea, samples of topics that might be more extensively presented and discussed in single books of the series.

The collection also has another aim: to disclose the inspiration for the new book series, which came from the topics discussed in the conference "Novel Optical Materials and Applications (NOMA)." This meeting, which has been recurring in Italy every alternate year for over three decades, provides an international forum to discuss the fundamentals of these materials and their roles in actual optical devices. Hence, before the chapters we present an introduction devoted to a brief description of the scientific and non-scientific characteristics of NOMA.

Iam Choon Khoo
Francesco Simoni
Cesare Umeton

Contents

Chapter 2. Optothermal Marangoni Effect: Phenomena and Applications   31

Andrzej Miniewicz, Stanisław Bartkiewicz, Monika Bełej,
Katarzyna Grześkiewicz, and Michalina Ślemp

Chapter 3. Molecular Alignment Patterning Enabled by Novel Photopolymerization with Structured Light and its Optical Applications   71

Sayuri Hashimoto, Kyohei Hisano, Miho Aizawa,
and Atsushi Shishido

**Chapter 4. Nonlinear Optical Propagation in
Heliconical Cholesteric Liquid Crystals   93**

Ashot H. Gevorgyan and Francesco Simoni

**Chapter 5. Liquid Crystals for Displays,
Smart Windows, Tunable Metamaterials,
Plasmonic Nanostructures, Micro-ring
Resonators, and Ultrafast Laser Manipulations   121**

I. C. Khoo and T.-H. Lin

Chapter 6.   Plasmonic-based Biosensors for the Rapid Detection of Harmful Pathogens   155

Francesca Petronella, Daria Stoia, Yasamin Ziai,
Federica Zaccagnini, Viviana Scognamiglio, Dana Maniu,
Chiara Rinoldi, Monica Focsan, Amina Antonacci,
Filippo Pierini and Luciano De Sio

Chapter 7.   A Double Plasmonic/Photonic Approach for Multilevel Anticounterfeit and Food Safety Applications   195

Antonio De Luca, Vincenzo Caligiuri, Aniket Patra,
Maria P. De Santo, Agostino Forestiero,
Giuseppe Papuzzo, Dante M. Aceti,
Giuseppe E. Lio, and Riccardo Barberi

Chapter 8. Laser-Assisted Micromachining and Applications

225

L. Criante, R. Ramos-García, and S. Bonfadini

Chapter 9. Novel Photo-sensitive Materials for Microengineering and Energy Harvesting 251

D. Sagnelli, A. Vestri, A. D'Avino, M. Rippa,
V. Marchesano, F. Ratto, A. De Girolamo Del Mauro,
F. Loffredo, F. Villani, G. Nenna, G. Ardila,
P. Meneroud, J. Gauthier, S. Duc, M. Thomachot,
F. Claeyssen, and L. Petti

Introduction: NOMA: Scientific and Human Uniqueness

Iam Choon Khoo[*,§], Francesco Simoni[†,¶], and Cesare Umeton[‡,||]

Pennsylvania State University, University Park, PA, USA
†*Università Politecnica delle Marche, Ancona, Italy*
‡*Università della Calabria, Rende, Italy*
§*ick1@psu.edu*
¶*f.simoni@photomat.it*
||*umeton.fis@gmail.com*

1. The Beginning

The intended year of publication of this book, 2023, coincides with the 16th edition of the Novel Optical Materials and Applications (NOMA) conference, which will be held from June 4 to June 9, 2023, in Cetraro — a small town in southern Italy, following a tradition of meeting every alternate year[1] since 1995. Actually, NOMA originated from the "School on Nonlinear Optics and Optical Physics"[2,3] organized by two of us (I.C.K and F.S.) in 1992 in Capri (Italy), which is designated as the first edition. This meeting was organized with the following ingredients: (a) outstanding invited lecturers with specialty covering a broad range of novel optical materials and their applications; (b) limited and selected number of participants from all over the world; (c) accommodation of all participants in the same hotel of limited size in a remote but idyllic setting conducive for localized interaction.

---

[1]The edition of 2021 was shifted to 2022 due to the COVID-19 pandemic.
[2]See Khoo, I. C., Lam, J. F., and Simoni, F. (Eds.). *Nonlinear Optics and Optical Physics*. Singapore: World Scientific (1994).
[3]See also Khoo, I. C., Simoni, F., and Umeton, C. (Ed.). *Novel Optical Materials and Applications*. NJ: Wiley (1997).

These features ensure high level of scientific interactions and exchanges in a secluded but very pleasant environment, and the limited number of participants enables the scientific exchanges and discussions to take place during the entire day spent together in a friendly atmosphere. Another important ingredient of NOMA is affordability. The cost of attendance has been kept at a minimal level by Grand Hotel San Michele, home of NOMA since 1995. It will take a treatise to elaborate on these special, arguably unique, features of NOMA and all the wonderful human experiences participants have appreciated in the three decades since its formation. We hope that the following brief writings can convey what we have called NOMA flavors — the special atmosphere, the sciences and friendships fostered, and the fond memories.

2. The Location

Grand Hotel San Michele is located in Cetraro, a small town on the Tyrrhenian Sea part of the Mediterranean in the north side of Calabria region, in southern Italy. Appearing as a big house surrounded by beautiful trees and flowers, it is about two kilometers outside the town on a nice green hill about 150 m above sea level. The border of the hotel property is a rocky cliff; an elevator going down through the rocks brings hotel guests to a private secluded beach and an open-air restaurant.

Besides the Mediterranean secluded beach for swimming and outdoor dining, the hotel features many facilities including a large, fully functional conference hall, golf course, vineyard and cellar, swimming pool, tennis court, bars, and restaurants. It usually provides full board to the hotel guests, starting with a very pleasant breakfast on the terrace or veranda. It offers many seats, sofas, and small tables both indoors and outdoors for guests to talk, work, or have a drink together. During the conference, lunch is usually held at the beach terrace restaurant, a family place with typical Mediterranean architecture providing a congenial atmosphere. Of course, being in Italy, maximum emphasis is placed on the quality and "quantity" of the food, with some typical dishes of the Calabria region.

All these amazing features are strengthened by the great hospitality extended by the working personnel and the owner, keeping the hotel guests comfortable with their kind and familial attitude. Thanks to the prestige of the sponsoring and cooperating Societies and to the efforts and connections involving also the Department of Physics of the University of Calabria and the Administration of the Grand Hotel San Michele, these wonderful experiences come at very low cost, and ensure that all meeting participants rather stay on site after the meeting sessions — the fundamental requirement to create the right social environment for fruitful scientific exchanges and relaxations.

A further opportunity for socialization is represented by the cultural trip. Conference attendees are offered the possibility of spending one day of "full immersion" in the Calabrian cultural tradition, with visits to castles, ancient villages, archaeological excavations of Magna Graecia, museums, and centuries-old pine forests, with lunches in high-level gastronomic restaurants where typical dishes of the Calabrian culinary tradition are offered in a landscape of enchanting beauty.

3. Scientific Excellence

The focus of the NOMA meeting series is on novel optical materials that possess unique characteristics for application in nonlinear- and electro-optics, communications, sensing, integrated-, nano-, and bio-photonics/phononics.[2,3] Needless to say, in the last 3 decades since its formation, optical materials of interest have evolved from many conventional types to new emergent ones, and participants, while limited in numbers, include many world-leading experts in wide ranging and rapidly developing optical and photonic sciences. NOMA is intended to be international in coverage, and participants come from all the countries of Europe, the Americas, Asia, and Australia.

Every edition of the NOMA Meetings has benefitted from many outstanding speakers on current hot topics in the broad field of *Optical Materials and Applications*, including the late Prof. Nicolas Bloembergen (Nobel Physics Laureate, 1981) — the Plenary speaker in the first NOMA held in the Grand Hotel St. Michele, followed by A. Yariv, E. Garmire, O. Svelto, C. Flytzanis, H. Eichler, T. Ikeda, E. Hanamura, R. Boyd, C. Tang, and many other outstanding scientists. We would like to pay special tributes to several colleagues who have passed away since: F. T. Arecchi, V. Degiogio, H. Gibbs, A. Kaplan, G. Stegeman, for their frequent participation and contribution to the technical excellence of NOMA.

The strong international character is testified by the average participation of 80/100 scientists (including 60/70 speakers) coming from 20/25 different countries.

Another ingredient that makes each NOMA Meeting unique from the scientific point of view is that while it is a topical meeting "at large," it actually does function as more comprehensive. Usually, a person participates in a topical conference focused on a particular topic of his/her own expertise or in a huge general conference comprising many topical subconferences among thousands of attendees. The result is that in any given case, rarely is one able to hear something not related to his work, and even when this is possible (e.g., at some plenary session of big conferences) it is difficult to have any meaningful scientific exchange with speakers in a different field. This limitation is overcome at NOMA Meetings where in a small environment it is possible to follow speeches on quite different topics, and this has the advantages of making possible the scientific exchanges between scientists working in different areas, and possibly for creating new ideas for each of the participants' own research in a highly international scientific environment.

In a span of nearly three decades, material sciences and optical physics have evolved as dramatically as other aspects of life, with once hot topics being replaced by emergent ones every so often, while many have withstood the passage of time and have become textbook materials.

The following is a partial list of topics covered by NOMA speakers in recent years:

Nanostructures	*Photonic Crystals*
Metamaterials	*Semiconductors*
Metasurfaces	*Optical Storage*
Nonlinear Optics	*Nanooptics*
Nonlinear dynamics	*Organic Material*
Biosensing	*Plasmonics*
Biophotonics	*Optical Manipulation*
Biomaterials	*Optical Tweezers*
Liquid crystals	*Photo-mobile materials*
Nanomaterials	*Optofluidics*
Holographic materials	*Silk Optics*
Smart Materials	*2-D materials*
Photo-sensitive Materials	*Topology Photonics*
Light–Matter Interaction	

Acknowledgments

Over the years, the following agencies and technical societies have provided valuable support to NOMA:

ANOMA, Association for Novel Optical Materials and Applications

UNICAL, Department of Physics — University of Calabria

AFOSR, Air Force Office of Scientific Research

CNR-Nanotec, Istituto di Nanotecnologia

BEAM, Engineering for Advanced Measurements Co.

OPTICA, Formerly OSA — Optical Society of America

IEEE Photonics Society

SIOF, Società Italiana di Ottica e Fotonica

SICL, Società Italiana Cristalli Liquidi

SIF, Società Italiana di Fisica

ITT, Istituto Italiano di Tecnologia

TOP Class, Tour Operator

Pubbliturco, Laboratorio Pubblicitario

https://doi.org/10.1142/9789811280603_0001

Chapter 1

Optical Properties and Emerging Phenomena of Two-Dimensional Materials

Kunyan Zhang*, Arpit Jain†, Wenjing Wu*, Jeewan Ranasinghe*,
Ziyang Wang*, and Shengxi Huang*,‡

**Rice University, Houston, TX 77005, USA*
†*Pennsylvania State University, University Park, PA 16802, USA*
‡*shengxi.huang@rice.edu*

Two-dimensional (2D) materials have attracted a great amount of interest because of their novel optical properties induced by the reduced dimension. In this chapter, we discuss different types of 2D materials and the related fabrication techniques for producing 2D monolayers and heterostructures. The obtained 2D materials host extremely strong light–matter interaction because of the weak dielectric screening. In addition to the fundamental properties, they demonstrate intriguing optical phenomena such as moiré exciton, single photon emission, and hybrid polaritons. The unique optical properties and tunable surface adsorption of 2D materials contribute to applications such as optical modulation, quantum optics, and biological sensing. The vast potential of 2D materials can be further explored with the aid of advanced machine-learning models.

1. Introduction to 2D Materials

A half-century ago, Richard Feynman brought up the idea of layered materials during his famous lecture.[1]

There's plenty of room at the bottom. by asking, *What could we do with layered structures with just the right layers?* With decades of efforts, scientists may now be able to answer Feynman's question, thanks to two-dimensional (2D) material research.

2D van der Waals material refers to a family of materials with strongly bonded 2D layers attached by weak van der Waals forces.[2] Individual layers can be easily separated by breaking the van der Waals bonds. The first successful isolation of single-layer graphene back in 2004[3] has brought 2D materials under the spotlight. Over the past two decades, tremendous efforts have been devoted by the scientific community to exploring the novel physics and applications of such materials.

1.1. *Characteristics of 2D materials*

Compared to their parental bulk counterparts, 2D materials process distinctive physical properties owing to their atomically thin nature. The electronic band structure of 2D materials strongly depends on the layer thickness. For example, monolayer MoS_2 becomes a direct bandgap semiconductor, while the bulk MoS_2 has an indirect bandgap.[4] This enables a new way of bandgap engineering via thickness control.

The electrical and optical properties of 2D semiconductor monolayers are also dominated by excitonic effects.[5,6] Thanks to the weak dielectric screening and strong charge confinement within the reduced dimension, an ultra-strong Coulomb interaction emerges, giving rise to intriguing many-body phenomena in such material systems. Charge carriers are tightly bounded within or across the layers as electron–hole pairs, called excitons. Excitons and their higher-order complexes (trion, biexciton, hexciton, etc.) have been observed in few-layer 2D semiconductors and their heterostructures. The reduced dielectric screening of the sub-nanometer layers results in the high sensitivity of the local environment to factors such as strain, defects, and temperature changes.[7-10] Electrons and holes can be trapped in local potential wells, resulting in localized excitons. Those trapped excitons can either become pronounced quantum emission sources or, in contrast, quench the intrinsic photoluminescence emission.

The merit of an "all-surface" crystal structure makes 2D materials highly favorable for surface-related applications. They have been widely studied for electrocatalysis and sensing purposes, such as hydrogen/oxygen evolution reaction (HER/OER)[11] and disease detection,[12] respectively. Furthermore, the atomic thickness together with high in-plane stiffness provides 2D materials with extremely flexible mechanical properties.

More excitingly, those layered 2D sheets have considerable freedom in layer-by-layer integration, creating a diversity of heterostructures without the constraints of lattice mismatching.[13] This merit offers the opportunity to design and realize new material platforms with desired properties. The easy-to-integrate nature also makes them promising for the fabrication of flexible and compact electronic/optoelectronic devices.

1.2. *Examples of 2D materials*

As of today, a large variety of 2D materials have been discovered and investigated as summarized in Fig. 1. This large family of materials can be categorized as elemental materials like graphene, compound materials like boron nitride, and structure-engineered materials, including heterostructures.

Elemental 2D materials have attracted considerable attention, despite their chemical simplicity.[14] There exist some elemental 2D materials that

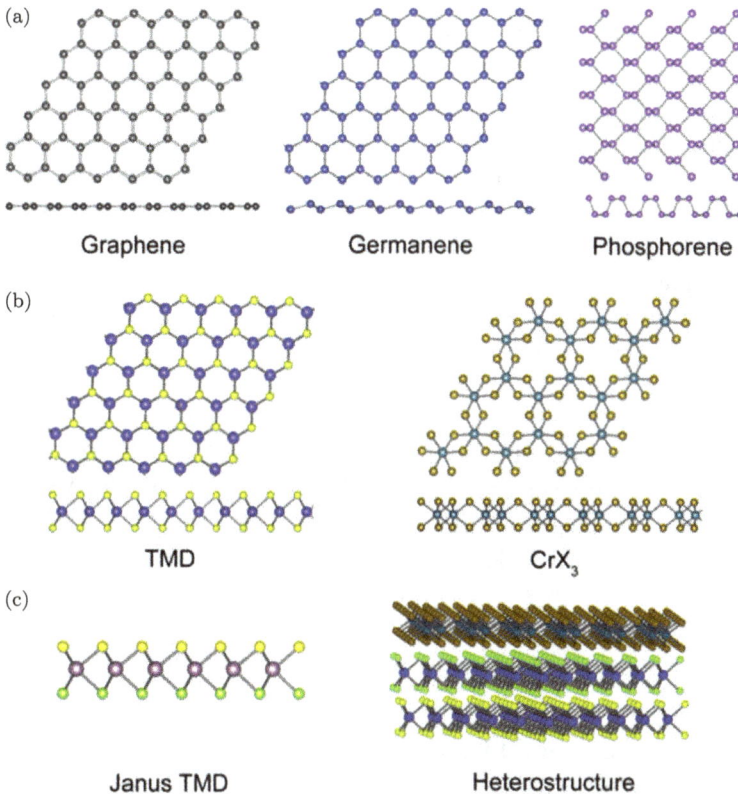

Figure 1. Examples of 2D materials. (a) Crystal structures of elemental 2D materials (top view and side view); (b) Two compound 2D materials: TMD and CrX_3 (X: I, Br); (c) Structure-engineered materials like Janus TMD and artificially assembled heterostructures.

have their bulk form in nature, such as graphene and phosphorene. Graphene is a single layer exfoliated from graphite and it is composed of a honeycomb lattice of sp^2-hybridized carbon.[15] It has a zero-gap electronic band structure with linear dispersion. Phosphorene is the monolayer form of phosphorus, exhibiting an orthorhombic layered structure with layers buckling out of the plane.[16] Its highly anisotropic physical properties come from this structure. Many other synthetic elemental 2D materials have been experimentally achieved,[17] including borophene (B),[18] silicene (Si),[19] germanene (Ge),[20] stanine (Sn),[21] and 2D tellurium (Te).[22]

Compound 2D materials can be further classified into many categories. Among them, transition metal dichalcogenides (TMDs) offer a wide range of electronic properties, from semiconductors to superconductors.[23] Semiconducting TMDs, including MoS_2 and WSe_2, have been exclusively studied for merits like strong light–matter interaction,[24] nonlinear properties from broken inversion symmetry,[25] strong spin-orbit coupling,[26] etc. Some metallic TMDs have been proven to host intriguing physical phenomena like charge density wave (CDW)[27] and superconductivity[28] under certain conditions. Another group of compound 2D layered materials has recently gained a lot of interest. These materials include transition metal halides (CrI_3, $CrBr_3$), $Cr_2Ge_2Te_6$, and XPS_3 (X: Ni, Mn, Fe). Magnetism is present in these materials even at the monolayer limit.[29]

Structure engineering of crystals emerges as an efficient pathway to modify and achieve desired novel properties. One exciting example of structure-engineered 2D materials is Janus TMDs,[30] which have different atomic species on their upper and lower facets. This symmetry breaking induces the formation of an out-of-plane electric dipole, which is capable of tuning the interlayer coupling.[30,31] Building novel van der Waals heterostructures by integrating 2D layers laterally or vertically provides access to programmable properties beyond their building blocks. When stacking layers with a finite twist angle, a moiré energy landscape can be created, resulting in a plethora of effective low-energy quantum Hamiltonians and further leading to the realization of many correlated phases.[32]

2. Fabrication of 2D Materials and Heterostructures

Various bottom-up and top-down synthesis methods have been developed for 2D materials. For example, vapor-based synthesis methods have enabled the wafer-scale synthesis of 2D materials for optoelectronic applications and electronic devices.[33] Chemical vapor deposition (CVD) involves using metal-organic precursors like $Mo(CO)_6$, chalcogen precursors like H_2S, and a carrier gas like Ar/H_2. The gases enter a tube furnace at a specific temperature (200–1,100°C) and pressure (1–760 Torr), dissociate, and react on the substrate placed downstream to form mono to a few layers of TMDs.[34] In contrast to CVD synthesis at relatively high temperatures, substituting the top surface atoms of TMD monolayers at room temperature can produce Janus TMD monolayers with broken mirror symmetry.[35] Powder vaporization is also very similar but involves using powder precursors kept inside the tube furnace during growth, leading to a loss of precise control over the precursor amounts. Metal transformation involves depositing thin films of either transition metals or chalcogen precursors and subsequent heat treatment to convert them to 2D TMDs.

Additional epitaxial growth techniques include molecular beam epitaxy and atomic layer deposition, which are important in specific applications and for understanding the precursor-substrate chemistries necessary for CVD growth. Graphene can be grown epitaxially on SiC substrates by subliming it at high temperatures with layer control depending on annealing time.[36] Hydrogenating this structure can lead to the formation of quasi-freestanding epitaxial graphene, which shows 2D electron gas properties and is an excellent template for epitaxial growth of hBN,[37] GaN,[38] and elemental 2D metals.[39] Chemical vapor transport is an essential technique for bulk crystal formation for exfoliation purposes and utilizes a halogen transport agent to grow centimeter-scale bulk crystals. These techniques are highlighted in Fig. 2.

Exfoliation is also a promising method used to fabricate 2D materials (Fig. 3). As the name implies, it involves removing one or a few van der Waals bonded layers from a bulk single crystal employing an exfoliating medium.[3] In the case of mechanical exfoliation, the exfoliating medium is a tape that is repeatedly peeled to reduce the thickness of 2D materials. Graphene, one of the first 2D materials to be discovered, was synthesized using the tape exfoliation method by Geim and his coworkers in 2004.[3] In liquid phase exfoliation, a solvent can be used as an exfoliation medium that enters the interlayer spaces in 2D materials chemically (oxidative liquid exfoliation) or through ultrasonication, resulting in large-area exfoliation of 2D materials.[40] In addition, large-area exfoliation that involves the use of thin metal films deposited on bulk 2D crystal can also exfoliate millimeter-scale mono to a few layers of 2D materials and transfer them to the substrate of our choosing.[41] 2D materials can also be precisely thinned

Bottom-up approach

Chemical Vapor Deposition (CVD)

Physical Vapor Deposition (PVD)

Wet Chemical Preparation

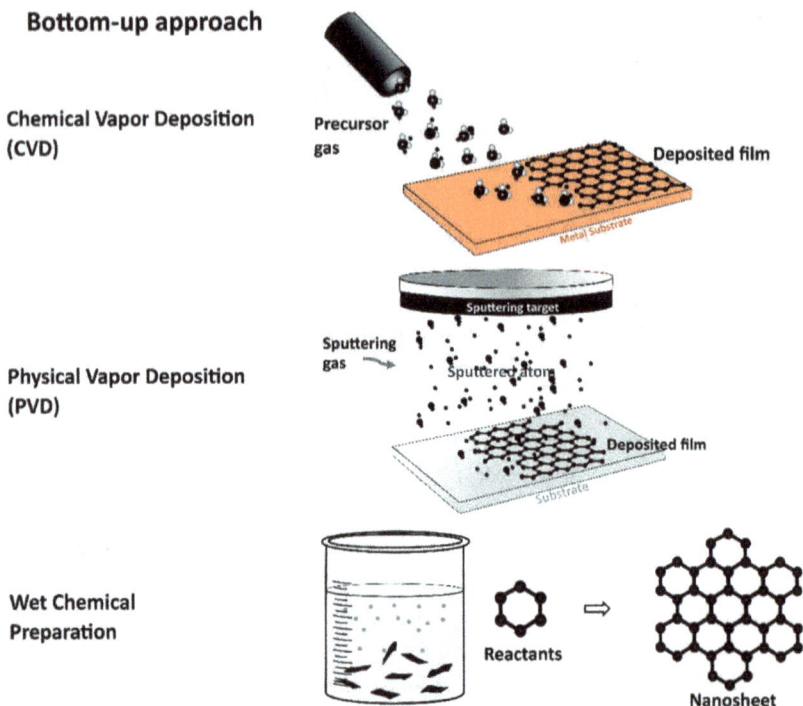

Figure 2. Primary bottom-up synthesis techniques for 2D materials. Adapted with permission from Ref. 17 © 2020 Royal Society of Chemistry.

Top-down approach

Mechanical Exfoliation

Liquid Exfoliation

Etching

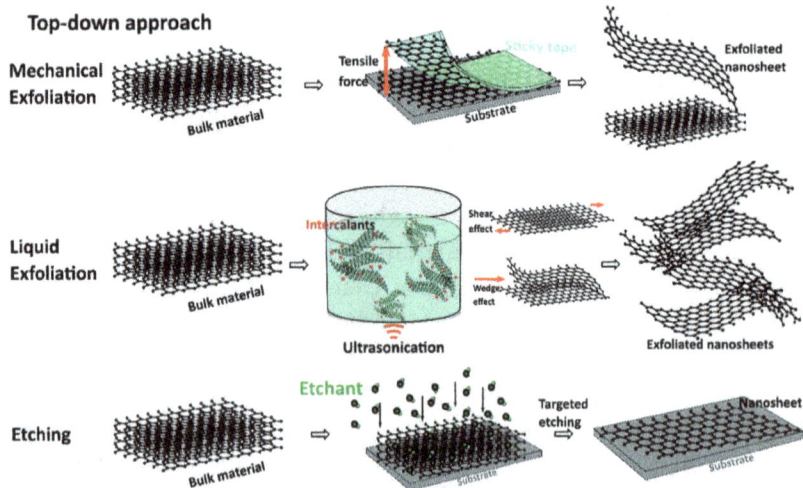

Figure 3. Primary top-down synthesis techniques for 2D materials. Adapted with permission from Ref. 17 © 2020 Royal Society of Chemistry.

down using laser etching to yield selective and damage-free layer control.[42] The obtained 2D layers can be stacked on top of each other to create van der Waals heterostructure through polymer-assisted deterministic transfer.[43] A polymer stamp like Polydimethylsiloxane/Polycarbonate (PDMS/PC) is used to pick up the desired 2D material flake from the substrate and then drop it onto another 2D material flake to create the heterostructure. This process can be repeated to achieve multiple-layer stacks. Researchers have also developed autonomous machines which can exfoliate, identify, and characterize the 2D materials and correspondingly create their van der Waal stacks, all inside a glove box.[44,45]

3. Optical Properties of 2D Materials

Two-dimensional materials possess novel optical properties due to their reduced dimensionality, unique electronic structure, and dielectric screening. Their unique properties can be utilized in next-generation flexible optoelectronic devices. In this section, we will discuss some of the fundamental optical properties of 2D materials.

3.1. *Light absorption*

Even though 2D material monolayers are less than one nanometer thick, they are excellent light absorbers compared to bulk crystals.[46] Absorbance in these materials can be measured using differential reflectance or transmittance measurements and can also be inferred from their dielectric constants.[12] For example, the electronic structure of graphene is analogous to a linear Dirac cone at the Fermi energy level. The optical response of graphene is defined by interband transitions occurring between the valence and conduction bands. It has been shown that for pristine monolayer graphene, its optical conductance is defined only by universal constants and is independent of frequency given by $\sigma(\omega) = \pi e^2/2h$.[47] The corresponding absorbance is given by $A(\omega) = (4\pi/c)\sigma(\omega) = \pi\alpha$, where α is the fine structure constant. This leads to a theoretically predicted absorption value of 2.29% for graphene in the visible range, which is very similar to the experimentally observed value of 2.3%.[47,48] However, there are some dispersions in the experimentally observed absorption data due to deviation from the linear Dirac cone of graphene.[48] Transmittance spectra for multilayer graphene show the reduction in transmittance by a factor of $\pi\alpha$ for every additional layer of graphene up to five layers in total.[48] Graphene can also be electrostatically doped by gating to shift its Fermi

level and correspondingly change the interband absorption, thereby leading
to the tunability of its optical response.

TMD monolayers are also excellent absorbers of light, achieving about
5–10% absorption of incident light in the visible range,[46] which is about ten
times higher than that of GaAs. Their high absorption can be explained
by the dipole transitions with a large density of states and oscillator
strengths between localized d states with strong spatial overlap on the
group VI transition metal atoms.[46] Excitonic effects in TMDs lead to a
highly constructive superposition of the oscillator strengths near the onset
of the absorption edge,[49] which has been utilized to create photovoltaic
devices using these materials.

3.2. *Excitons*

Monolayer MoS_2 is a direct gap semiconductor with a bandgap of about
1.8 eV,[50] making it ideal for optoelectronic applications. When MoS_2
absorbs a photon, an electron leaves the valence band for the conduction
band, leaving behind a positively charged hole in the valence band. This
hole is strongly bonded to the electron by Coulombic forces and is called
an exciton, a neutral quasiparticle. If the exciton also has an associated
charge, it is called a trion. 2D TMDs are unique because of highly stable
excitons with sizeable binding energy, which exist even at room temperature
due to their reduced dimensionality and low dielectric screening. These
excitons can also be converted to photons and can be easily measured using
temperature-dependent photoluminescence spectroscopy.[51]

Figure 4(a) describes the various types of excitons which can exist in
a 2D TMD material.[52] Bright excitons consist of bonded electrons and
holes with the same spin and position in momentum space. Momentum-
forbidden dark excitons consist of bonded electrons and holes with the
same spin but different momentum space positions. Spin-forbidden dark
excitons consist of bonded electrons and holes with the opposite spin but the
same momentum space. Localized excitons occur due to electron–hole pairs
trapped in a defect-induced potential. Therefore, defect engineering can be
used to manipulate the exciton dynamic in 2D monolayers.[53] Figure 4(b)
describes interlayer excitons in heterolayers in which the electrons are
present in the conduction band of one material and holes in the valence
band of another.

Excitonic binding energy is defined as the difference between the
bandgap energy (E_0) and the energy of the observed excitonic transition
(E_x).[52] For $MoSe_2$, E_0 and E_x were experimentally measured to get an

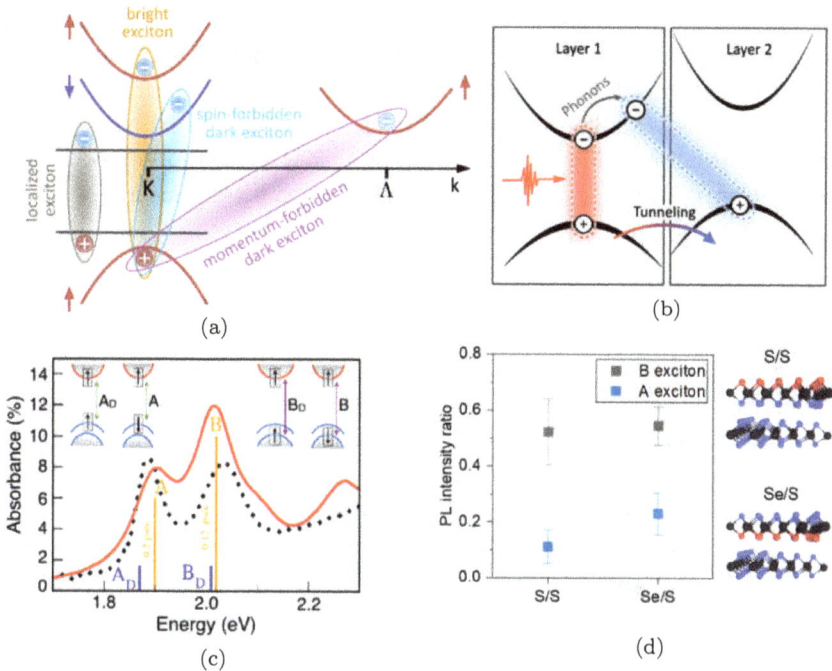

Figure 4.   Excitons in 2D materials. (a) Illustration of different types of excitons. The arrows represent the spin. Adapted with permission from Ref. 52 © 2018 Springer Nature. (b) Illustration of interlayer excitons in heterostructure of 2D materials. Adapted with permission from Ref. 52 © 2018 Springer Nature. (c) The optical absorption spectra of MoS$_2$. Adapted with permission from Ref. 55 © 2015 American Chemical Society. (d) The intensity of intralayer exciton of MoS$_2$ is modulated by the intrinsic electric field of MoSSe at the heterostructure interface. Adapted with permission from Ref. 31 © 2021 American Chemical Society.

exciton binding energy of 0.55 eV, which is significantly higher than other bulk semiconductors.[54] The absorption spectra of MoS$_2$ in Fig. 4(c) show the two prominent bright excitons,[55] A and B excitons. These two peaks arise due to a vertical transition from the spin-split valence band to the conduction band, as seen in the inset. Both the peaks also have a satellite peak coming from the associated dark exciton transition. A and B excitons resemble the 1s state in the 2D hydrogen model[56] and their higher excited excitonic states with decreasing oscillator strength are observed in the optical spectra similar to the hydrogen Rydberg series.[57] However, there is a deviation in the energy spacing between the transitions due to non-local dielectric screening in TMDs. An absorption peak called C peak appears around 2.7 eV, marking the onset of the continuum regime with excitons of low binding energies. The intensity of the intralayer A exciton of MoS_2 can be effectively modulated by the intrinsic dipole of Janus TMD materials as shown in Fig. 4(d).[52] The intrinsic dipole in Janus MoSSe affects the charge transfer between MoSSe and MoS_2 and therefore the PL intensity of the intralayer exciton of MoS_2. Trions (charged excitons) are observed at low temperatures in monolayer TMDs with significant binding energies.[58]

3.3. *Optical phonons*

Phonon is the quantum of the vibrations in a crystal lattice. These vibrations are unique to the chemistry and the structure of the material. Phonons and phonon dispersions have been widely studied for 2D materials with first-principles theoretical prediction[59] and experimental observations.[60] Raman spectroscopy is a powerful experimental tool to measure and probe these phonons and characterize these 2D materials. In this section, we will focus on phonons in a model TMD with few layers.

TMDs can be represented by an X–M–X structure, with M being a transition metal sandwiched between two X chalcogen atoms arranged in a hexagonal lattice termed 2H. The primitive unit cell for this structure has six different atomic positions; hence, there are 18 different phonons (15 optical and 3 acoustic) at the Γ point in the Brillouin zone.[61] These modes have been highlighted for a monolayer 2H-MoS_2 film in Fig. 5(a). The phonon dispersion for monolayer MoS_2 is shown in Fig. 5(b). Experimentally, E_{2g} and A_{1g} modes for MoS_2 are observed using 532 nm laser excitation and can be used to determine the number of layers in the 2D materials by probing the difference in the peak positions of these two peaks.[62] The layer-dependent shift occurs due to the blue shifting of the A_{1g} mode and red shifting of the E_{2g} mode on increasing thickness. Similar phonon dispersion calculations and layer-dependent Raman spectra have also been evaluated for other TMDs.[61]

The phonons discussed above correspond to vibrations occurring within a single layer of 2D material. However, if there are two or more layers of 2D materials, coupled vibrations between those layers can occur, leading to interlayer phonons (Fig. 5(c)).[63] These interlayer phonons typically have very low energy as interlayer bonding in 2D materials is inherently weak and is observed at low wavenumbers in Raman spectra, as seen in Fig. 5(d). E_{2g}^2 is the interlayer shear mode whose energy increases with an increase in layer number. The dotted line in Fig. 5(d) represents layer breathing mode, which corresponds to the simultaneous up and down movement

(a)

$E^{'}$ (Acoustic) $A^{''}_2$ (Acoustic) $E^{''}$ (Raman) $A^{'}_1$ (Raman) $E^{'}$ (IR+Raman) $A^{''}_2$ (IR)

(c)

B^2_{2g} breathing mode E^2_{2g} shear mode

(b)

(d)

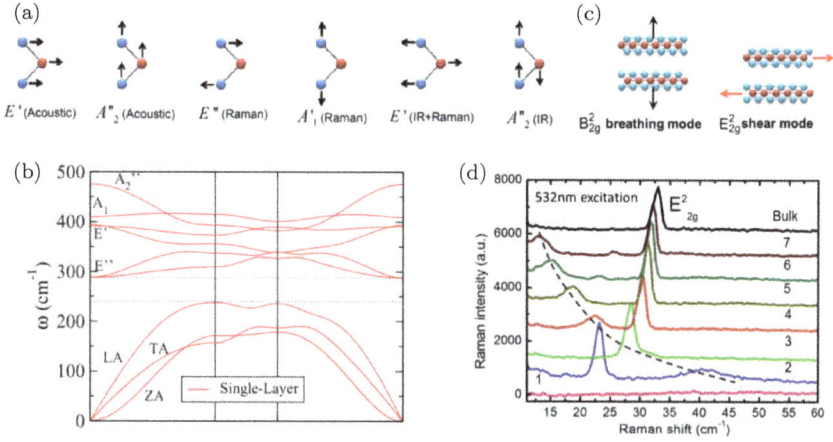

Figure 5.   Optical phonon modes of MoS_2. (a) The vibrational modes for monolayer MoS_2. Adapted with permission from Ref. 60 © 2013 American Chemical Society. (b) Calculated phonon dispersion for monolayer MoS_2. Adapted with permission from Ref. 59 © 2011 American Physical Society. (c) Interlayer phonon modes for bilayer MoS_2. Adapted with permission from Ref. 60 © 2013 American Chemical Society. (d) Layer-dependent low-frequency Raman spectra of MoS_2 showing shear and layer breathing modes. Adapted with permission from Ref. 63 © 2012 American Physical Society.

of the entire layer. Its energy decreases with an increase in the layer number. Hence, these low-frequency modes also become an essential tool for characterizing layer numbers. 2D material heterostructures also exhibit these interlayer phonons, including both the shear and layer breathing modes whose peak position depends on the type of 2D material and the coupling strength between them.[30] The positions of interlayer phonons are also dependent on the relative orientation of the stacked heterolayers.[31]

4. Emerging Optical Phenomena

The unique optical characteristics of 2D monolayers and few layers distinguish them from their bulk counterparts. In addition to the strong optical absorption, excitonic effect, and phonon behaviors, 2D materials possess unconventional optical phenomena induced by moiré stacking. The exciton and phonon properties can be modulated by the periodic moiré potential. In addition to the narrow emission lines generated by moiré lattice, defect engineering can also be used to generate single photons from the defect states in 2D materials. The strong light–matter interaction of

2D materials also constitutes an ideal platform for studying polaritons, a quasiparticle stemming from the interaction between photons and other excited states.

4.1. *Moiré exciton*

Stacking van der Waals materials with interlayer rotation or lattice mismatch forms moiré lattice with a unit cell of a few nanometers, which gives rise to intriguing optical phenomena. The moiré lattice creates a period potential as shown in Fig. 6, whose depth is determined by the stacking configuration, material species, as well as doping levels. This moiré potential can trap the exciton in certain high symmetry points.[64] The radiative recombination of interlayer exciton in such a moiré lattice is typically demonstrated as a series of narrow emission lines at low temperatures.[64] The localization and long valley lifetime of the moiré exciton also enable direct visualization of the interlayer exciton using time- and angle-resolved photoemission spectroscopy (Tr-ARPES).[65] Unlike intralayer exciton in the monolayers, the circular polarization of the moiré exciton depends on the symmetry of the local atomic registry. For example, a heterobilayer of transition metal dichalcogenides has two high-symmetry locations with opposite symmetry selection rules in the moiré supercell at the A and B points as shown in Fig. 6(a).[64] Through interlayer stacking, the local potential minimum in the supercell can be tuned to be at A or B points. This allows for co- or cross-circular polarized emissions, meaning that the circular polarization of the trapped interlayer exciton is the same or the opposite as the excitation.[66] Additionally, the Zeeman splitting of these trapped interlayer excitons is governed by the valley pairing at different twist angles. When the twist angle corresponds to a commensurate pattern such as $\vartheta = 21.8°$, the recombination of the interlayer excitons at the K–K or K–K' transitions are facilitated by the electron–phonon Umklapp scattering and becomes momentum-direct in the second Brillouin Zone.[67,68]

2D moiré superlattices also constitute a promising platform to study correlated electronic states and many-body interaction. When the hopping between different moiré superlattices becomes negligible, the system hosts a series of unconventional phenomena described by the Hubbard model. The interaction between these strongly correlated electronic states with excitons enables studying electron correlation and quantum phase transition through optical spectroscopies. For a moiré lattice with a large superlattice constant of 24 nm and a weak interlayer interaction, the interlayer exciton can be

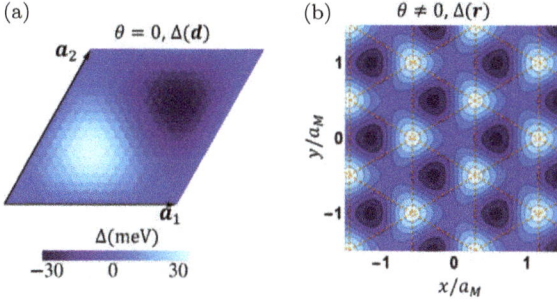

Figure 6.   Valence band maximum of AA stacked $WSe_2/MoSe_2$. (a) No twist angle. (b) With twist angle. Adapted with permission from Ref. 71 © 2018 American Physical Society.

dynamically screened by itinerant electrons to form the exciton-polaron.[69] This is demonstrated by the hybridization of interlayer exciton and intralayer exciton in the Stark effect. The many-body physics revealed by the strong light–matter interaction provides a versatile paradigm to study the Bose–Einstein condensation (BEC) in 2D materials.[70]

4.2. *Moiré phonon*

The quantum confinement in the momentum space folds the dispersion into the mini-Brillouin zone and forms excitonic minibands at the crossing points. Similar to the electronic bands, the zone folding of the moiré supercell also folds the phonon dispersion and activates additional phonons at the Γ points which are otherwise not observable. Such phonons induced by the moiré patterns are named moiré phonons. For a moiré superlattice formed by interlayer twist, the crystallographic superlattice is denoted by a pair of prime numbers (m, n) as shown in Fig. 7. The twist angle is then given by $\cos\theta = [m^2 + 4mn + n^2/2(m^2 + mn + n^2)]$ and the lattice constant is given by $L = [a|m - n|/2\sin(\theta/2)]$ in which a is the in-plane lattice constant of the monolayer. Additionally, the lattice constant of the moiré superlattice is defined as $L^M = [a|m - n|/2\sin(\theta/2)]$. In the reciprocal space, the basic vectors of the moiré reciprocal lattice are written as $g = b_2 - b_1$ and $g' = b'_1 - b'_2$ in which b_1 and b_2 are the reciprocal lattice vectors. The absolute value of the moiré reciprocal lattice vectors are $|g| = |g'| = 2b\sin(\vartheta/2)$. The measured phonons can be mapped to different locations in the moiré reciprocal lattice based on the relationship between g and θ. They can be used to map out the phonon dispersion

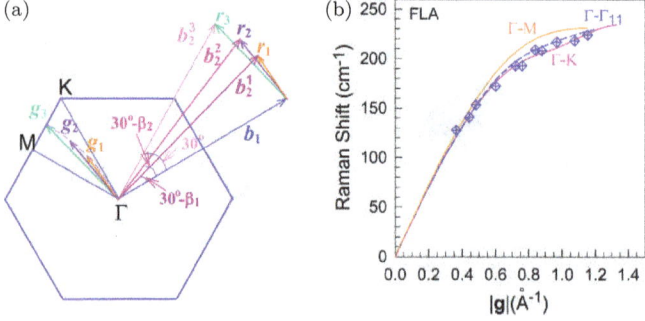

Figure 7. Moiré phonons in twisted bilayer TMDs. (a) Schematic diagram of moiré basic vectors. (b) Raman shift of the folded longitudinal acoustic (FLA) modes as a function of the reciprocal lattice vector. Adapted with permission from Ref. 77 © 2018 American Physical Society.

of the constituting monolayer by varying the twist angle.[72] In addition to the zone folding of phonon dispersion, the moiré pattern also affects the low-frequency phonon modes that represent interlayer vibration. As the moiré superlattice switches from a relaxed regime to a rigid regime at around $\vartheta = 2\text{--}4°$, interlayer phonon modes exhibit a sharp transition in frequency because of the changed strain condition.[73] Instead of a simple model such as phonon dispersion folding, this phonon renormalization can be described by the low-energy continuum model.[74,75] In the small angle limit, the optical phonon modes near the Γ point are calculated from local crystalline dynamical matrices $\bar{D}(q'|d)$ as a function of the displacement d between twisted layers. The moiré dynamical matrix $\bar{D}_m(q, q')$ is obtained from the Fourier transform of the local crystalline dynamical matrices for each local stacking,

$$\bar{D}_m(q, q') = \sum_{G,d} \delta(q - q' - \tilde{G}(\theta, G)) \bar{D}(q'|d) e^{id \cdot G}$$

q' is the reciprocal space variable, $\tilde{G}$ is the reciprocal space vector. This model takes into account the layer separation and lattice reconstruction and can reproduce the interlayer phonon modes in the moiré lattice. The van der Waals stacking not only breaks the point group symmetry, but also the time-reversal symmetry of phonon modes. In systems like magic-angle twisted bilayer graphene and compound materials like hexagonal boron nitride (hBN), time-reversal breaking chiral phonons are predicted to emerge, providing additional perspectives to the origin of the correlated states.[76,77]

4.3. *Single photon emission*

The strong light–matter interaction in 2D materials offers versatile routes to engineer quantum states for quantum photonics. For example, quantum defects in WSe_2 contribute to weakly bound excitons at cryogenic temperatures, which host single photon emissions.[78] In hBN, the single photon emission stems from the defect levels residing in the bandgap.[79,80] The emission of the single photon in hBN can be achieved at room temperature by excitation with energy smaller than the 6 eV bandgap of hBN. In addition to defects, local strain around nanostructures, such as nanopillars, is also an effective approach to realize the spatial localization of a single photon. In this case, the photons are trapped by the trapping potential created by the local strain gradient. Besides the tunability in generating defects and strain gradients, another advantage of 2D materials for single photon emission is the capability to integrate them with a cavity, photonic waveguides, and plasmonic structures, which are rather difficult to achieve for bulk emitters. By coupling a quantum emitter in an optical cavity, the Purcell effect can lead to a strong enhancement in the light–matter interaction.[81] This would facilitate the quantum efficiency of the photon emission defined by a $g^2(0)$ function. In a quantum network, the photonic qubits supported by single photons can function as interlinks that entangle distant qubit notes.[82]

4.4. *Polariton*

2D materials serve as a promising platform to study polaritons, a quasiparticle originating from the interaction between photons and electric dipole. Different types of polaritons exist in 2D materials including plasmon-polariton, phonon-polariton, exciton-polariton, Cooper-pair-polariton, and magnon-polariton.[83] Exciton- and phonon-polaritons are associated with the high dispersive permittivity of 2D materials because of weak dielectric screening. The optical anisotropy of 2D materials leads to the permittivity of different signs in the in-plane and out-of-plane direction, contributing to hyperbolic phonon-polaritons.[84] This allows phonon-polaritons to propagate within the top and bottom surface of 2D materials as in a waveguide. The hyperbolic phonon-polaritons of hBN can enable applications such as slow light[85] and sub-diffraction imaging.[86,87] In contrast, the hyperbolic phonon-polariton of Bi_2Se_3 and Bi_2Te_3 would lie in the terahertz (THz) range.[88] The phonon-polaritons can be measured by a range of techniques including scanning near-field optical microscopy, Fourier-transform infrared spectroscopy, peak force infrared microscopy, and photothermal-induced resonance microscopy. In comparison, the observation of exciton-polaritons in near-field spectroscopy is very recent in exfoliated WSe_2 thin films.[89] Experimental observation of exciton-polariton is typically demonstrated by the anti-crossing of the exciton and cavity modes with a Rabi splitting as large as 46 meV in MoS_2.[90]

5. Optical Applications of 2D Materials

These extraordinary properties of 2D materials give birth to a wide range of optical and biological applications. During the past few decades, the scientific community witnessed rapid growth in the semiconductor industry demanding compact optoelectronic devices at the nanoscale. On the other hand, the biomedical field requires unique materials with fascinating properties to fill the gaps in personalized medicine. Therefore, designing unique 2D material-based devices is important for harnessing light–matter interactions for optical applications.

5.1. *Applications in optoelectronics*

The advancement in nanotechnology attracts tremendous interest in exploring various 2D materials. The optical responses of these layered materials could span a broad region of the electromagnetic spectrum. Their bandgaps could be further manipulated by various means such as doping, stacking heterostructures, alloy composition, and layer numbers.[91] The extraordinary properties of many 2D materials make them available for a range of optoelectronic applications.

5.1.1. *Photodetectors*

Ultrathin 2D materials have drawn significant research interest for photodetectors due to their excellent properties such as tunable optoelectronic properties, dangling bond-free surface, wide optical absorption, and high carrier mobility.[92] Over the past decade, graphene as a material for photodetectors ranging from visible to THz has been widely reported.[93] However, expanding the capabilities of graphene in optoelectronics is restricted by its zero bandgap.[91–94] In addition to graphene, black phosphorus (BP) has also been demonstrated as a functional material in photodetectors. For instance,

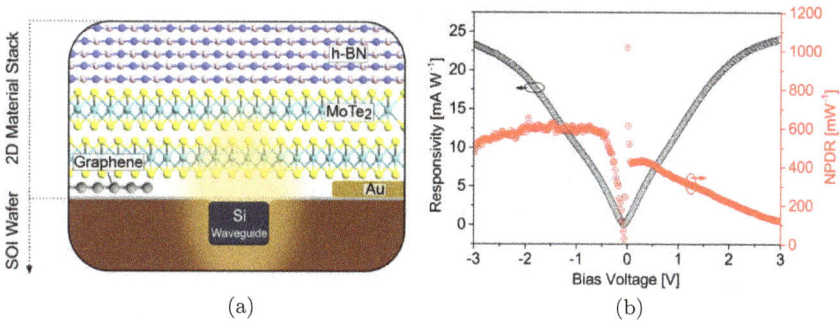

Figure 8.   MoTe$_2$ waveguide photodetector. (a) Schematic illustration of a graphene-MoTe$_2$-Au photodetector stacked onto a silicon waveguide. (b) Photoresponsivity and normalized photocurrent-to-dark-current ratio (NPDR) as a function of applied bias voltages. Adapted with permission from Ref. 97 © 2018 American Chemical Society.

2D BP was used to build a mid-infrared spectrometer enabled by the strong photoresponse and Stark effect.[95] Besides, BP-based polarimeters are also reported to show potential in high-speed polarization-division-multiplexed imaging.[96] Another interesting avenue is waveguide-integrated 2D photodetectors based on the photoconductivity of 2D materials. Leuthold and coworkers fabricated a few-layer MoTe$_2$ waveguide photodetector that can be used in O-band (1.31 μm) optical communications.[97] The MoTe$_2$ waveguide photodetector showed a photoresponsivity of 23 mA/W under 3 V bias at a wavelength of 1.31 μm. The normalized photocurrent-to-dark current ratio was up to 600 mW^{-1}, outperforming those of graphene-based photodetectors (Fig. 8). The eye diagram with a bit error ratio of 3.5×10^{-3} at 1 Gbit s^{-1} indicates that the MoTe$_2$ waveguide photodetector can be used in optical communications. Furthermore, strain engineering has been utilized to enhance the performances of MoTe$_2$-based photodetectors.[98] With this approach, a responsivity of 0.5 A/W and responsive wavelength in the range of 1,500 nm can be achieved. The strain engineering can effectively modulate the bandgap from 1.04 eV for pristine MoTe$_2$ to 0.8 eV for strained MoTe$_2$. Overall, 2D material-based photodetectors are being continuously investigated and upgraded for better device performance.[99]

5.1.2. *Light-emitting diodes*

Low-dimensional light sources play an important role in photonic-integrated circuits. TMDs have become a promising candidate for light emitting due to their direct bandgap, large quantum confinement effect, and diversified heterobilayer structures.[100] Pan *et al.* proposed a light emitting diode (LED) formed by p-type monolayer WSe$_2$ and *n*-type CdS nanoribbon allowing efficient optical routing of WSe$_2$ photoluminescence and electroluminescence (EL) emission.[100] Additionally, valleytronics can be used to create valley-light emitting diodes (vLED). In this regard, Yang *et al.* demonstrated electrically tunable circularly polarized EL from monolayer WS$_2$.[101] In this device, EL shows distinct circular polarization, where the right-handed circularly polarized light is stronger than the left-handed one. In another report, Novoselov and coworkers presented an innovative approach to design LED by a vertical heterostructure based on insulating hBN, conducting graphene, and semiconducting WS$_2$.[102] This engineered structure results in a large light emitting area, high current density, and low contact resistance.

5.1.3. *Quantum cascade lasers*

Van der Waals materials with semiconductor photonics were used to realize quantum cascade lasers with tunable emission characteristics. For example, graphene has a complex optical dispersion that can be used to modulate the emission of THz quantum cascade lasers.[103] Graphene can also be used as an active tunable gate material to modulate the emission of quantum cascade lasers.[103] The highly doped graphene can be used instead of gold in the quantum cascade laser to achieve wavelength regulation. Other researchers have also investigated this topic improving the practicality of the THz laser in the real environment.[104]

5.2. *Applications in healthcare*

The strong light–matter interactions as well as surface and structural engineering pathways enabled by 2D materials are highly preferred for targeted measurements. In general, optical sensing platforms measure the analyte-induced changes in signals, such as absorbance, fluorescence, surface plasmon resonance, and Raman scattering. A variety of van der Waals materials have been investigated and examined for a range of biological applications such as photothermal cancer therapy, bioimaging, and biomarker sensing of neurodegenerative diseases.

5.2.1. *Photothermal cancer therapy*

The versatile physicochemical properties of 2D materials provide great potential for therapeutic applications such as photothermal cancer therapy. These materials are influential candidates for photothermal cancer therapy due to tunable optical properties and the large surface area available for functionalization and drug loading. In photothermal cancer therapy, an external light source with strong tissue penetration is absorbed by a photothermal agent accumulated in the diseased site to convert light to energy resulting in cancer cell death by increased local temperature. Graphene-based materials are widely explored as potential photothermal cancer therapy agents, such as the PEG-modified graphene oxide nanosheets.[105] Additionally, ultrathin BP nanosheets also show high efficiency in converting light to heat. For example, BP nanosheets decorated with gold nanoparticles exhibit higher photothermal conversion efficiency as compared to BP nanosheets.[106] MXenes and TMDs have also been investigated as promising materials with superior photo-absorption characteristics for photothermal conversion.[91] For instance, Ti_3C_2, Nb_2C, and Ta_4C_3 have demonstrated photothermal conversion efficiencies of over 30%.[107–109]

5.2.2. *Computed tomography imaging*

The stability in physiological environments of 2D materials enables their use as an imaging contrast agent. MXenes and TMDs are two particular candidates for computed tomography (CT) bioimaging due to the presence of high atomic-number elements that can provide excellent X-ray attenuation.[91] Another class of promising 2D materials in bioimaging is graphene oxide.[110] For instance, gold nanoparticles with strong X-ray attenuation can be loaded onto graphene oxide to realize CT imaging, as seen in Ref. 110. In that report, gold nanoparticles were integrated with poly(lactic acid) followed by graphene oxide adsorption via electrostatic technique. The obtained microcapsule served as a contrast agent for CT imaging allowing accurate identification of the size and location of a tumor as well as a guide for photothermal cancer therapy. Dai *et al.* further demonstrated that $MnO_x/Ta_4C_3T_x$ and $Fe_3O_4/Ta_4C_3T_x$ composites can be used as CT imaging agents and magnetic resonance imaging agents as shown in Fig. 9.[109,111]

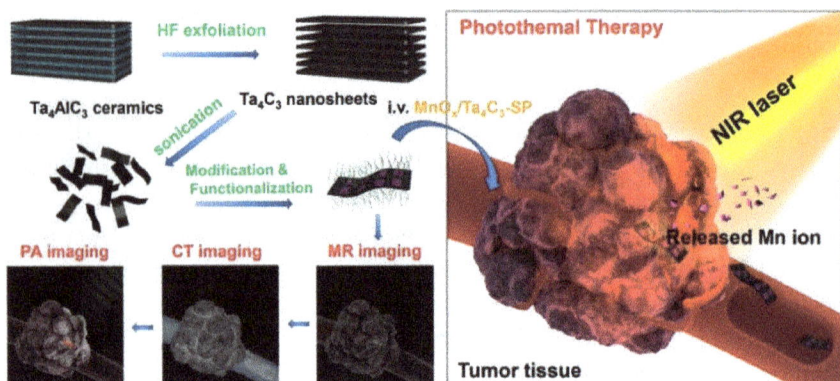

Figure 9.   Schematic illustration of theranostic functions of MnO_x/Ta_4C_3–soybean phospholipid composite nanosheets. Adapted with permission from Ref. 111 © 2017 American Chemical Society.

5.2.3. *Biological sensing*

Surface-enhanced Raman scattering (SERS) has been considered a rapid, sensitive, and label-free technique that allows even single-molecule detection.[112] Typically, metallic nanoparticles are used as SERS substrates through electromagnetic field enhancement.[113,114] SERS substrates that integrate 2D materials with plasmonic nanoparticles can produce a much stronger Raman signal and they benefit from both electromagnetic and chemical enhancements. Chandra Ray *et al.* reported graphene oxide and nanoparticle-based hybrid SERS substrates for biomolecular fingerprinting and label-free disease marker diagnosis.[115] The hybrid SERS probe was utilized to detect the deoxyribonucleic acid (DNA) of the human immunodeficiency virus (HIV) and bacteria.[115] The hybrid system enabled the detection of HIV DNA with 500 fM concentration and methicillin-resistant Staphylococcus aureus (MRSA) bacteria with a limit of detection of 10 CFU/mL. The rapid diagnosis of the COVID-19 virus is highly essential to minimize the spread and treat patients in a timely manner.[116] Several 2D material-based sensors have been developed and proposed to detect components of the COVID-19 virus such as SARS-CoV-2 antibody and protein components.[117,118]

The biosensing application of Raman spectroscopy combined with 2D materials extends to the diagnostics of Alzheimer's disease. Wang *et al.* utilized graphene to enhance the classification accuracy and signal-to-noise ratio of acquired spectra due to the high thermal conductivity and

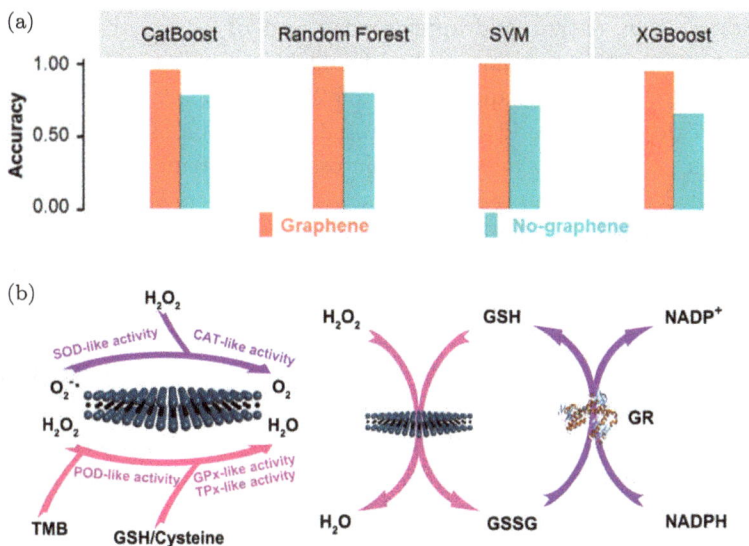

Figure 10. Biosensing and treatment of neurodegenerative disease enabled by 2D materials. (a) The classification accuracy of Raman spectra of mice brain with Alzheimer's disease with and without graphene on the cortex region. Adapted with permission from Ref. 119 © 2022 American Chemical Society. (b) Schematic illustration of enzyme-mimicking activities of MXenzyme and glutathione recycling by glutathione reductase. Adapted with permission from Ref. 122 © 2021 Springer Nature.

fluorescence quenching ability of graphene (Fig. 10(a)).[119] Guo and coworkers presented a nanohybrid based on BP and Au nanoparticles for label-free SERS analysis of the mouse brain.[120] In addition to Alzheimer's disease, sensors and therapeutics based on 2D materials for neurodegenerative diseases have been achieved.[121] The biocompatibility of 2D materials can further allow the treatment of neurodegenerative diseases. Feng *et al.* discovered that 2D vanadium carbide nanoenzyme, also known as MXenzyme, can mimic up to six naturally occurring enzymes (Fig. 10(b)).[122] Synthesized MXenzyme possesses excellent biocompatibility and the potential to rebuild the redox homeostasis and relieve reactive oxygen species (ROS)-stimulated damages opening a new avenue for applying MXene-based nanoplatforms in the treatment of neurodegenerative disease.

6. Outlooks on Machine Learning

Over the last decade, the number of new 2D materials has rapidly grown. New automated approaches are needed to accelerate the production, characterization, design, and application stages of 2D materials. Machine learning (ML) is a promising tool to facilitate the preparations, property predictions, and sensing applications of 2D materials.[12,91,123,124] In 2D material exfoliation, the identification and localization are manually done by human experts with an optical microscope. To reduce the amount of time needed for 2D material identification, researchers have implemented different deep-learning methods to rapidly identify and characterize the locations and thicknesses of 2D materials with optical images and hyperspectral images.[125-127] The neural networks can identify the types and thicknesses of 2D materials with high prediction accuracy and real-time processing capability. With more computational power available, further developments can be made to increase the identification accuracy and preparation efficiency in the exfoliation of 2D materials. Furthermore, with novel interpretable ML methods, the ML methods also have the potential to extract deep graphical features including edges, shapes, and flake sizes of the 2D materials from the optical images.

Prediction and measurement of properties of 2D materials are also critical tasks. Studies have been made to develop ML algorithms to predict material strength, thermodynamic stability, synthesizability, exciton binding energy, etc.[128-131] One of the most important properties is the refractive index, which can be measured by ellipsometry and reflectometry. Analysis of ellipsometry requires heavy model fitting and needs to be operated by experienced researchers. Different approaches have been made and enabled the automated fitting of ellipsometry.[132-134] With ML methods, researchers will be able to predict and measure the properties of 2D materials in an automated and rapid way with high accuracy. With more data available, the prediction and measurement accuracies can be further improved. The prediction and measurement algorithms can also be combined with the identification and localization algorithms to form a robust and efficient pipeline for 2D materials production, measurement, and design.

In biosensing, 2D materials have unique properties that can be used as substrates to enhance the sensing signal. Chen et al. have reviewed and summarized the enhancement of different types of novel 2D materials applied in Raman Spectroscopy, including graphene, hBN, and TMDs.[135] Cui et al. summarized the ML advanced applications on biosensing.[136] Without any modification, the ML approaches can be easily extended to 2D-enhanced biosensors for neurodegenerative diseases.[119] The combination of ML and 2D enhancement biosensing has the potential to accelerate the study of various diseases, tissues, biofluids, and human samples.

ML can be an essential tool in 2D material preparation, in-operando characterization, and sensing applications. As a future prospect, novel deep-learning models can be trained and used to automatedly identify 2D materials in exfoliation with high accuracy. The property predictions and measurements implemented with ML approaches can also be further improved to accelerate the development of novel 2D materials. In 2D-enhanced biosensing, ML methods can be deployed to enable rapid, accurate, and interpretable diagnosis of different diseases.

References

1. R. Feynman, *Feynman and Computation*, CRC Press (2018).
2. K. S. Novoselov, A. Mishchenko, A. Carvalho, and A. H. Neto Castro, 2D materials and van der Waals heterostructures, *Science* **353**, aac9439 (2016).
3. K. S. Novoselov *et al.* Electric field effect in atomically thin carbon films, *Science* **306**, 666–669 (2004).
4. K. F. Mak, C. Lee, J. Hone, J. Shan, and T. F. Heinz, Atomically thin MoS_2: A new direct-gap semiconductor, *Phys. Rev. Lett.* **105**, 136805 (2010).
5. K. F. Mak *et al.*, Tightly bound trions in monolayer MoS_2, *Nat. Mater.* **12**, 207–211 (2013).
6. J. S. Ross *et al.*, Electrical control of neutral and charged excitons in a monolayer semiconductor, *Nat. Commun.* **4**, 1474 (2013).
7. S. Kumar, A. Kaczmarczyk, and B. D. Gerardot, Strain-induced spatial and spectral isolation of quantum emitters in mono- and bilayer WSe_2, *Nano Lett.* **15**, 7567–7573 (2015).
8. T. P. Darlington *et al.*, Imaging strain-localized excitons in nanoscale bubbles of monolayer WSe_2 at room temperature, *Nat. Nanotechnol.* **15**, 854–860 (2020).
9. Y. Bai *et al.*, Excitons in strain-induced one-dimensional moiré potentials at transition metal dichalcogenide heterojunctions, *Nat. Mater.* **19**, 1068–1073 (2020).
10. A. Arora *et al.*, Dark trions govern the temperature-dependent optical absorption and emission of doped atomically thin semiconductors, *Phys. Rev. B* **101**, 241413 (2020).
11. X. Liu *et al.*, The critical role of electrolyte gating on the hydrogen evolution performance of monolayer MoS_2, *Nano Lett.* **19**, 8118–8124 (2019).
12. Z. Wang, Y. Cosmi Lin, K. Zhang, W. Wu, and S. Huang, EllipsoNet: Deep-learning-enabled optical ellipsometry for complex thin films. arXiv:2210.05630 (2022).
13. C. Jin *et al.*, Ultrafast dynamics in van der Waals heterostructures, *Nat. Nanotechnol.* **13**, 994–1003 (2018).
14. A. J. Mannix, B. Kiraly, M. C. Hersam, and N. P. Guisinger, Synthesis and chemistry of elemental 2D materials, *Nat. Rev. Chem.* **1**, 0014 (2017).

15. A. K. Geim, Graphene: Status and prospects, *Science* **324**, 1530–1534 (2009).
16. X. Ling, H. Wang, S. Huang, F. Xia, and M. S. Dresselhaus, The renaissance of black phosphorus, *Proc. Natl. Acad. Sci.* **112**, 4523–4530 (2015).
17. S. B. Mujib, Z. Ren, S. Mukherjee, D. M. Soares, and G. Singh, Design, characterization, and application of elemental 2D materials for electrochemical energy storage, sensing, and catalysis, *Mater. Adv.* **1**, 2562–2591 (2020).
18. H. Liu, J. Gao, and J. Zhao, From boron cluster to two-dimensional boron sheet on Cu (111) surface: Growth mechanism and hole formation, *Sci. Rep.* **3**, 1–9 (2013).
19. A. J. Mannix, B. Kiraly, B. L. Fisher, M. C. Hersam, and N. P. Guisinger, Silicon growth at the two-dimensional limit on Ag (111), *ACS Nano* **8**, 7538–7547 (2014).
20. L. Li *et al.*, Buckled germanene formation on Pt (111), *Adv. Mater.* **26**, 4820–4824 (2014).
21. Zhu, F.-F. *et al.*, Epitaxial growth of two-dimensional stanene, *Nat. Mater.* **14**, 1020–1025 (2015).
22. Y. Wang *et al.*, Field-effect transistors made from solution-grown two-dimensional tellurene, *Nat. Electron.* **1**, 228–236 (2018).
23. K. F. Mak and J. Shan, Photonics and optoelectronics of 2D semiconductor transition metal dichalcogenides, *Nat. Photon.* **10**, 216–226 (2016).
24. K. F. Mak, D. Xiao, and J. Shan, Light–valley interactions in 2D semiconductors, *Nat. Photon.* **12**, 451–460 (2018).
25. Y. Li *et al.*, Probing symmetry properties of few-layer MoS_2 and h-BN by optical second-harmonic generation, *Nano Lett.* **13**, 3329–3333 (2013).
26. Z. Wang, J. Shan, and K. F. Mak, Valley- and spin-polarized Landau levels in monolayer WSe_2, *Nat. Nanotechnol.* **12**, 144–149 (2017).
27. M. M. Ugeda *et al.*, Characterization of collective ground states in single-layer $NbSe_2$, *Nat. Phys.* **12**, 92–97 (2016).
28. Y. Saito, T. Nojima, and Y. Iwasa, Highly crystalline 2D superconductors, *Nat. Rev. Mater.* **2**, 16094 (2016).
29. B. Huang *et al.*, Emergent phenomena and proximity effects in two-dimensional magnets and heterostructures, *Nat. Mater.* **19**, 1276–1289 (2020).
30. K. Zhang *et al.*, Enhancement of van der Waals interlayer coupling through polar Janus MoSSe, *J. Am. Chem. Soc.* **142**, 17499–17507 (2020).
31. K. Zhang *et al.*, Spectroscopic signatures of interlayer coupling in Janus $MoSSe/MoS_2$ heterostructures, *ACS Nano* **15**, 14394–14403 (2021).
32. Q. Tong *et al.*, Topological mosaics in moiré superlattices of van der Waals heterobilayers, *Nat. Phys.* **13**, 356–362 (2016).
33. D. Dumcenco *et al.*, Large-area epitaxial monolayer MoS_2, *ACS Nano* **9**, 4611–4620 (2015).
34. S. Das, J. A. Robinson, M. Dubey, H. Terrones, and M. Terrones, Beyond graphene: Progress in novel two-dimensional materials and van der Waals solids, *Ann. Rev. Mater. Res.* **45**, 1–27 (2015).

35. Y. Guo *et al.*, Designing artificial two-dimensional landscapes via atomic-layer substitution, *Proc. Natl. Acad. Sci.* **118**, e2106124118 (2021).
36. C. Berger *et al.*, Ultrathin epitaxial graphite: 2D electron gas properties and a route toward graphene-based nanoelectronics, *J. Phys. Chem. B* **108**, 19912–19916 (2004).
37. P. C. Mende, J. Li, and R. M. Feenstra, Substitutional mechanism for growth of hexagonal boron nitride on epitaxial graphene, *Appl. Phys. Lett.* **113**, 031605 (2018).
38. Z. Y. Al Balushi *et al.*, Two-dimensional gallium nitride realized via graphene encapsulation, *Nat. Mater.* **15**, 1166–1171 (2016).
39. N. Briggs *et al.*, Atomically thin half-van der Waals metals enabled by confinement heteroepitaxy, *Nat. Mater.* **19**, 637–643 (2020).
40. E.-M. Kirchner and T. Hirsch, Recent developments in carbon-based two-dimensional materials: Synthesis and modification aspects for electrochemical sensors, *Microchimica Acta* **187**, 441 (2020).
41. F. Liu *et al.*, Disassembling 2D van der Waals crystals into macroscopic monolayers and reassembling into artificial lattices, *Science* **367**, 903–906 (2020).
42. A. Nipane *et al.*, Damage-free atomic layer etch of WSe_2: A platform for fabricating clean two-dimensional devices, *ACS Appl. Mater. Interfaces* **13**, 1930–1942 (2021).
43. Y. Liu, S. Zhang, J. He, Z. M. Wang, and Z. Liu, Recent progress in the fabrication, properties, and devices of heterostructures based on 2D materials, *Nano-Micro Lett.* **11**, 13 (2019).
44. S. Masubuchi *et al.*, Autonomous robotic searching and assembly of two-dimensional crystals to build van der Waals superlattices, *Nat. Commun.* **9**, 1413 (2018).
45. A. J. Mannix *et al.*, Robotic four-dimensional pixel assembly of van der Waals solids, *Nat. Nanotechnol.* **17**, 361–366 (2022).
46. M. Bernardi, M. Palummo, and J. C. Grossman, Extraordinary sunlight absorption and one nanometer thick photovoltaics using two-dimensional monolayer materials, *Nano Lett.* **13**, 3664–3670 (2013).
47. R. R. Nair *et al.*, Fine structure constant defines visual transparency of graphene, *Science* **320**, 1308 (2008).
48. K. F. Mak *et al.*, Measurement of the optical conductivity of graphene, *Phys. Rev. Lett.* **101**, 196405 (2008).
49. L. Britnell *et al.*, Strong light-matter interactions in heterostructures of atomically thin films, *Science* **340**, 1311–1314 (2013).
50. K. F. Mak, C. Lee, J. Hone, J. Shan, and T. F. Heinz, Atomically thin MoS_2: A new direct-gap semiconductor, *Phys. Rev. Lett.* **105**, 136805 (2010).
51. J. S. Ross *et al.*, Electrical control of neutral and charged excitons in a monolayer semiconductor, *Nat. Commun.* **4**, 1474 (2013).
52. T. Mueller and E. Malic, Exciton physics and device application of two-dimensional transition metal dichalcogenide semiconductors, *npj 2D Mater. Appl.* **2**, 29 (2018).

53. Q. Qian *et al.*, Defect creation in WSe$_2$ with a microsecond photoluminescence lifetime by focused ion beam irradiation, *Nanoscale* **12**, 2047–2056 (2020).

54. M. M. Ugeda *et al.*, Giant bandgap renormalization and excitonic effects in a monolayer transition metal dichalcogenide semiconductor, *Nat. Mater.* **13**, 1091–1095 (2014).

55. M. Palummo, M. Bernardi, and J. C. Grossman, Exciton radiative lifetimes in two-dimensional transition metal dichalcogenides, *Nano Lett.* **15**, 2794–2800 (2015).

56. D. Y. Qiu, F. H. da Jornada, and S. G. Louie, Optical spectrum of MoS$_2$: Many-body effects and diversity of exciton states, *Phys. Rev. Lett.* **111**, 216805 (2013).

57. A. Chernikov *et al.*, Exciton binding energy and nonhydrogenic Rydberg series in monolayer WS$_2$, *Phys. Rev. Lett.* **113**, 076802 (2014).

58. S. Tongay *et al.*, Defects activated photoluminescence in two-dimensional semiconductors: Interplay between bound, charged and free excitons, *Sci. Rep.* **3**, 2657 (2013).

59. A. Molina-Sánchez and L. Wirtz, Phonons in single-layer and few-layer MoS$_2$ and WS$_2$, *Phys. Rev. B* **84**, 155413 (2011).

60. Y. Zhao *et al.*, Interlayer breathing and shear modes in few-trilayer MoS$_2$ and WSe$_2$, *Nano Lett.* **13**, 1007–1015 (2013).

61. X. Zhang *et al.*, Phonon and Raman scattering of two-dimensional transition metal dichalcogenides from monolayer, multilayer to bulk material, *Chem. Soc. Rev.* **44**, 2757–2785 (2015).

62. C. Lee *et al.*, Anomalous lattice vibrations of single- and few-layer MoS$_2$, *ACS Nano* **4**, 2695–2700 (2010).

63. H. Zeng *et al.*, Low-frequency Raman modes and electronic excitations in atomically thin MoS$_2$ films, *Phys. Rev. B* **86**, 241301 (2012).

64. H. Yu, G.-B. Liu, J. Tang, X. Xu, and W. Yao, Moiré excitons: From programmable quantum emitter arrays to spin-orbit coupled artificial lattices, *Sci. Adv.* **3**, e1701696 (2017).

65. O. Karni *et al.*, Structure of the moiré exciton captured by imaging its electron and hole, *Nature* **603**, 247–252 (2022).

66. K. L. Seyler *et al.*, Signatures of moiré-trapped valley excitons in MoSe$_2$/WSe$_2$ heterobilayers, *Nature* **567**, 66–70 (2019).

67. H. Baek *et al.*, Highly energy-tunable quantum light from moiré-trapped excitons, *Sci. Adv.* **6**, eaba8526 (2020).

68. H. Yu, Y. Wang, Q. Tong, X. Xu, and W. Yao, Anomalous light cones and valley optical selection rules of interlayer excitons in twisted heterobilayers, *Phys. Rev. Lett.* **115**, 187002 (2015).

69. Y. Shimazaki *et al.*, Strongly correlated electrons and hybrid excitons in a moiré heterostructure, *Nature* **580**, 472–477 (2020).

70. Z. Wang *et al.*, Evidence of high-temperature exciton condensation in two-dimensional atomic double layers, *Nature* **574**, 76–80 (2019).

71. F. Wu, T. Lovorn, E. Tutuc, and A. H. MacDonald, Hubbard model physics in transition metal dichalcogenide moiré bands, *Phys. Rev. Lett.* **121**, 026402 (2018).

72. M.-L. Lin *et al.*, Moiré phonons in twisted bilayer MoS_2, *ACS Nano* **12**, 8770–8780 (2018).

73. J. Quan *et al.*, Phonon renormalization in reconstructed MoS_2 moiré superlattices, *Nat. Mater.* **20**, 1100–1105 (2021).

74. J. Jung, A. Raoux, Z. Qiao, and A. H. MacDonald, Ab initio theory of moiré superlattice bands in layered two-dimensional materials, *Phys. Rev. B* **89**, 205414 (2014).

75. R. Bistritzer and A. H. MacDonald, Moiré bands in twisted double-layer graphene, *Proc. Natl. Acad. Sci.* **108**, 12233–12237 (2011).

76. N. Suri, C. Wang, Y. Zhang, and D. Xiao, Chiral Phonons in Moiré Superlattices, *Nano Lett.* **21**, 10026–10031 (2021).

77. X. Liu, R. Peng, Z. Sun, and J. Liu, Moiré phonons in magic-angle twisted bilayer graphene, *Nano Lett.* **22**, 7791–7797 (2022).

78. M. Koperski *et al.*, Single photon emitters in exfoliated WSe_2 structures, *Nat. Nanotechnol.* **10**, 503–506 (2015).

79. G. Grosso *et al.*, Tunable and high-purity room temperature single-photon emission from atomic defects in hexagonal boron nitride, *Nat. Commun.* **8**, 705 (2017).

80. T. T. Tran, K. Bray, M. J. Ford, M. Toth, and I. Aharonovich, Quantum emission from hexagonal boron nitride monolayers, *Nat. Nanotechnol.* **11**, 37–41 (2016).

81. T. Vogl, R. Lecamwasam, B. C. Buchler, Y. Lu, and P. K. Lam, Compact cavity-enhanced single-photon generation with hexagonal boron nitride, *ACS Photon.* **6**, 1955–1962 (2019).

82. X. Liu and M. C. Hersam, 2D materials for quantum information science, *Nat. Rev. Mater.* **4**, 669–684 (2019).

83. T. Low *et al.*, Polaritons in layered two-dimensional materials, *Nat. Mater.* **16**, 182–194 (2017).

84. Y. Wu *et al.*, Manipulating polaritons at the extreme scale in van der Waals materials, *Nat. Rev. Phys.* **4**, 578–594 (2022).

85. E. Yoxall *et al.*, Direct observation of ultraslow hyperbolic polariton propagation with negative phase velocity, *Nat. Photon.* **9**, 674–678 (2015).

86. L. Lu *et al.*, Experimental observation of Weyl points, *Science* **349**, 622 (2015).

87. S. Dai *et al.*, Subdiffractional focusing and guiding of polaritonic rays in a natural hyperbolic material, *Nat. Commun.* **6**, 6963 (2015).

88. J.-S. Wu, D. N. Basov, and M. M. Fogler, Topological insulators are tunable waveguides for hyperbolic polaritons, *Phys. Rev. B* **92**, 205430 (2015).

89. Z. Fei *et al.*, Nano-optical imaging of WSe_2 waveguide modes revealing light-exciton interactions, *Phys. Rev. B* **94**, 081402 (2016).

90. X. Liu *et al.*, Strong light–matter coupling in two-dimensional atomic crystals, *Nat. Photon.* **9**, 30–34 (2015).

91. J. C. Ranasinghe *et al.*, Engineered 2D materials for optical bioimaging and path toward therapy and tissue engineering, *J. Mater. Res.*, 1–25 (2022).
92. J. Wu *et al.*, High-performance waveguide-integrated Bi_2O_2Se photodetector for Si photonic integrated circuits, *ACS Nano* **15**, 15982–15991 (2021).
93. X. Cai *et al.*, Sensitive room-temperature terahertz detection via the photothermoelectric effect in graphene, *Nat. Nanotechnol.* **9**, 814–819 (2014).
94. Y. Cao *et al.*, Ultra — Broadband photodetector for the visible to terahertz range by self-assembling reduced graphene oxide-silicon nanowire array heterojunctions, *Small* **10**, 2345–2351 (2014).
95. S. Yuan, D. Naveh, K. Watanabe, T. Taniguchi, and F. Xia, A wavelength-scale black phosphorus spectrometer, *Nat. Photon.* **15**, 601–607 (2021).
96. Y. Xiong *et al.*, Twisted black phosphorus–based van der Waals stacks for fiber-integrated polarimeters, *Sci. Adv.* **8**, eabo0375 (2022).
97. P. Ma *et al.*, Fast $MoTe_2$ waveguide photodetector with high sensitivity at telecommunication wavelengths, *ACS Photon.* **5**, 1846–1852 (2018).
98. R. Maiti *et al.*, Strain-engineered high-responsivity $MoTe_2$ photodetector for silicon photonic integrated circuits, *Nat. Photon.* **14**, 578–584 (2020).
99. J. Jiang *et al.*, Recent advances in 2D materials for photodetectors, *Adv. Electron. Mater.* **7**, 2001125 (2021).
100. X. Yang *et al.*, A waveguide-integrated two-dimensional light-emitting diode based on p-type WSe_2/n-type CdS nanoribbon heterojunction, *ACS Nano* **16**, 4371–4378 (2022).
101. W. Yang *et al.*, Electrically tunable valley-light emitting diode (vLED) based on CVD-grown monolayer WS_2, *Nano Lett.* **16**, 1560–1567 (2016).
102. F. Withers *et al.*, Light-emitting diodes by band-structure engineering in van der Waals heterostructures, *Nat. Mater.* **14**, 301–306 (2015).
103. M. Polini, Tuning terahertz lasers via graphene plasmons, *Science* **351**, 229–231 (2016).
104. Y.-M. Bahk *et al.* Plasmon enhanced terahertz emission from single layer graphene, *ACS Nano* **8**, 9089–9096 (2014).
105. K. Yang *et al.*, Graphene in mice: Ultrahigh in vivo tumor uptake and efficient photothermal therapy, *Nano Lett.* **10**, 3318–3323 (2010).
106. G. Yang *et al.*, Facile synthesis of black phosphorus–Au nanocomposites for enhanced photothermal cancer therapy and surface-enhanced Raman scattering analysis, *Biomater. sci* **5**, 2048–2055 (2017).
107. H. Lin, X. Wang, L. Yu, Y. Chen, and J. Shi, Two-dimensional ultrathin MXene ceramic nanosheets for photothermal conversion, *Nano Lett.* **17**, 384–391 (2017).
108. H. Lin, S. Gao, C. Dai, Y. Chen, and J. Shi, A two-dimensional biodegradable niobium carbide (MXene) for photothermal tumor eradication in NIR-I and NIR-II biowindows, *J. Am. Chem. Soc.* **139**, 16235–16247 (2017).
109. C. Dai *et al.*, Biocompatible 2D titanium carbide (MXenes) composite nanosheets for pH-responsive MRI-guided tumor hyperthermia, *Chem. Mater.* **29**, 8637–8652 (2017).

110. Y. Jin, J. Wang, H. Ke, S. Wang, and Z. Dai, Graphene oxide modified PLA microcapsules containing gold nanoparticles for ultrasonic/CT bimodal imaging guided photothermal tumor therapy, *Biomaterials* **34**, 4794–4802 (2013).

111. C. Dai *et al.*, Two-dimensional tantalum carbide (MXenes) composite nanosheets for multiple imaging-guided photothermal tumor ablation, *ACS Nano* **11**, 12696–12712 (2017).

112. J. Langer *et al.*, Present and future of surface-enhanced Raman scattering, *ACS Nano* **14**, 28–117 (2019).

113. R. A. Khoury *et al.*, Monitoring the seed-mediated growth of gold nanoparticles using in situ second harmonic generation and extinction spectroscopy, *J. Phys. Chem. C* **122**, 24400–24406 (2018).

114. J. C. Ranasinghe *et al.*, Monitoring the growth dynamics of colloidal gold-silver core-shell nanoparticles using in situ second harmonic generation and extinction spectroscopy, *J. Chem. Phys.* **151**, 224701 (2019).

115. Z. Fan, R. Kanchanapally, and P. C. Ray, Hybrid graphene oxide based ultrasensitive SERS probe for label-free biosensing, *J. Phys. Chem. Lett.* **4**, 3813–3818 (2013).

116. K. Zhang *et al.*, Understanding the excitation wavelength dependence and thermal stability of the SARS-CoV-2 receptor-binding domain using surface-enhanced Raman scattering and machine learning, *ACS Photon.* **9**, 2963–2972 (2022).

117. P. Ranjan, V. Thomas, and P. Kumar, 2D materials as a diagnostic platform for the detection and sensing of the SARS-CoV-2 virus: A bird's-eye view, *J. Mater. Chem. B* **9**, 4608–4619 (2021).

118. C. Ménard-Moyon, A. Bianco, and K. Kalantar-Zadeh, Two-dimensional material-based biosensors for virus detection, *ACS Sens.* **5**, 3739–3769 (2020).

119. Z. Wang *et al.*, Rapid biomarker screening of Alzheimer's disease by interpretable machine learning and graphene-assisted Raman spectroscopy, *ACS Nano* **16**, 6426–6436 (2022).

120. T. Guo *et al.*, Full-scale label-free surface-enhanced Raman scattering analysis of mouse brain using a black phosphorus-based two-dimensional nanoprobe, *Appl. Sci.* **9**, 398 (2019).

121. C. Tapeinos, Graphene — Based nanotechnology in neurodegenerative disorders, *Adv. NanoBiomed Res.* **1**, 2000059 (2021).

122. W. Feng *et al.*, 2D vanadium carbide MXenzyme to alleviate ROS-mediated inflammatory and neurodegenerative diseases, *Nat. Commun.* **12**, 1–16 (2021).

123. B. Ryu, L. Wang, H. Pu, M. K. Y. Chan, and J. Chen, Understanding, discovery, and synthesis of 2D materials enabled by machine learning, *Chem. Soc. Rev.* **51**, 1899–1925 (2022).

124. Y. Jin and K. Yu, A review of optics-based methods for thickness and surface characterization of two-dimensional materials, *J. Phys. D: Appl. Phys.* **54**, 393001 (2021).

125. B. Han *et al.*, Deep-learning-enabled fast optical identification and characterization of 2D materials, *Adv. Mater.* **32**, 2000953 (2020).

126. X. Dong *et al.*, 3D deep learning enables accurate layer mapping of 2D materials, *ACS Nano* **15**, 3139–3151 (2021).

127. X. Dong *et al.*, Deep-learning-based microscopic imagery classification, segmentation, and detection for the identification of 2D semiconductors, *Adv. Theory Simul.* **5**, 2200140 (2022).

128. J. Yang and H. Yao, Automated identification and characterization of two-dimensional materials via machine learning-based processing of optical microscope images, *Extreme Mech. Lett.* **39**, 100771 (2020).

129. G. R. Schleder, C. M. Acosta, and A. Fazzio, Exploring two-dimensional materials thermodynamic stability via machine learning, *ACS Appl. Mater. Interfaces* **12**, 20149–20157 (2020).

130. N. C. Frey *et al.*, Prediction of synthesis of 2D metal carbides and nitrides (MXenes) and their precursors with positive and unlabeled machine learning, *ACS Nano* **13**, 3031–3041 (2019).

131. J. Liang and X. Zhu, Phillips-inspired machine learning for band gap and exciton binding energy prediction, *J. Phys. Chem. Lett.* **10**, 5640–5646 (2019).

132. E. Simsek, Determining optical constants of 2D materials with neural networks from multi-angle reflectometry data, *Mach. Learn. Sci. Technol.* **1**, 01LT01 (2020).

133. Y. Li, Y. Wu, H. Yu, I. Takeuchi, and R. Jaramillo, Deep learning for rapid analysis of spectroscopic ellipsometry data, *Adv. Photon. Res.* **2**, 2100147 (2021).

134. Y. Li, Y. Wu, H. Yu, I. Takeuchi, and R. Jaramillo, High-speed analysis of spectroscopic ellipsometry data using deep learning methods. in *OSA Adv. Photon. Congress 2021*. JW3D.4 (Optica Publishing Group).

135. M. Chen *et al.*, 2D materials: Excellent substrates for surface-enhanced Raman scattering (SERS) in chemical sensing and biosensing, *TrAC Trends Anal. Chem.* **130**, 115983 (2020).

136. F. Cui, Y. Yue, Y. Zhang, Z. Zhang, and H. S. Zhou, Advancing biosensors with machine learning, *ACS Sens.* **5**, 3346–3364 (2020).

Chapter 2

Optothermal Marangoni Effect: Phenomena and Applications[*]

Andrzej Miniewicz[†,§], Stanisław Bartkiewicz[†], Monika Bełej[†],
Katarzyna Grześkiewicz[†], and Michalina Ślemp[†,‡]

[†]*Institute of Advanced Materials, Faculty of Chemistry,
Wroclaw University of Science and Technology,
Wybrzeże Wyspiańskiego 27, 50-370 Wrocław, Poland*
[‡]*NonLinear Optical and Interfaces Group (ONLI),
Institut Lumière Matière ILM, UMR CNRS 5306,
Université Claude Bernard Lyon 1, 10 Rue Ada Byron,
69622 Villeurbanne cedex, France*
[§]*andrzej.miniewicz@pwr.edu.pl*

Marangoni effect is ubiquitous in nature and plays a vital role in the
interface between two or more liquid phases. Any change of surface tension
due to concentration and/or thermal gradients is accompanied by directed
liquid flows. Local thermal gradients can be remotely induced by laser light,
just opening the field to many interesting phenomena that have recently
found novel and exciting applications in optofluidics and nanotechnology.
We describe several laser-induced thermocapillary effects and explain them
with the help of numerical simulations in 2D and 3D using the COMSOL
Multiphysics programming platform. We report on optothermal Marangoni
effect in thin layers of dyesolution enabling efficient light absorption and
heat production. The differently designed experiments allowed for direct
visualization of Marangoni flows at the interface causing its bending around
gas bubbles, allowing for their trapping, actuating droplets by laser light,
enriching content of solute in a droplet confined inside a gas bubble, unique
crystal growth inside a bubble, etc. We describe the method of laser light-
aided manipulation of Marangoni swimmers and rotors, surface deformation
due to Marangoni effect, and formation of novel type hydrodynamic trapping.
The experiments and simulation methodology described in this chapter come
from our laboratory, however, the literature data on various aspects of

---

[*]This chapter is dedicated to Professor Juliusz Sworakowski on his 80th birthday.

Marangoni effect are very rich, as this phenomenon can be used in metallurgy, crystal growth, surface cleaning, micro- and nanoparticles translation and rotation, lithography, trapping, and for various purposes in biology.

1. Introduction

Marangoni phenomenon is observed at the interface between two phases, mainly in fluids (liquids, soft matter, molten metals, etc.) and manifests itself as near interface liquid flows from the areas of lower to higher values of surface tension.[1] The necessary condition for occurrence of the Marangoni effect is therefore a presence of surface tension differences at a given liquid–gas or liquid–liquid interface,[2] i.e., gradients of surface tension $\nabla\gamma$. Generally, these gradients can arise either via change in solute concentration, surfactant concentration at free liquid surface, or via generation of temperature gradients. The former case is known as *solutocapillary Marangoni* and the latter, as *thermocapillary Marangoni effect*.[3] In fact, any local non-uniformity in interface temperature, chemical composition, presence of light, electric or magnetic field modulation, evaporation in binary mixtures, or unbalanced distribution of colloidal particles can stimulate the appearance of surface tension gradients. These gradients are relaxed by mass flow, heat flow, and surfactants redistribution and usually take place simultaneously. The produced convective motions are relatively strong, allowing their use in various applications, mostly in micro-scale events, but not limited to this, and are also observed in nano- and macroscales.[4–8] The shear stress at the surface due to viscosity is transferred into the liquid bulk and usually a fluid mixing is observed.[9–11]

The main reason to study different aspects of Marangoni effect is the aspiration to deeply understand the mechanisms of fluid convection in liquids under conditions of strong temperature or concentration gradients. By controlling them, one can perform translocation of small objects (Marangoni swimmers), induce their rotation (Marangoni rotors), deposit nanoparticles on a substrate (bubble pen lithography), control crystal growth by laser light, produce and steer gas bubbles in various microfluidic systems, initiate chemical reactions in nanoliter volumes, etc.

The goal of this chapter is to introduce readers into the world of thermocapillary Marangoni effect launched by laser irradiation of different liquids or interfaces between them. The use of laser light opens new and exciting possibilities in studying thermocapillary Marangoni effect due to easily controlled local heating and addressing the position of laser light under microscope, as well as its frequency, intensity modulation, and incident angle. Due to broad use of laser light in the experiments for heating purposes the optothermal Marangoni effect term is frequently used. Optothermal Marangoni effect can be treated as a tool for remote control over small objects' movements at the liquid surface and bulk, like nanocrystals, nanoparticles, or formation of micro assemblies of crystallites of a given geometry. Marangoni effect-assisted gas bubble trapping or hydrodynamic small objects trapping were already demonstrated in the literature. All the above-mentioned phenomena fall within an emerging field of optofluidics that involves the use of optical devices and light to induce and detect flowing media and the use of fluids to modify optical properties in devices. Ultimately, its value is highly dependent on the successful integration of photonic integrated circuits with microfluidic or nanofluidic systems.

In this chapter, the focus is directed toward our own experience of the study of optothermal and related Marangoni effects. Presenting several examples of these exciting phenomena, we would like to show their beauty, complexity, and application perspectives. We are exploiting Computer Fluid Dynamics (CFD) simulation approach for better understanding of the underlying complex physics. Giving a description of such complex and coupled physical mechanisms of heat and mass transport is a challenge and cannot be done so analytically. Furthermore, the phenomena are located at the edge of hydrodynamics, heat transport, thermodynamics, light–matter interaction, and cohesive forces of involved liquids and phases. They require solving time-dependent coupled differential equations in three dimensions (3D) with Marangoni effect defined at the interface. Thus, we used programming and a simulation platform designed for solving complex physical systems, COMSOL Multiphysics.[12] Together with presentation of our own results, the state of the art of Marangoni-related phenomena and applications known in literature will be presented and discussed when appropriate.

2. Physics of Marangoni Effect

2.1. *Brief history of Marangoni effect and its impact*

J. Thomson in 1855 described interesting behavior of droplets of alcohol climbing up and down over a glass vessel, a phenomenon known today as "tears of wine".[13] Ten years later Italian physicist Carlo Giuseppe

Marangoni (1840–1925) explained this phenomenon. Marangoni had found that surface tension along a liquid–gas interface may vary due to temperature or chemical composition gradients.[14,15] These gradients generate tangential forces that produce flows in liquids known as Marangoni forces or Marangoni stresses. In this phenomenon, evaporation of alcohol creates a gradient of surface tension and Marangoni stresses force the wine to stream up the wall of the glass against gravity. At some point when gravitational forces start to dominate over surface tension forces "tears of wine" are released flowing down the glass. The consequences of this discovery are extremely wide and the phenomenon is still discussed in the literature.[16] Nowadays the Marangoni effect is observed in various fluid flow processes including metals,[17,18] polymers,[19,21] semiconductors,[22] biological materials,[23–25] and surface chemistry.[26,27] It is employed in industrial crystal growth, in semiconductor industry, electron beam melting where large thermal gradients are generated as well as in welding processes. However, recently its importance has been shifted toward nanotechnology, optofluidics, and remote translational or rotational movement of small objects.[28,29]

2.2. *Surface tension and Marangoni effects*

The surface tension of liquids denoted as γ can be observed in the spherical shape of small drops of liquids and in many other properties of wetting phenomena. It is believed that it depends upon the attraction forces between the molecules constituting the liquid.[30] Almost equal attraction forces inside the liquid volume compensate each other, but not at the surface. At the surface, molecules are attracted toward the liquid bulk. The energy required to remove the surface layer of molecules in a unit area defines the surface tension γ [J$\cdot$m^{-2}]. It is also a measure of forces acting in the surface plane tending to minimize its area and therefore γ is alternatively expressed in units [N$\cdot$m^{-1}]. Phenomenological correlation of surface tension, γ, with bulk properties of liquids, has been proposed in 1886 by the equation known as Eötvös rule[31]:

$$\gamma(T) \cdot V_m^{2/3} = k \cdot (T - T_c), \tag{1}$$

where T_c is the critical temperature of the liquid, V_m is the molar volume, and the constant $k \approx 2.1 \cdot 10^{-7}$ J$\cdot$K$^{-1}\cdot$mol$^{-2/3}$. This equation has been modified by several research groups (cf. Ref. 32, and references therein) tending to correlate surface tension with molar Gibbs surface energies Δg, surface entropies Δs, and surface enthalpies Δh. Such an approach can better describe microscopic molecular configurations at the surface and their influence on the surface tension values.

An increase in temperature lowers the average force of attraction among the molecules, and in the majority of liquids, in which cohesive forces are described by van der Waals potentials, results in a decrease in surface tension. These potentials weakly depend on temperature and therefore surface tension $\gamma(T)$ decreases almost linearly with temperature increase[33]:

$$\gamma(T) = \gamma(T_0) + \gamma_T \cdot (T - T_0), \tag{2}$$

where $\gamma_T = \frac{\partial \gamma}{\partial T}$ is the temperature coefficient of surface tension and $\gamma(T_0)$ is the surface tension at the reference temperature T_0. For the majority of liquids, the γ_T coefficients are negative. For liquids correctly described with London dispersion forces (e.g., organic liquids) surface tension values are similar[33] ranging from 10–100 mN$\cdot$m^{-1}. However, the presence of any other interactions in liquids comprising polar interactions, like Coulomb interactions, dipole–dipole, H-bonds, charge transfer (CT), or magnetic, metallic, and any type of interactions tending to molecular or particle self-organization may produce deviation from the linear dependence predicted by Equation (2). Strong deviations of surface tension and its temperature coefficient from typical values were observed for some liquid metal alloys,[34] and liquid crystals[35] close to their isotropic-to-nematic phase transition temperatures. The classic behavior can also severely be influenced by the type of interface between the liquid and other phases: gaseous, liquid, or solid. The surface tension can be dramatically changed by the presence of surfactants at the surface. Several methods of surface tension measurements have been elaborated, including: (i) capillary rise method, (ii) stalagmometer method — drop weight method, (iii) Wilhelmy plate or ring method, (iv) maximum bulk pressure method, (v) methods analyzing shape of the sessile or hanging liquid drop or gas bubble, (vi) contact angle measurements, and (vii) dynamic methods. F. M. Fowkes and his followers came to the conclusion that surface tension of a liquid or solid can be expressed as a sum of several components[36]:

$$\gamma = \gamma^d + \gamma^p + \gamma^i + \gamma^H + \gamma^\pi + \gamma^{DA} + \gamma^e, \tag{3}$$

where the superscripts denote the following interactions: d — dispersion interaction, p — dipole–dipole interaction, i — dipole-induced dipole interactions, H — hydrogen bond, $\pi - \pi$ electron interactions, DA — donor-acceptor, and e — electrostatic interactions. Depending on the particular case, some of the above interactions may not be present or some additional should be included to correctly predict the surface tension of the free surface and surface tension between two or three phases. Interfacial energy/tension between two liquid phases diminishes if more interactions are possible between the phases.

In many practical situations, the infinitesimal change in the surface tension might come from thermal and composition effects:

$$d\gamma = \frac{\partial \gamma}{\partial T}dT + \frac{\partial \gamma}{\partial c}dc, \tag{4}$$

where the first term is responsible for the temperature T induced change of γ while the second, for the concentration c one. The coefficient before dc in Equation (4) describes the concentration dependence of the surface tension and is frequently denoted as $\gamma_c = \frac{d\gamma}{dc}$. This, frequently met in practice, dependence opens a new door for light–matter interaction with liquids. The most commonly used substances influencing surface tension of liquids are surfactants such as amphiphilic (cationic, anionic, or neutral) compounds that lower the surface tension between two phases, e.g., CTAB (hexadecyl trimethyl ammonium bromide, cationic), SDS (sodium dodecyl sulphate, anionic), or alkyl phenol ethoxylate (non-ionic).[37] For example, if the surface of water is covered with the photochromic surfactant (e.g., azobenzene derivative molecule), both thermocapillary Marangoni effect as well as solutocapillary one can be observed upon localized laser light absorption. Absorbed photons can increase solution surface temperature but also change the surface tension by inducing local change in the initial concentration of azobenzene *trans* and *cis* photoisomers. It has been noticed that different light wavelengths of used light can reversibly transform surface tension due to its dependence on molecular conformation *trans* or *cis*. The movement of sessile liquid droplet illuminated on both sides with different color beams has been observed over liquid or solid substrates.[38,39] In the literature, this effect has been named the *chromocapillary Marangoni effect*.[40]

The thermocapillary Marangoni effect can be characterized by Marangoni number *Ma*, a non-dimensional parameter that compares advective heat transport rate due to surface tension gradient and diffusive heat transport rate initiated by heating source[41]:

$$Ma = \frac{|\gamma_T|L\Delta T}{\mu\kappa}, \tag{5}$$

where L is the characteristic length of the system, ΔT is the maximum temperature difference in the system, μ is the dynamic viscosity in $[\mathrm{N} \cdot \mathrm{s}^{-1} \cdot \mathrm{m}^{-2}]$, and κ is the coefficient of thermal diffusion in $[\mathrm{m}^2 \cdot \mathrm{s}^{-1}]$.

When both solutocapillary and thermocapillary effects are present, the Marangoni number includes both the heat transfer and concentration transfer rates and compares these motive forces to damping ones[42]:

$$Ma_c = -\frac{\frac{d\sigma}{dc} \cdot \frac{dT}{dL} \cdot L^2}{D\mu}, \tag{6}$$

where D is diffusion coefficient, L is thickness of a layer. In the geometries different from the classic, Marangoni number depends greatly on the system geometry and L is the characteristic length.

In order to address different phenomena and scales, several non-dimensional numbers were defined, such as capillary number Ca indicating the ratio of viscosity to capillarity, Bond number Bo describing the gravity to capillarity ratio, Biot number Bi referring to the ratio of the convective heat transfer from liquids to solids, Prandtl number Pr indicating viscous diffusion rate to thermal diffusion rate, and many others.[43,44] In a majority of the phenomena observed in the microfluidics domain, capillarity and viscosity are the dominant factors, however, other system properties (e.g., evaporation, phase separation, etc.) should not be neglected in many cases.[28] For instance, Reynolds number Re that corresponds to the ratio of inertial forces to viscous forces is used for the flow patterns forecast in fluid mechanics:

$$\mathrm{Re} = \sqrt{Ma/Pr}, \tag{7}$$

where $Pr = \nu/\kappa$, with ν denoting kinematic viscosity. The laminar flow is expected for low Re, while turbulent flow will be formed for large Re.

2.3. *Navier–Stokes and heat transfer equations*

Marangoni flows induced at the surface produce the surface shear stress $\vec{\tau}_s$ in $[\mathrm{N} \cdot \mathrm{m}^{-2}]$:

$$\vec{\tau}_s = \mu \cdot \left. \frac{d\vec{u}}{dz} \right|_{z=0}, \tag{8}$$

where $\vec{u}$ is the surface flow velocity and the z-axis is normal to the free surface.

Motion in fluids is governed by the Navier–Stokes equations that describe conservation of momentum by Newton's second law of motion for fluids[45]:

$$\rho\left(\frac{\vartheta\vec{u}}{\vartheta t} + \vec{u}\cdot\vec{\nabla}\vec{u}\right) = -\vec{\nabla}p + \vec{\nabla}\cdot(\mu[(\vec{\nabla}\vec{u}) + (\vec{\nabla}\vec{u})^T] - \frac{2}{3}\mu(\vec{\nabla}\cdot\vec{u})\boldsymbol{I}) + \vec{F},$$

$$(9)$$

where ρ is fluid density $[\mathrm{kg\cdot m^{-3}}]$, $\vec{u}$ is the fluid velocity $[\mathrm{m\cdot s^{-1}}]$, p is the fluid pressure $[\mathrm{N\cdot m^{-2}}]$, $\boldsymbol{I}$ is the unitary tensor, and the superscript T connected to velocity gradient describes a component tangential to the surface. The term on the left-hand side of Equation (9) corresponds to inertial forces, then the first term on the right-hand side represents pressure forces, second, viscous forces, and the last, external forces applied to the fluid. These equations are solved together with the continuity equation representing conservation of mass:

$$\frac{\partial\rho}{\partial t} + \vec{\nabla}\cdot(\rho\vec{u}) = 0.$$

$$(10)$$

The heat transfer, from an external source Q, into liquid may be included by solving the following equation[46]:

$$\rho C_p\frac{\partial T}{\partial t} + \rho C_p\vec{u}\cdot\vec{\nabla}T = \vec{\nabla}\cdot(k\cdot\vec{\nabla}T) + Q,$$

$$(11)$$

where T is the temperature in $[\mathrm{K}]$, C_p is the heat capacity in $[\mathrm{J\cdot kg^{-1}\cdot K^{-1}}]$, k denotes the thermal conductivity of the fluid in $[\mathrm{W\cdot m^{-1}\ K^{-1}}]$, ρ is the liquid density in $[\mathrm{kg\cdot m^{-3}}]$, and Q is the heat produced by laser light in $[\mathrm{W\cdot m^{-3}}]$.

Laser heating is preferred among the other possible heat sources, because the large localized and dynamic (using continuous or pulsed laser light) temperature gradients can be realized that boost the thermocapillary convective flows. Laser heating of a liquid layer of thickness h by a collimated Gaussian laser beam of a power P_{laser} and radial beam extension ω_0 can be approximated by the expression[47]:

$$Q_{\mathrm{laser}}(r, z) = \frac{2P_{\mathrm{laser}}\alpha}{\pi\omega_0^2}\exp\left(\frac{-2r^2}{\omega_0^2} - \alpha(h - z)\right),$$

$$(12)$$

where α is absorption coefficient of a liquid at a given wavelength in $\mathrm{m^{-1}}$, r is the radial coordinate, and z is the axial coordinate. Depending on the value of the absorption coefficient α, the expression describes the exponential decrease in heat delivered to the liquid volume. The gradients can be easily enlarged by focusing the beam. Depending on the configuration of an experiment, the laser beam can be directed from above the liquid surface, from the bottom, and at oblique incidences. The important factor for producing thermocapillary flows is to adjust laser power and liquid absorption coefficient (alternatively heated objects' properties, e.g., Marangoni surfers, plasmonic substrates, etc.) to obtain large gradients of temperature without extensive heating of the whole liquid volume. The boundary conditions and different regimes of the applicability of the above-presented equations to the studied liquids, volumes, and thermal conditions must be applied to obtain the agreement between simulations and experiments. All the above-presented equations comprise the physical foundations of a majority of Marangoni experiments presented in this chapter.

3. Numerical Simulations and Experiments on Marangoni Effect in 2D

The complexity of the equations describing Marangoni thermocapillary effects coupling derivatives in time and space leads naturally to the extensive use of numerical tools. Numerical solutions of equations verified by experiments constitute a synergy in solving and understanding complex situations met in real experiments. The predictive power of simulations coupled with the ease of parameter sweeping is a great advantage in experimental work. Results frequently help to determine the best experimental conditions and to select the best materials suitable for obtaining the desired functionalities.

Full three-dimensional (3D) simulations are time consuming, therefore for simplification the axis-symmetric (applied when symmetry of the system allows for two-dimensional (2D) solutions that are next revolved around the symmetry axis to 3D view) or 2D simulations are used. They usually serve as a beginning step before undertaking 3D simulations. All the simulations presented in this chapter were done within the COMSOL Multiphysics platform.[12]

3.1. *Simple 2D simulation of thermocapillary Marangoni effect*

Two-dimensional simulations of Marangoni thermocapillary effect are performed in a two-dimensional geometry (x, y) assuming that the liquid is forming a uniform thin layer. This considerably reduces the time necessary to calculate time-dependent coupled Navier–Stokes and thermal transfer equations especially when the mesh is dense. The thermocapillary Marangoni effect force is introduced as a weak contribution in the laminar flow module for a given interface. The Marangoni tangential stress $\vec{\tau}_s$ is balanced by bulk viscous stress, hence in the adjacent to interface phase momentum transfer occurs:

$$\rho \nu \vec{n} \nabla \vec{u} = \vec{\tau}_s \tag{13}$$

giving rise to liquid flows in the whole simulated area. In Equation (13), ν is the kinematic viscosity coefficient. Results of simulations depend on the specific boundary conditions applied to the system. Exemplary conditions are:

- No-slip or slip conditions can be set at the given liquid boundaries.
- *Initial values for velocities*: $|\vec{u}| = 0\,m \cdot s^{-1}$ and pressure difference $\Delta p = 0$ atm.
- Thermal insulation can be assumed on all the liquid boundaries.
- The Marangoni effect overdrives the no-slip conditions.
- *The initial parameters*: reference temperature $T_0 = 293.15\,\mathrm{K}$, pressure $p_0 = 1$ atm.
- The heat can be released from the beam center positioned at (x_c, y_c) in the form of excess temperature ΔT_{exc}.

In Figure 1, we show exemplary results of numerical simulations performed for 1,4-dioxane where the maps of velocity and pressure demonstrate the main features of thermocapillary effect, i.e., surface escape of fluid from the heated region, formation of vortices in the liquid volume, and generation of negative pressure at the central part of the heated region.

Parameters of 1,4-dioxane that have been used in simulations are listed in Table 1.

Fast liquid flow velocity reaching at maximum $0.6\,\mathrm{m} \cdot \mathrm{s}^{-1}$ (see Figure 1(b)) is observed in this system. Prandtl number Pr that characterizes the ratio between molecular diffusivity of momentum and molecular diffusivity of heat is given be the following equation:

$$Pr = \frac{\mu \cdot C_p}{k}. \tag{14}$$

and calculated for 1,4-dioxane amounts to $Pr = 14$. For 1,4-dioxane, heat diffuses slowly relative to momentum diffusion. Typically, Prandtl number

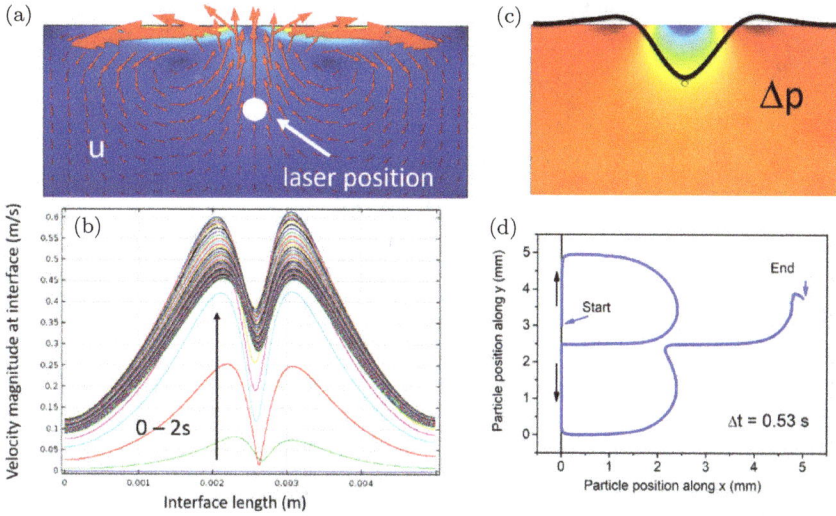

Figure 1. Main features of Marangoni capillary effect in 2D Newtonian liquid heated by laser light, represented here by an excess temperature of $\Delta T_{exc} = 10$ K. (a) Red arrows show proportional velocity vectors together with colors, blue — the lowest velocity, red — the highest. (b) Evolution with time of the interface velocity within the time range of 0 to 2 s. (c) Pressure difference Δp [Pa] induced in the liquid layer due to laser heating. The negative pressure is denoted by blue color and overpressure, by red one. The black line suggests possible interface deformation due to generated pressure differences. (d) Simulation of particle with mass trajectory. Particle was injected at t = 1 s and position marked with an arrow (Start), terminating (End) at t = 1.53 s.

Table 1. Substantial 1,4-dioxane parameters used in COMSOL Multiphysics simulations.[48]

Parameter	Description	Value
$\gamma(T)$	Surface tension	$33 \cdot 10^{-3}$ N $\cdot$ m^{-1} (at 293 K)
		$28.83 \cdot 10^{-3}$ N $\cdot$ m^{-1} (at 323 K)
$\gamma_T = \frac{\partial \gamma}{\partial T}$	Temperature coefficient of surface tension	$-0.1391 \cdot 10^{-3}$ N $\cdot$ m$^{-1} \cdot$ K^{-1}
ρ	Liquid density	1033 kg $\cdot$ m^{-3}
μ	Dynamic viscosity coefficient	$1.3125 \cdot 10^{-3}$ Pa $\cdot$ s (at 293,15 K)
		$0.9603 \cdot 10^{-3}$ Pa $\cdot$ s (at 313,15 K)
$\nu = \frac{\mu}{\rho}$	Kinematic viscosity coefficient	$1.139 \cdot 10^{-6}$ m$^2 \cdot$ s^{-1} (at 298 K)
k	Thermal conductivity coefficient	0.159 W $\cdot$ m$^{-1} \cdot$ K^{-1} (at 298 K)
		0.147 W $\cdot$ m$^{-1} \cdot$ K^{-1} (at 323 K)
$\alpha_T = \frac{k}{\rho C_p}$	Thermal diffusion coefficient	$8.94 \cdot 10^{-6}$ m$^2 \cdot$ s^{-1} (at 298 K)
C_p	Heat capacity coefficient at constant pressure	1700 J $\cdot$ kg$^{-1} \cdot$ K^{-1}
α	Heat expansion coefficient	$1.03 \cdot 10^{-3}$ K^{-1}

$Pr < 1$ for liquid metals (0.004–0.03) and gasses (0.7–1), (1.7–13.7) for water, and (5–50) for light organic fluids, and Pr takes much large values for oils and glycerin $Pr \gg 50$.

3.2. 2D Marangoni effects viewed experimentally

Most of the experiments presented in this chapter were performed in solution of para-nitroaniline ($C_6H_6N_2O_2$) dissolved in an organic solvent 1,4-dioxane ($C_4H_8O_2$). The molecular structures of these compounds are presented in Scheme 1. The solution was either sandwiched between two glass plates separated by spacers or deposited directly on a glass plate, forming a layer. For excitation the focused continuous wave (cw) laser operating at wavelength 405 nm was used.

This system has been found to be a perfect one, demonstrating the main features of 2D thermocapillary Marangoni effect shown in Fig. 1.[49,50] In work by Miniewicz et al.,[50] Marangoni effect was visualized by tracers (carbon particles) introduced to the solution. The observed bending of the interface could be easily observed, as illustrated in Fig. 2.

Frame coloration technique prepared from a movie with frame rate of 1,200 fps allows for measurement of particles' velocities, and results reported in Ref. 50 gave value for $|u_{max}| \approx 90 \, \text{mm} \cdot \text{s}^{-1}$, and in Ref. 48, $|u_{max}| \approx 60 \, \text{mm} \cdot \text{s}^{-1}$. Numerical experiments gave larger values for fluid flow velocity. The discrepancy between experiment and simulations can be rationalized if one takes into account the proximity of two glass plates which seriously limit the fluid velocity and the fact that tracers are slowed down by Stokes drag force. This effect has not been addressed by us in 2D simulations.

Moving the laser beam closer to the interface results in a jump of the interface bending toward the laser spot position. This stabilizes the interface position with respect to the heat source. Gentle movement of the laser in (x, y) plane, into the interior of the liquid layer, allowed for increasing of

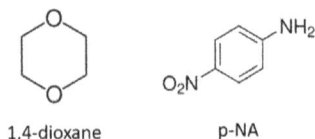

Scheme 1. Molecular structures for solvent 1,4-dioxane and solute para-nitroaniline (p-NA).

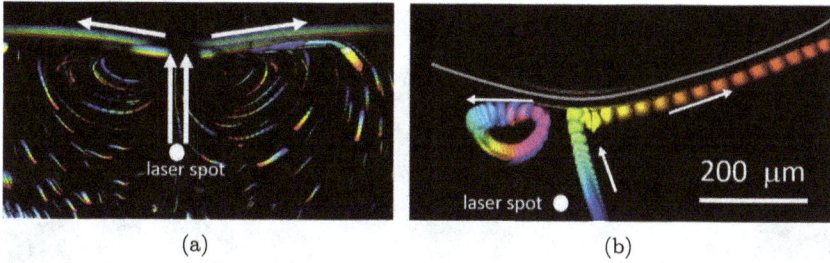

Figure 2. Two-dimensional laser-induced thermocapillary effect measured in 1,4 dioxane with dissolved para-nitroaniline thin layer confined between two glass plates. Soot particles are added as traces allowing observation of fluid flows. Frame coloration technique has been used. (a) Vortices of fluid flow are clearly seen on both sides. The inward bending in the direction of laser spot position (white circle) is clearly visible. (b) Left side carbon particle makes a whirling route and the right particle moves close to the interphase line with average velocity of $60 \, \text{mm} \cdot \text{s}^{-1}$. A total of 46 movie frames, each 0.833 ms, were colored with rainbow sequence.

interface bending until, at some critical distance, the liquid collapses and the air bubble is formed. In Fig. 3, a composite figure showing interface bending and the formation of a gas bubble is presented.

3.3. *Creation and manipulation of gas bubbles with optothermal Marangoni effect*

Different mechanisms and applications of light-driven bubbles were the subject of several reviews and articles.[29,51–54] Nowadays the microbubbles are used in biology,[55,56] nanolithography,[57,58] bubble printing,[59,60] sensing,[61,62] etc.

Gas (vapor) bubble suspended in a liquid undergoes the Young–Laplace equation representing a mechanical equilibrium condition between two fluids separated by an interface, as follows:

$$p_{\text{gas}} - p_{\text{liq}} = \gamma \left(\frac{1}{r_1} + \frac{1}{r_2} \right), \tag{15}$$

where p_{gas} and p_{liq} are the pressures of gas (inside) and liquid (outside) on the interface, γ is surface tension of the interface, and r_1 and r_2 are the radii defining the curvature of a two-dimensional surface. In the case of the sphere-like section of the bubble, the above equation reduces to

$$p_{\text{gas}} - p_{\text{liq}} = \frac{2\gamma}{R_b}, \tag{16}$$

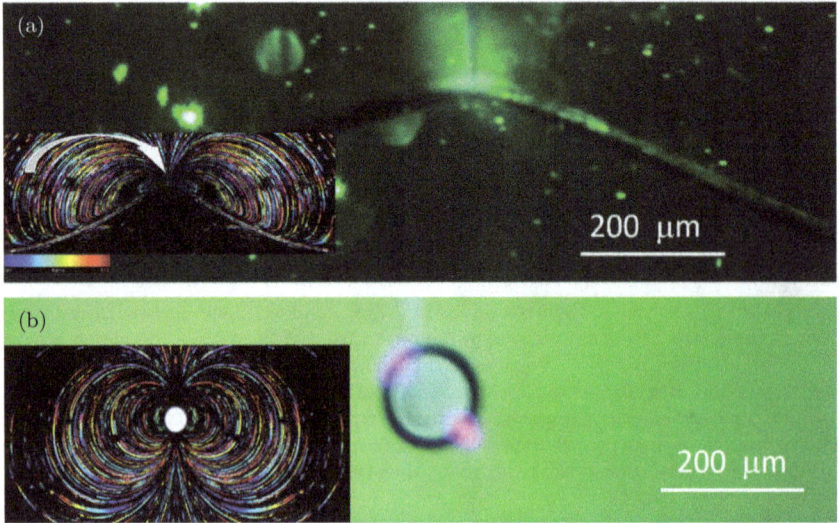

Figure 3. Photographs taken from the experiment of 1,4-dioxane with para-nitroaniline confined between two glass plates and irradiated with 532 nm cw laser. Observations were made under polarizing optical microscope. (a) Thermocapillary Marangoni effect induced interface bending viewed under crossed polarizer, enabling to see the tracer movements. Inset shows the trajectories of traces (streamlines) with vortices. Laser spot touches the interface. (b) Movement of laser into the layer interior results in detachment of a cylindrical air bubble. This bubble is shown with non-crossed polarizers and a trajectory of tracer around it is pasted as inset to this photograph. Remark: Insets showing the flows via trajectories of carbon particles where symmetrized. In reality, this symmetry is frequently broken.

where R_b is the bubble radius. The Young–Laplace equation can predict the droplet and bubble shapes. A more elaborate Young–Laplace equation valid for describing the dynamics of a spherical bubble in an infinite volume of incompressible fluid is the Rayleigh–Plesset equation.[63,64]

By making a very thin layer of liquid ($d \sim 70\,\mu m$) with dissolved organic substances, it is possible to create, transport, and release gas bubbles, in the form of approximate cylinders, in a desired place using suitable laser beams. We demonstrate this using 1,4-dioxane with para-nitroaniline in Fig. 4. In this case, one may say that the bubble is hydrodynamically trapped by the heat source (laser beam intensity $210\,W \cdot cm^{-2}$) due to thermocapillary Marangoni effect operating at the edge of the gas bubble. In Fig. 4, the above-described process is shown in a sequence of nine photographs taken from the movie.

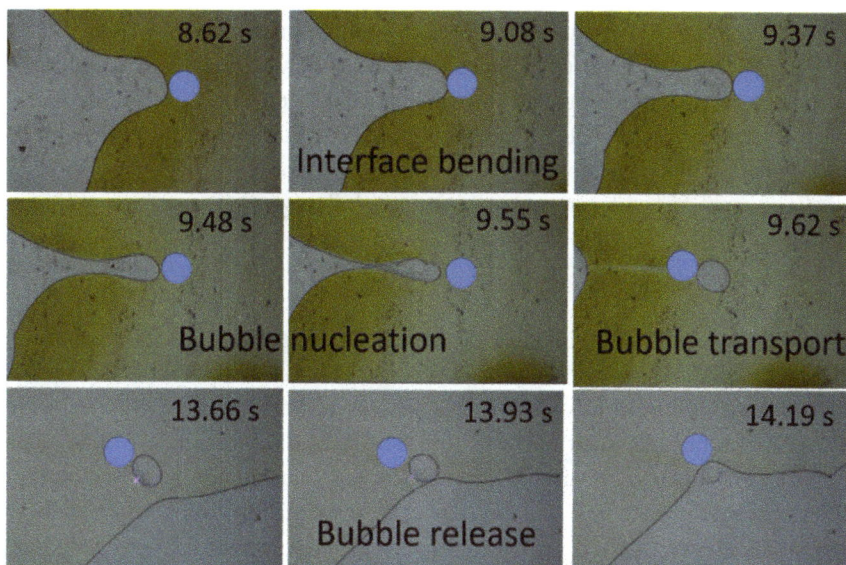

Figure 4. Process of laser-induced generation, transport, and release of gas bubble in 1,4-dioxane with para-nitroaniline. Times given for each photograph come from the real-time movie. A blue circle is added to each photograph to indicate the position of a laser beam spot, which is shifted using translation of the microscope table on which the sample is mounted. The microscope objective used in this case has 10× magnification.

We started with the creation of large interfacial deformation of liquid, the large deformation is easier for very thin layers and at places with high enough concentration of absorbing dye in the solution (see the yellowish color in Fig. 4). By moving the laser beam spot from the edge of the liquid finely, interface bending can be made so large that the liquid collapses, leaving gas bubbles trapped by the beam. The capillary forces evidently help to achieve this formation of bubbles. Then, the slow movement of laser beam spot can translate the bubbles to another destination, for example, to the other side of the liquid layer (see Fig. 4). The investigated solution confined between two glass plates can be treated as a 2D membrane separating two different gases. In principle, one can use two controlled laser beams to generate bubbles on both sides of the liquid membrane, trap them, and next join them in the middle to perform the specific ~300 picoliter gaseous reaction. Alternatively, the system allows for demonstration of Marangoni bubble trapping, which cannot be accomplished using light gradient force in optical tweezers experiments[65] as in this case the gradient force will be positive, i.e., pushing the gas bubble out of focus. However, thermal

trapping could be possible.[66] The idea was experimentally verified by us
in Ref. 48 and the theoretical considerations show that the trapping force
comes from the under-pressure appearing at the edge of the bubble when
heat from the laser reaches the bubble. The driving force is the difference
between the Marangoni force F_M and the drag force F_d, the total force is
directed toward the laser spot position, as follows:

$$|\vec{F}_{\text{total}}(r)| = \vec{F}_M - \vec{F}_d = |\vec{p}_u|(1 - \cos\theta(r)) \cdot \pi \cdot R_b^2 - 4\pi\mu R_b u(r), \quad (17)$$

where $\vec{p}_u$ represents the integral of pressure distribution around the bubble
being at the distance r from the position of the laser beam, R_b is the
cylindrical bubble radius, and $u(r)$ is the local value of fluid flow velocity
toward the bubble position (see Ref. [48] for more details). Experiment
scheme and results of simple simulations of thermocapillary Marangoni
effect acting on a cylindrical bubble in a layer of liquid heated by a laser
beam are shown in Fig. 5.

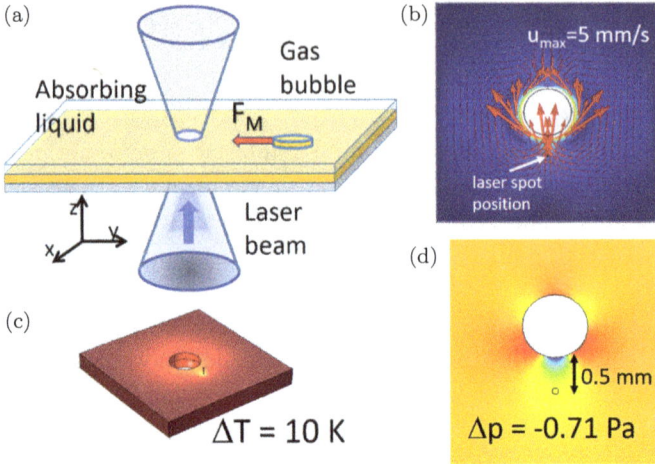

Figure 5.   Experimental setup and illustration of the results of simulation of thermocapillary Marangoni effect acting on a bubble of gas confined between two glass plates and
filled with liquid absorbed laser light. (a) Scheme of an experiment. (b) Top view cross-
section of fluid flow velocity marked by arrows whose length is proportional to the local
velocity values. (c) Side view of position of laser beam delivering heat and its convection
around the bubble. (d) Cross-section of pressure difference field that is responsible for
the attraction force acting on a bubble and finally trapping it at the laser spot position.
The force acts toward the position of laser spot shown by a circle at a distance of 0.5 mm
from the bubble edge.

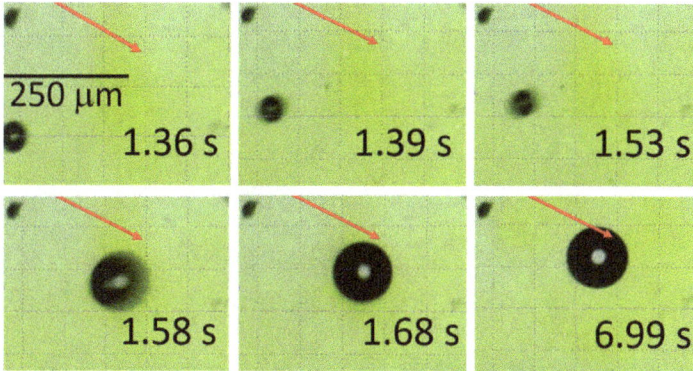

Figure 6.   The optothermal Marangoni trapping of a gas bubble viewed by a sequence of events observed under an optical microscope. Red arrow shows the position of a laser spot of a diameter of $4\,\mu m$ (not visible here due to filtering of the light of laser). Within a time of $320\,ms$, the bubble travels a distance of about $300\,\mu m$ ($u_{av} \approx 0.937\,mm \cdot s^{-1}$) and next it is trapped at the laser beam position, staying there for 5 s. When trapped, the bubble can be translated to a desired destination with laser via movement of the microscope stage.

Fluid flows form a kind of a clamp around a gas bubble (see Fig. 5(b)). The bubble moves toward the laser spot position due to hydrodynamic flow generated by Marangoni effect, resulting in an asymmetric pressure difference $\Delta p = -0.71$ Pa (see Fig. 5(d)). Marangoni force acting on a bubble is at the maximum when laser spot touches the gas-fluid interface, then $\Delta p_{max} = -6$ Pa, for similar other simulation parameters. In Fig. 6, we show the Marangoni trapping of a gas bubble in 2D case (the absorbing liquid layer is confined between two glass plates).

It should be noted that the trapping distance is larger than $300\,\mu m$ where there is no intensity of laser light. So, the effect is purely optothermal. In fact, from the theory of optical trapping of dielectric particles[65] it is evident that the trapping of a particle with refractive index lower than refractive index of surrounding liquid is not possible because the gradient force becomes positive, pushing the object out of focus.

Experiments similar to ours were already described in the literature. Study of optically actuated thermocapillary movements of microbubbles was reported in a silicone oil, but laser light was absorbed in a thin film of amorphous silicon.[53] The 3D trapping of gas bubbles was reported by R. Ramos-Garcia group,[67] where 5 ns pulsed laser with repetition frequency of $10\,kHz$ coupled to optical fibers was used to generate bubbles via light absorption on Ag nanoparticles photo-deposited at the distal end of a multimode optical fiber. Authors obtained quasi-stable 3D trapping of bubbles. Later the same group reported on the 3D trapping and manipulation of 100 μm bubbles[68] and explained it on the grounds of Marangoni optothermal effect.

4. Experiments on Marangoni Effect in 3D

4.1. *Droplets and crystal growth in gas bubbles*

One of the most spectacular new Marangoni-driven phenomena is related to spontaneous organic crystal growth in a droplet confined inside a gas bubble. The pNA crystal has been grown in the hottest place in the layer of solution of pNA dissolved in 1,4-dioxane due to laser-induced concentration increase in a droplet. This may occur because the stream of liquid, created by Marangoni optothermal effect, hits the bubble with quite large velocity (cf. Fig. 5(b)). This phenomenon has intriguing consequences. Frequently, the stream of solution breaks the liquid–gas interface and enters into the empty area of the cylindrical bubble. Initially the small droplet, due to constant supply of solution, increases with time, as is shown in Fig. 7 in the sequence of five photographs taken from the movie. Depending on the concentration of the dissolved substance in the solution and whether the bubble is staying or moving across the layer, the droplet size rises during the time interval of a second up to several seconds. Due to continuous laser heating the temperature increase promotes solvent evaporation from the droplet and consequently increases the solution concentration, which

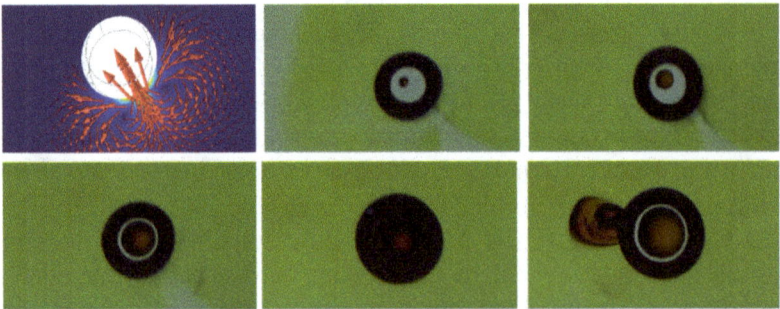

Figure 7.   Sequence of photographs taken from the movie showing the process of droplet formation in a bubble trapped by laser, its growth, and the partial ejection of its mass.[49] The first figure shows the liquid flow direction due to Marangoni thermocapillary effect occurring around the gas bubble.

momentarily becomes much larger than in the surrounding layer. The larger concentration of a dye increases the surface tension of the droplet and its contact angle with the substrate. The droplet growth enlarges the pressure inside the bubble (notice the increase of bubble diameter). At some moment, when the droplet is too large with respect to the bubble diameter, it touches the bubble edge and a certain amount of its mass is vigorously ejected outside. Controlling the laser power, the removal of excess droplet content may be forced to occur periodically.

Ejected saturation solution is either dispersed in the surrounding solution or forms outside spontaneously growing microcrystals due to the lowering of temperature.

Sometimes, the oversaturated solution may give rise to the unique process of microcrystal growth inside the gas bubble. Repeating experiments several times, we succeeded in registering a nucleation and growth of a single crystal inside a droplet trapped in a bubble. To show this in a more spectacular fashion, we used polarizing microscope with nearly crossed polarizers aimed to detect any birefringent material that potentially may appear in an optically isotropic droplet of solution. In Fig. 8, we show this process on few photographs taken from the movie. The crystal growing inside the saturated solution appears as the bright yellow crystal freely moving inside the droplet that is continuously fed from outside with the solvent with dissolved substance. Judging from the bright-colored uniform rectangular shape appearing in the droplet and seen under crossed polarizers (see Fig. 8), the para-nitroaniline micro-crystal grown from the supersaturated solution of 1,4-dioxane was nearly a single crystal as checked by polarizing optical microscope.

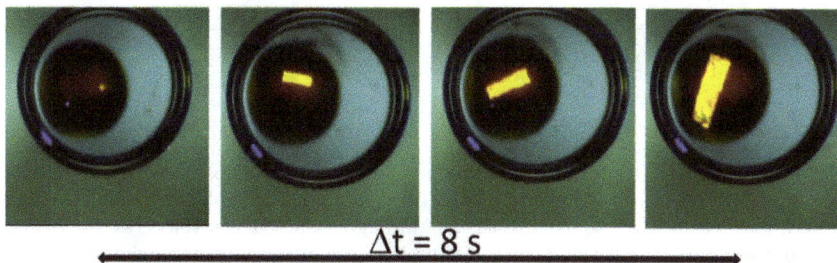

Figure 8. An example of single crystal growth of para-nitroaniline from a droplet of 1,4-dioxane saturated solution formed due to laser trapping of quasi-cylindrical gas bubble in a layer of solution confined in between two glass plates.

4.2. Manipulation of droplets via optothermal Marangoni effect

Mechanisms of droplet manipulations are really versatile. The literature devoted to this subject is numerous and was already described in several review papers as well as original publications.[29,39,51,69-74].

Based on the similar mechanism to that described in Subchapter 3.3, i.e., the flow of the solution through liquid–gas interface of a bubble, we observed a nice mechanism of conversion from optical to mechanical energy. In reference to Fig. 3(a), where an initially flat interface gas–liquid bending occurs due to laser-induced thermocapillary effect, it has been observed that some part of the solution can be ejected toward the empty space when the laser spot is close to but not touching the interface. In Fig. 9(a), we demonstrate the controlled large amplitude interface bending that is

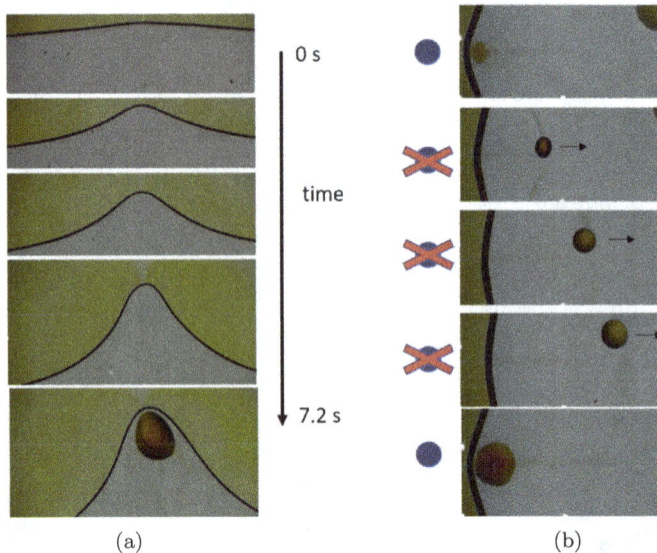

(a)   (b)

Figure 9.   Laser-induced liquid–gas interface bending in 1,4-dioxane–para nitroaniline system confined between two glass plates. (a) Continuous controlled interface bending toward laser beam spot due to optothermal Marangoni effect. Layer thickness $d = 70\,\mu$m. Part of the saturated solution penetrates the interface, making a thin liquid layer. With time a droplet is formed close to the interface edge. (b) Similar experiment for layer thickness $d = 250\,\mu$m. Interface bending is qualitatively smaller and droplet formation, easier. Shutting off the laser for a moment (marked by a cross placed on the laser spot) facilitates release of its mechanical energy and pushes the nearby droplet in forward direction. The Marangoni "catapult" for droplets can be used several times.

characteristic of the relatively thin liquid layer (here $d = 70\,\mu$m). At some moment, there appears a droplet near the center of interface that is enlarged with time due to supply of solution caused by the leakage in the interface. The similar experiment has been repeated for a thicker layer ($d = 270\,\mu$m). Bending has much smaller amplitude, but the ejected solution immediately forms a sessile droplet close to the interface in the air zone. Momentary blocking of laser light causes the optothermal Marangoni flow to cease and the previously bent interface to returning to its less bent form. The elastic energy of deformation is released, and the moving interface pushes forward the droplet just formed (see Fig. 9(b)), which escapes from its previous position. Periodic repeating of opening and blocking the laser beam leads to periodic formation of the droplet and its actuation due to interface elastic energy. We named this effect Marangoni "catapult".[50] The ejected droplets glide over the bottom glass surface which is wetted with the very thin layer of the solution. The traces of this layer can be seen as two wings formed aside the moving droplet (cf. second row in Fig. 9(b)). After being pushed by the interface, the droplets slow down due to drag force and finally increase their diameter and disappear.

Droplets from near-saturated solution of para-nitroaniline in 1,4-dioxane can be formed directly by laser heating of a free layer on a glass plate from below.[49] The formation of a droplet is immediate and starts from the escape of a solution from the region heated by laser due to the Marangoni thermocapillary effect (see Fig. 10(a,b)). The intense evaporation of dioxane causes the droplet with higher concentration of p-NA appear at the center of the heated region. However, the contact of this droplet with the liquid layer is maintained via a thin layer of solution that allows for increase in droplet volume and contact angle with the substrate. This happens due to increase of p-NA concentration in a droplet which increases the surface tension of the solution. When the laser beam is blocked, the droplet remains on the surface for some short time ($\sim$1 s) and then disappears, leaving the layer as in its initial condition. Subtly tuned laser power to the optimum can sustain steady-state condition for a droplet and the movement of laser beam to another destination is accompanied with the respective movement of the droplet.

This feature makes possible the laser-assisted enrichment of a solution via droplet formation, then its movement toward the seed crystal and crystal growth[75] (see also Fig. 10(b)). Repeated transport of a new solution rich in pNA helps in crystal growth and decides on the direction of its growth. By this method, we were able to grow single crystals of pNA under

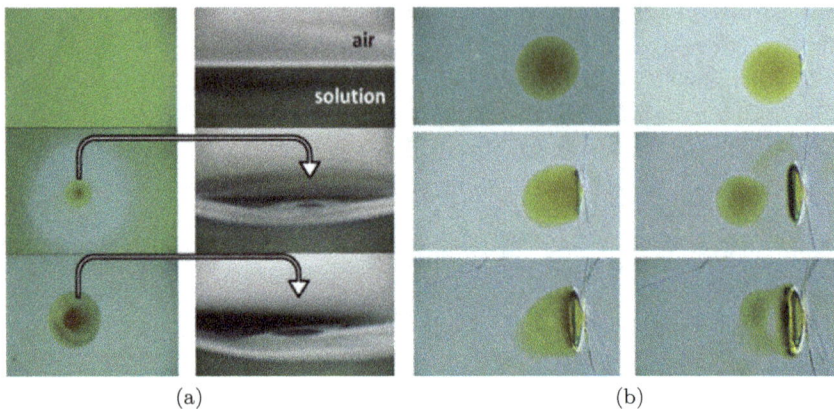

Figure 10. (a) Photographs showing the process of droplet formation due to localized laser-induced heating of a solution of p-NA in dioxane. First column — top view, second column — side view. (b) Example of using droplet formation and translation in laser-assisted crystal growth.[49]

optical microscope.[75] Laser-assisted crystal growth using optical tweezers setup for lysozyme protein crystal was reported in 2004,[76] Femtosecond laser pulses and laser ablation process were used for the same purpose by the Masuhara group.[77-79]

In Fig. 11(a), we demonstrate how rapid the droplet formation and its movement over the glass plate can be.

This fast movement is possible due to the thin layer of solution that remains at the glass surface and allows for gliding of a droplet when the laser changes its position. The droplet growth is possible due to: (i) Marangoni effect-induced flows in an initially flat absorbing laser light liquid, (ii) formation of a droplet and evaporation of dioxane that increases the surface tension of the droplet making it stable, and (iii) the connection with the reservoir of solution by a thin layer enabling liquid flows two ways: toward and outward from the droplet (see Figs. 11(b) and 11(c)).

4.3. Marangoni swimmers

Driving of micro-objects in a contactless fashion can be realized by conversion of light energy into mechanical work, either by direct optical momentum transfer (light pressure, gradient force in optical tweezing)[80] or indirectly by optothermal effects. Many examples of light-actuated objects' movements[81-86] show that thermocapillary propulsion mechanism is one of

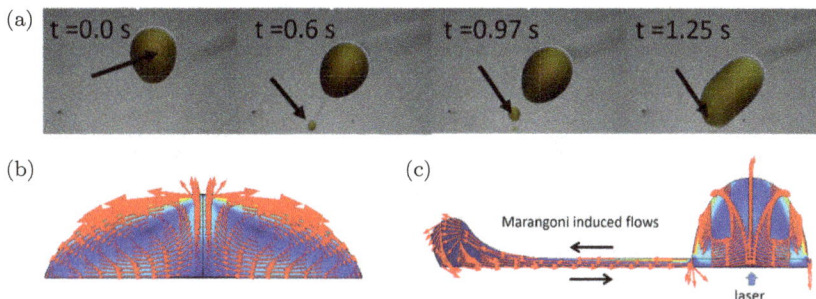

Figure 11. (a) Small pNA-dioxane droplet was formed by a laser absorbed by a solution layer. Then rapid movement of laser spot started to form a new droplet (see arrow pointing to the laser spot position). The bigger droplet left aside is attracted to the smaller one and forms a new large droplet rich in pNA. (b) The velocity of flow inside the sessile droplet irradiated from below. Arrows are proportional to the velocity. (c) Simulated mechanism of droplet growth and enrichment in pNA due to solvent evaporation. The thin layer of liquid connects the droplet generated by laser via Marangoni thermocapillary effect with the solution — the reservoir of solvent and solute. Observe the flow outside and inside the droplet in a thin liquid layer. Velocity arrows are normalized.

the strongest acting at the micron scale. At this scale, surface tension keeps objects at the surface, while any change in surface temperature generates liquid flows inducing capillary forces typically tens of nanonewton per micrometer. Thermally induced surface tension gradients $\partial T/\partial x$ produce thermally driven convective flows. On heating the liquid upper surface by laser light, which energy is partially absorbed, the temperature gradient is generated directly on it, so there is no critical or threshold temperature gradient value to start the flow. Surface flows then can be used to manipulate objects that are floating over it. When laser light is not directly absorbed by the liquid, the object can be heated directly by laser light and it can transfer the produced heat to the liquid surface generating in this way the localized temperature gradients. Using micro-sized chiral symmetry gears absorbing incident light of few mW, the rotation rates ranging 300 r.p.m. were demonstrated.[87] Since then several simple methods have been elaborated for manufacturing the swimmers and rotors of different sizes and different absorbing coatings.[88,89]

Here, we show how much bigger, millimeter size, floating objects can be rotated or translated by light. For remote non-contact actuation of Marangoni swimmers with laser light, we used the SCANcube scanning head equipped with two coupled galvano-mirrors sending the laser beam at wide angle destinations. Using the dedicated software (LaserDesk program), the trajectory and velocity of the laser beam motion at the target surface can be designed and controlled with very high accuracy. The system allows high reproducibility of traces and laser power, keeping exactly the same parameters that can be used for different liquids then allowing for easy comparison of the swimmers' movements in different environments. The obvious advantage in using thermocapillary Marangoni effect for actuated swimmers and not solutocapillary ones is that heat is dissipated relatively easily, while in solutocapillary effect the surface is contaminated with surfactants or other chemicals. For successful actuation, the generation of large temperature gradients is of primary importance as most of the liquids have similar surface tension temperature coefficients γ_T. Also beneficial is the high value of Peclet number in mass transportation Pe $\gg 1$, informing that the mass transport in a liquid should be dominated by advective transport. The thermal gradients should be estimated from thermal camera measurements and in function of time. The movement of the laser beam cannot be too fast with respect to heat dissipation. An example of the problem is shown in Fig. 12(a). To induce the object rotation, one can use the laser beam moving on a circular trajectory. The rotation rate must be slower than the heat dissipation rate, otherwise no gradients will be induced at the beam position. The floating objects can be relatively easily moved by heating the surface of a liquid in an asymmetric way with respect to the position of the floating object, e.g., beam is moving slowly around the rectangular swimmer (cf. Fig. 12(b)) each cycle pushing the ends of an object in the same clockwise manner. Beam rotation that is too fast around the object has no effect on momentum transfer.

In Fig. 12(c), the difficult movement of the rectangular object being perpendicular to its long axis is realized. One can accomplish that movement only having a laser spot oscillating along the long axis of the rectangle and at the same time moving in the orthogonal direction (a zig-zag-like motion, cf. Fig. 13(a)). An attempt to move the laser along a straight line after the object results in its movement aside. However, having a ship-like floating object (see Fig. 13(b)), the laser beam movement along the straight line is sufficient to obtain a corresponding straight-line trajectory.

The galvano-mirror system controlled by computer can be programmed to translate, rotate, and move floating objects in every complex trajectory by laser heating of a liquid. Alternatively, flat objects floating over non-absorbing laser light liquid can be heated by laser beam if they are covered by substance showing high absorption coefficient. In the latter case, the irradiation can be realized either from the top or from below the liquid layer.

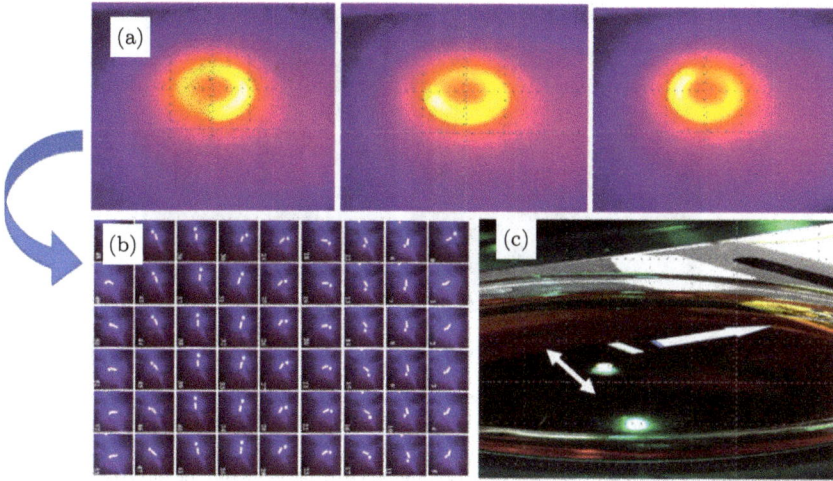

Figure 12. Examples of performing floating objects rotation and translation over rapeseed oil surface. Laser light heats the upper layer of oil due to light absorption by dissolved azo-dye. Laser beam is moved over the surface according to preprogrammed trace with velocity enabling the induction of highest possible temperature gradients. (a) Laser beam makes circles as monitored by FLIR 96 thermal camera. (b) The rectangular Marangoni swimmer (1 × 8 mm) floating over the surface rotates while the laser spot circulates clockwise around it. (c) Millimeter size swimmers (1 × 8 mm) can be translated in desired direction remotely by predesigned movements of the laser beam, here along the long axis of the rectangle accompanied with small movement perpendicular to it (see the white arrow showing translation direction of the rectangle).

Special asymmetric patterns of light, when projected on liquids able to effectively absorb this light, can induce their rotation or mixing (cf. Fig. 14). For liquids with small heat diffusion coefficient and strong absorption of used light, the heat produced in illuminated pattern generates temperature gradients which, according to the Marangoni thermocapillary effect, launch the liquid rotation. Examples of simulations performed for rapeseed oil containing a dye in a Petri dish are shown in Fig. 14.

Analysis of the obtained simulation results is encouraging as it demonstrates that the use of asymmetric (chiral) patterns of light projected on highly absorbing liquid surfaces is able to remotely mix a liquid. The maximum fluid velocity according to the numerical experiment with $\Delta T = 10\,\mathrm{K}$ reaches the value $|u|_{\max} \approx 0.2\,\mathrm{m \cdot s^{-1}}$. Due to momentum transfer from the center of a Petri dish toward its periphery and heat convection, the inertial forces increase and the temperature gradients decrease, resulting in decreasing of liquid speed, which is clearly observed by change of symmetry in the

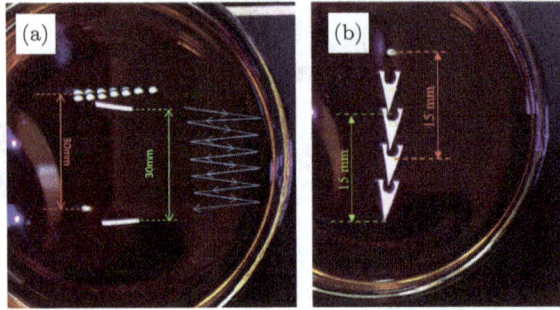

Figure 13.   Experiments with Marangoni swimmers floating over the rapeseed oil with an azo-dye and actuated remotely by laser heating. (a) Translational movement of rectangular swimmers propelled by laser light making a zig-zag trajectory. White dots show the laser spot movement captured every 20 s. (b) The boat-like swimmer pushed by laser beam moving along a straight line. The distance between laser spot and swimmer was constant and equal to 2.8 mm. The velocity of the swimmer in this case was $u = 125\,\mu\text{m s}^{-1}$. The marked red arrow shows the distance traveled by the laser beam, while the green one shows the distance traveled by the swimmer.

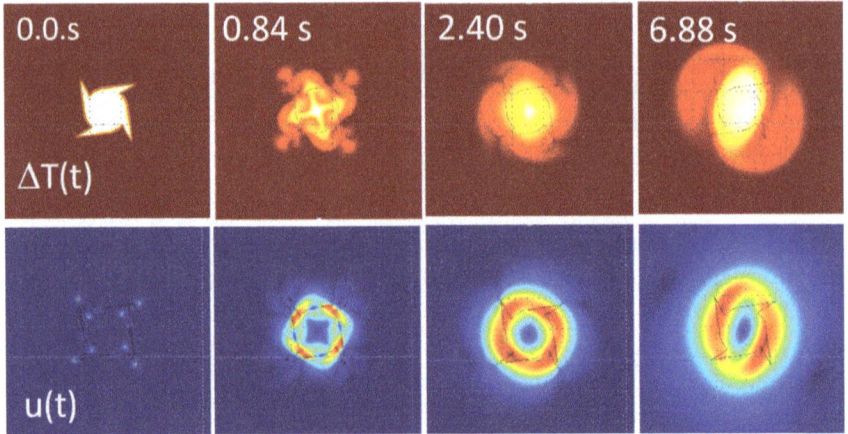

Figure 14.   Simulation of temperature field $\Delta T(t)$ (upper row) and velocity field u(t) (lower row) in function of time. 2D liquid is heated by projection of an asymmetric (chiral) four-winged figure on highly absorbing liquid surface, increasing its temperature by 10 K. Light energy that is transferred to liquid induces temperature gradients that, due to Marangoni thermocapillary effect, generate anticlockwise fluid rotation.

velocity map and temperature map from the chiral-four to the chiral-two axis pattern. For successful mixing, strong light absorption is required. Experiments confirm the mixing in thin 1–2 mm thick liquid layers.

4.4. *3D liquid-free surface actuation by laser beam*

Marangoni thermocapillary effect induced by laser in 2D was responsible for interface air/liquid bending (cf. Fig. 3(a)). Similar deformation of liquid surface induced by laser beam incident on a liquid layer has been studied since 1979 by Da Costa *et al.*[90,91] Later, several publications devoted to measurement and modeling of this phenomenon emerged.[47,92–94] We repeated the measurements of this effect experimentally and numerically in 3D using a rapeseed oil pool with dissolved azo-dye to guarantee strong light absorption of 514.5 nm Ar$^+$ ion cw excitation laser.[95] The schematic representation of the experiment is shown in Fig. 15(a). Due to the Marangoni optothermal effect induced by normally incident strong laser beams, the surface of the upper oil layer is deformed. This deformation is relatively weak, therefore, to observe its magnitude, we employed the second coherent laser source of 650 nm, incident obliquely. Light of this wavelength was not absorbed by the dye and the purpose of using it was to observe light interference pattern in the far field due to difference in phase between the beam reflected from the bottom of the vessel and the liquid surface. Without the excitation of the beam, the red beam reflection was not distorted, while when the excitation beam was open, the interference pattern appeared in a far field. This pattern was either photographed from a white screen or measured with the help of light detectors positioned at the center and the periphery of the interference pattern. In such a manner, we were able to estimate not only the liquid depletion amplitude and shape (see Fig. 15(b)) but also the dynamics of the bending process in the function of heat delivered to the oil. We systematically studied the optothermal Marangoni process and simulated it by solving 3D time-dependent Navier–Stokes and heat transfer equations.[95] The difficulty in simulations was related to coupling of the pressure changes at the liquid surface with mechanical deformation of the liquid surface under conditions of buoyancy and gravitational forces as well as Marangoni flows. To accomplish them, we used a moving mesh procedure available in COMSOL 5.6, allowing to study time-dependent changes in incompressible liquid due to action of the

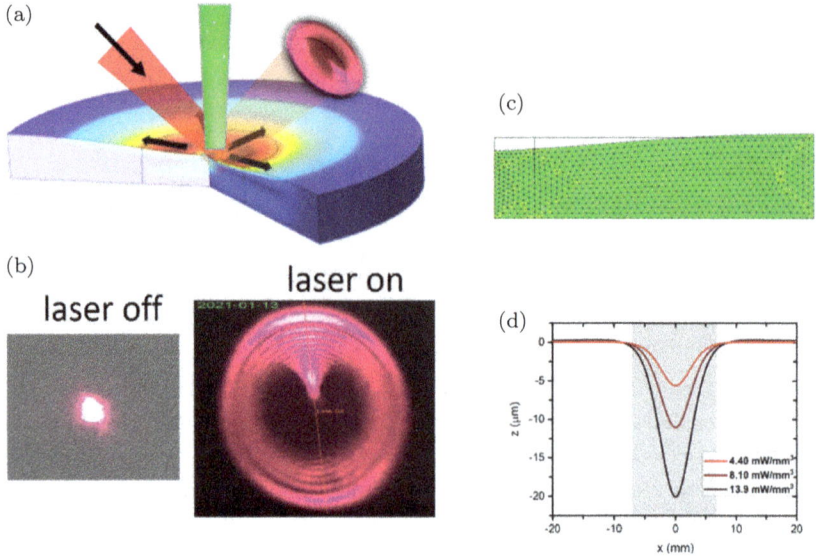

Figure 15. (a) Schematic representation of laser-induced oil surface deformation based on thermocapillary Marangoni effect. Laser beam of 514.5 nm hits the free oil surface and produces radial outward flow of liquid. The depletion of the surface is measured by interference method using oblique incident red light of 650 nm wavelength, which forms interference fringes proportional to the bending profile in far field. (b) Photographs of red laser spot when green laser irradiation is blocked and interference fringes arise due to laser green light excitation. The correction of beam interference pattern from elliptical to circular has been applied. (c) Axisymmetric geometry of simulation of solution of Navier–Stokes equations in Newtonian fluid with Marangoni effect. Bending at the central irradiated area was simulated using dynamic moving mesh method in COMSOL Multiphysics. (d) Numerical results of surface bending shown in the plot agree well with profile calculations based on the analysis of interference fringe pattern.

excitation laser. The results are described in detail in the recent paper.[95] In Figs. 15(c) and 15(d) we show the results of simulations of liquid bending profiles in function of heat delivered by laser beam with a Gaussian intensity profile. To accomplish that we have used a moving mesh method in an axisymmetric geometry.

In numerical simulations of laser-induced free surface deformation phenomenon, we used parameters of rapeseed oils given in Table 2. Remarkably, the temporal response of oil bending and its shape are in agreement with results obtained from real measurements, just proving that not the light pressure, but the optothermal Marangoni mechanism is a dominant one in this case.

Table 2. Physical properties of rapeseed oil as used in COMSOL Multiphysics simulations. Heat capacity at constant pressure C_p, and thermal conductivity κ were assumed to be independent of temperature. However, the density $\rho(T)$, the dynamic viscosity $\mu(T)$, and the surface tension $\gamma(T)$ were used in simulations according to the following approximated formula.

Properties of rapeseed oil	Value or formula
Density, $\rho(T) =$	$1111 - 0.6691 \cdot T(K)$ $[kg \cdot m^{-3}]$
Heat capacity at constant pressure, C_p	2030 $[J \cdot kg^{-1} \cdot K^{-1})]$
Dynamic viscosity, $\mu(T) =$	$0.00977 + 3.23 \cdot 10^{-6}$ $\exp(-0.0573 \cdot T)$ $[mPa \cdot s]$
Thermal conductivity, κ	0.16 $[W \cdot m^{-1} \cdot K^{-1}]$
Surface tension at 273.15 K, γ_0	0.03519 $[N \cdot m^{-1}]$
Temperature coefficient of surface tension, γ_T	$-0.0758 \cdot 10^{-3}$ $[N \cdot m^{-1} \cdot K^{-1}]$
Refractive index n at $\lambda = 589\,nm$	1.468
Refractive index n at $\lambda = 650\,nm$	1.465
Diffusion coefficient D	$D = 3.4 \cdot 10^{-12}\,m^2 \cdot s^{-1}$
Péclet number in mass transfer	$Pe = \frac{u \cdot L}{D} = 1.2 \times 10^6$
(mass transportation is determined by advective transport)	$(u = 0.081\,cm \cdot s^{-1},\ L = 0.5\,cm)$
Péclet number in heat transfer (momentum diffusivity dominates the heat transfer)	$Pe = \frac{u \cdot L}{\alpha} = \frac{uL\rho C_p}{\kappa} = 46.8$

4.5. *Hydrodynamic trap*

The numerical simulations of oil surface deformation due to laser heating allowed us to determine the vortex nature of fluid flow under the free surface. The quiet zone (a zone where the velocity of fluid is nearly zero) is clearly build-up close to the laser spot at some depth from the surface (see Fig. 16(a)). In the central part of the vortex, flow of liquid is very slow and outside it is increasing. This opens the possibility for hydrodynamic trapping of objects some tens of micrometers in size. We succeeded in conducting the experiment, proving that such trapping is possible.[95] In Fig. 16, we demonstrate the cross-section of vortices build up under the free surface and the photograph of elongated crystal in this trap. The trapped crystal rotates around its long axis, despite the fact that it can be translated together with slow translation of laser spot over the oil surface in Petri dish.

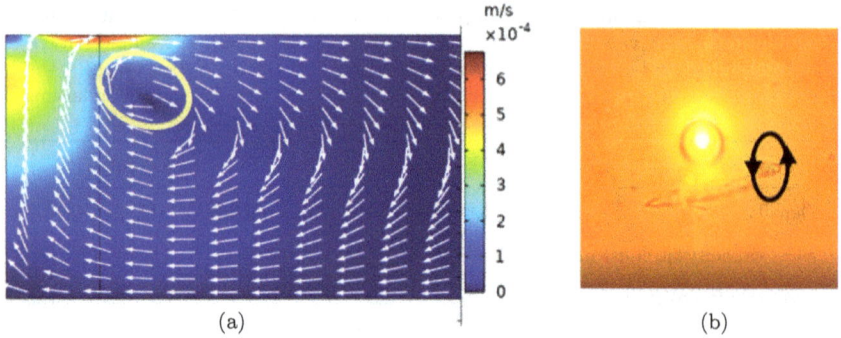

Figure 16. (a) Cross-section of axis-symmetric view of velocity field in oil layer irradiated by Gaussian laser beam from the top. At the periphery of laser beam spot (here marked by vertical black line) a stable vortex quiet zone (marked by ellipse) is formed. (b) The photograph of elongated crystal trapped in the quiet zone. The crystal cannot leave the trap until the laser supplies heat. Due to friction forces between crystal and the circular flow around it, the crystal rotates around its long axis several times per second.

The trap strength is proportional to the laser power. Trapping is accidental and it requires some patience to trap a crystal as shown in Fig. 16(b), similarly to optical tweezers, but here the Marangoni thermocapillary effect and not a gradient force of light is enabling the formation of this trap.

4.6. Two-phase liquid systems including liquid crystals and polymers

Optothermal Marangoni effect is not limited to liquid/air interface only. It can be observed also at the interface between two immiscible or partially immiscible fluids. In this case, multiphase flows are characterized by the ratio of viscous to surface forces. In Fig. 17, the 2D system of solution of dioxane with para-nitroaniline is in contact with the pure nematic liquid crystal (LC) pentyl-cyano-biphenyl (5CB). Irradiation with laser of the solution dioxane/pNA close to the air/solution interface generates Marangoni optothermal effect at this interface. The Marangoni flows form vortices close to this interface, but also break the interface solution/nematic LC and pull the liquid crystal toward the center of the position of the laser beam (cf. Fig. 17(a)).

The vortices of liquid crystal are formed with a characteristic funnel shape structure with characteristic side lobes (cf. Fig. 17(b)). The use of quasi-crossed polarizers reveals the phase transition from the optically

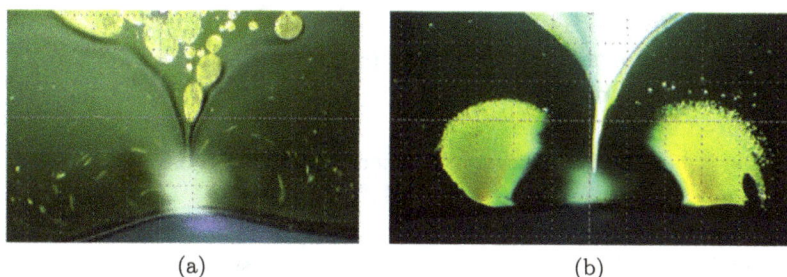

(a)   (b)

Figure 17.   Optothermal Marangoni effect observed in a layer of two immiscible liquid phases: solution of pNA in 1,4-dioxane and nematic liquid crystal 5CB. Laser is positioned close to the interface between air and solution. (a) Initially, a Marangoni flow produces vortices close to the air/solution interface, the fluid flow toward the surface pulls the nematic LC in the same direction due to momentum transfer between solution and LC. (b) View under crossed polarizers of funnel formation and flow of LC. Heating of 5CB over its nematic-to-isotropic phase transition causes the birefringent LC at the hottest place to transform into isotropic phase and becomes invisible under crossed polarizers.

birefringent nematic phase to the optically isotropic phase (which gives the black color) due to the local temperature increase.

Marangoni effect has been observed also in thin azo-polymer films due to the optofluidization process. In this process, laser-focused irradiation melts the upper polymer layer and induces Marangoni flows due to surface tension gradients based on photostationary equilibrium between trans and cis isomers in polymeric matrix. This phenomenon allows for shaping the azo-polymer surfaces with nanometric resolution.[21,96–99]

5. Applications of Marangoni Effect

Thermocapillary and solutocapillary Marangoni effects have found several applications. Particularly, microfluidic motion is currently a hot issue. Marangoni surfers, swimmers, boats, and rotors actuated by light, heat, composition, or surfactant concentration have attracted a lot of attention expressed in dedicated reviews[27–29,100] and in hundreds of publications. The concern was focused on the efficient conversion of light energy, heat, or chemical composition into mechanical work. It has been demonstrated that these stimuli movable objects might play important roles in sensing, imaging, biomedicine, and manufacturing. Here, we briefly present examples of the prospective, cutting-edge applications that attracted our attention:

- Mini-generators of electrical energy exploiting the "camphor engines" and floating polymer rotors bearing permanent magnets.[101,102]
- Highly efficient mechanism of maze solving based on Marangoni effect produced by surfactant.[103,104]
- Single-light-actuated dual-mode (thermal and composition) propulsion surfers.[105]
- Development of an extremely low-power optical tweezing technique through optically heating a thermoplasmonic substrate, termed opto-thermoelectric nanotweezers (OTENT).[106,107]
- Microactuators and microrobots which can transform external stimuli into mechanical motion at microscale.[108–112]
- Liquid micromotors exhibiting both positive and negative phototaxis, which relies on the molecular photochromism.[113]
- Marangoni-assisted patterning of nanoparticles.[114,115]
- Non-contact fuel-free light-driven micro/nanorotors. Controlling of Marangoni surfers and rotors.[116–120]

6. Conclusions

Enormous amount of work has recently been published about various aspects of solutal- and thermocapillary Marangoni effects. This is closely related with the progress in optofluidics, nanoplasmonics, nonlinear optics, nanotechnology of surfaces, nanophotonics, biology, etc. Light used as a remote and precise supplier of energy to the matter in micro and even nanoscale enables realization of advanced and smart applications. In this chapter, we provided basic knowledge about optothermal Marangoni effect, and presented some examples of non-intuitive experimental results and simple simulations that help to understand the observed, frequently very complex, phenomena occurring in liquids and interfaces of different phases. The solution of Navier–Stokes equation is still the oldest unsolved mathematical problem among seven Clay Mathematical Institute Millennium Prize Problems set in 2000. The prospects for the future of the Marangoni effects-related applications are quite encouraging and can be found in, cited in this book chapter, recent reviews and original publications.

Acknowledgment

This work was supported by the National Science Centre, Poland [grant number UMO-2018/29/B/ST3/00829].

References

1. L. E. Scriven and C. V. Sternling, The Marangoni effects, *Nature* **187**, 186–188, (1960).
2. P.-G. de Gennes, F. Brochard-Wyart, and D. Querena, *Capillarity and Wetting Phenomena- Drops, Bubbles, Pearls, Waves*, New York: Springer-Verlag (2004).
3. M. G. Velarde and R. K. Zeytounian, *Interfacial Phenomena and the Marangoni Effect*. Vienna: Springer (2014).
4. J. Chen, J. F.-C. Loo, D. Wang, Y. Zhang, S.-K. Kong, and H.-P. Ho, Thermal optofluidics: Principles and application, *Adv. Opti. Mater.* **8**, 1900829 (2020).
5. D. Baigl, Photo-actuation of liquids for light-driven microfluidics: State of the art and perspectives, *Lab Chip* **12**(19), 3637–3653 (2012).
6. Z. Chen, J. Li, and Y. Zheng, Heat-mediated optical manipulation, *Chem. Rev.* **122**(3), 3122–3179 (2022).
7. S. Ghosh, A. D. Ranjan, S. Das, R. Sen, B. Roy, S. Roy, and A. Banerjee, Directed self-assembly driven mesoscale lithography using laser-induced and manipulated microbubbles: Complex architectures and diverse applications, *Nano Lett.* **21**, 10–25 (2021).
8. B. D. Edmonstone and O. K. Matar, Simultaneous thermal and surfactant-induced Marangoni effects in thin liquid films, *J. Colloid Interf. Sci.* **274**(1), 183–199 (2004).
9. M. Schmitt and H. Stark, Marangoni flow at droplet interfaces: Three-dimensional solution and applications, *Phys. Fluids* **28**(1), 12106 (2016).
10. H. Kim, J. Lee, T.-H. Kim, and H.-Y. Kim, Spontaneous marangoni mixing of miscible liquids at a liquid–liquid–air contact line, *Langmuir* **31**, 8726–8731 (2015).
11. J. Park, J. Ryu, H. J. Sung, and H. Kim, Control of solutal Marangoni-driven vortical flows and enhancement of mixing efficiency, *J. Colloid Interf. Sci.* **561**, 408–415 (2020).
12. *COMSOL Multiphysics®* *V5.5, COMSOL AB*, Sweden: Stockholm (2018), www.comsol.com.
13. J. Thomson, On certain curious motions observable at the surfaces of wine and other alcoholic liquors, *Lond. Edinburgh Dublin, Philos. Mag. J. Sci.* **10**, 330 (1855).
14. C. Marangoni, Sull'espansione Delle Gocce D'un Liquido Galleggianti Sulla Superfice di Altro Liquido, *Tipografia dei Fratelli Fusi*, Pavia (1865).
15. C. Marangoni, On the principle of the surface viscosity of liquids, established by J. Plateau, Italian, *Il Nuovo Cimento* Series **2**(5/6), 239 (1872).
16. D. C. Venerus and D. Nieto Simavilla, Tears of wine: New insights on an old phenomenon, *Sci. Rep.* **5**, 16162 (2015).
17. K. C. Mills, B. J. Keene, R. F. Brooks, and A. Shirali, Marangoni effects in welding, *Philos. Trans. R. Soc. Lond. Ser. A* **356**, 911–925 (1998).
18. L. Aucott *et al.*, Revealing internal flow behavior in arc welding and additive manufacturing of metals, *Nat. Comm.* **9**(1), 5414 (2018).

19. T. A. Arshad, C. B. Kim, N. A. Prisco, J. M. Katzenstein, D. W. Janes, R. T. Bonnecaze, and C. J. Ellison, Precision Marangoni-driven patterning, *Soft Matter* **40**, 8043–8050 (2014).

20. I. Kitamura, K. Oishi, M. Hara, S. Nagano, and T. Seki, Photoinitiated Marangoni flow morphing in a liquid crystalline polymer film directed by super-inkjet printing patterns, *Sci. Rep.* **9**, 2556 (2019).

21. A. Miniewicz, A. Sobolewska, W. Piotrowski, P. Karpinski, S. Bartkiewicz, and E. Schab-Balcerzak, Thermocapillary Marangoni Flows in Azopolymers, *Materials* **13**(11), 2464 (2020).

22. S. B. Lee, S. Lee, D. G. Kim, S. H. Kim, B. Kang, and K. Cho, Solutal-Marangoni-flow-mediated growth of patterned highly crystalline organic semiconductor thin film via gap-controlled bar coating, *Adv. Funct. Mater.* **31**(28), 2100196 (2021).

23. A. Z. Stetten, S. V. Iasella, T. E. Corcoran, S. Garoff, T. M. Przybycien, and R. D. Tilton, Surfactant-induced Marangoni transport of lipids and therapeutics within the lung, *Curr. Opin. Colloid Interf. Sci.* **36**, 58–69 (2018).

24. M.-H. Hsieh, H.-J. Wei, K.-H. Chen, H.-C. Wang, C.-H. Yu, T.-H. Lu, Y. Chang, and H.-W. Sung, A fast and facile platform for fabricating phase-change materials-based drug carriers powered by chemical Marangoni effect, *Biomaterials* **271**, 120748 (2021).

25. B. A. Nerger, P.-T. Brun, and C. M. Nelson, Marangoni flows drive the alignment of fibrillar cell-laden hydrogels, *Sci. Adv.* **6**(24), 1–11 (2020).

26. M. Maillard, L. Motte, and M. P. Pileni, Rings and hexagons made of nanocrystals, *Adv. Mater.* **13**, 200–204 (2001).

27. M. Xiao, Y. Xian, and F. Shi, Precise macroscopic supramolecular assembly by combining spontaneous locomotion driven by the marangoni effect and molecular recognition, *Angew. Chem. Int. Ed.* **54**, 8952–8956 (2015).

28. A. Karbalaei, R. Kumar, and H. J. Cho, Thermocapillarity in Microfluidics — A review, *Micromachines* **7**, 13, (2016).

29. S.-Y. Park and P.-Y. Chiou, Light-driven droplet manipulation technologies for lab-on-a-chip applications, *Adv. OptoElectron* **2011**, 909174, 1–12 (2011).

30. W. C. Thoburn, The origin of surface tension in liquids, *Phys. Teach* **15**, 234 (1977).

31. L. Eötvös, Ueber den Zusammenhang der Oberflächenspannung der Flüssigkeitenmit ihrem Molecularvolumen, *Annalen der Physik* **27**(3), 448–459 (1886).

32. L. Glasser, Volume-based thermodynamics of organic liquids: Surface tension and the Eötvös equation, *J. Chem. Thermodynamics* **157**, 106391 (2021).

33. J. J. Jasper, The surface tension of pure liquid compounds, *J. Phys. Chem. Ref. Data* **1**(4), 841–1010 (1972).

34. S. G. Cook, Surface tension of mercury in the presence of gas under varying pressures, *Phys. Rev.* **34**, 513 (1929).

35. M. G. Gannon and T. E. Faber, The surface tension of nematic liquid crystals, *Philos. Mag. A* **37**(1), 117–135 (1978).
36. F. M. Fowkes, Attractive forces at interfaces. *Ind. Eng. Chem. Res.* **56**(12), 40–52 (1964).
37. M. J. Rosen and J. T. Kunjappu, *Surfactants and Interfacial Phenomena* (4th edn.). Hoboken, New Jersey: John Wiley & Sons (2012).
38. K. Ichimura, S.-K. Oh, and M. Nakagawa, Light-driven motion of liquids on a photoresponsive surface, *Science* **288**(5471), 1624–1626 (2000).
39. A. Diguet, R. Guillermic, N. Magome, A. Saint-Jalmes, Y. Chen, K. Yoshikawa, and D. Baigl, Photomanipulation of a droplet by the chromocapillary effect, *Angew. Chem. Int. Ed.* **48**(49), 9281–9284 (2009).
40. A. Kidess, S. Kenjereš, and C. R. Kleijn, The influence of surfactants on thermocapillary flow instabilities in low Prandtl melting pools, *Phys. Fluids* **28**(6), 062106 (2016).
41. M. J. Block, Surface tension as the cause of Benard cells and Surface deformation in a liquid film, *Nature* **178**(4534), 650–651 (1956).
42. J.-M. Wang, G.-H. Liu, Y.-L. Fang, and W.-K. Li, Marangoni effect in nonequilibrium multiphase system of material processing, *Rev. Chem. Eng.* **32**(5), 551–585 (2016).
43. D. Green and R. Perry, *Perry's Chemical Engineers' Handbook* (8th edn.), McGraw-Hill Professional (2007).
44. Y. Cengel and J. Cimbala, *Fluid Mechanics: Fundamentals and Applications*, Boston: McGraw Hill Higher Education (2006).
45. W. L. Hosch, Navier-Stokes equation, *Encyclopedia Britannica*, 28 May (2020). https://www.britannica.com/science/Navier-Stokes-equation.
46. D. M. McEligoi, Fundamentals of momentum, heat and mass transfer, *Int. J. Heat Mass Transf.* **13**(10), 1641 (1970).
47. H. Chraibi and J.-P. Delville, Thermocapillary flows and interface deformation produced by localized laser heating in confined environment, *Phys. Fluids* **24**(3), 032102 (2012).
48. A. Miniewicz, C. Quintard, H. Orlikowska, and S. Bartkiewicz, On the origin of the driving force in the Marangoni propeller gas bubble trapping mechanism, *Phys. Chem. Chem. Phys.* **19**(28), 18695–18703 (2017).
49. S. Bartkiewicz and A. Miniewicz, Whirl-enhanced continuous wave laser trapping of particles, *Phys. Chem. Chem. Phys.* **17**(2), 1077–1083 (2015).
50. A. Miniewicz, S. Bartkiewicz, H. Orlikowska, and K. Dradrach, Marangoni effect visualized in two-dimensions optical tweezers for gas bubbles, *Sci. Rep.* **6**, 34787 (2016).
51. B. Selva, V. Miralles, I. Cantat, and M. C. Jullien, Thermocapillary actuation by optimized resistor pattern: Bubbles and droplets displacing, switching and trapping, *Lab Chip* **10**(14), 1835–1840 (2010).
52. J. Rodríguez-Rodríguez, A. Sevilla, C. Martínez-Bazán, and J. M. Gordillo, Generation of microbubbles with applications to industry and medicine, *Annu. Rev. Fluid Mech.* **47**, 405–429 (2015).

53. A. T. Ohta, A. Jamshidi, J. K. Valley, H. Y. Hsu, and M. C. Wu, Optically actuated thermocapillary movement of gas bubbles on an absorbing substrate, *Appl. Phys. Lett.* **91**(7), 074103 (2007).

54. K. Namura, K. Nakajima, K. Kimura, and M. Suzuki, Photothermally controlled Marangoni flow around a micro bubble, *Appl. Phys. Lett.* **106**, 043101 (2015).

55. Y. Kim, H. Ding, and Y. Zheng, Enhancing surface capture and sensing of proteins with low power optothermal bubbles in a biphasic liquid, *Nano Lett.* **20**, 7020–7027 (2020)

56. S. Moon, Q. Zhang, D. Huang, S. Senapati, H. C. Chang, E. Lee, and T. Luo, Biocompatible direct deposition of functionalized nanoparticles using shrinking surface plasmonic bubble, *Adv. Mater. Interf.* **7**, 2000597 (2020).

57. E. H. Cgoyao, C. Goldmann, C. Hamon, and M. Herber, Laser-driven bubble printing of plasmonic nanoparticle assemblies onto nonplasmonic substrates, *J. Phys. Chem. C* **126**(17), 7622–7629 (2022).

58. Y. Zheng, H. Liu, Y. Wang, C. Zhu, S. Wang, J. Cao, and S. Zhu, Accumulating microparticles and direct-writing micropatterns using a continuous-wave laser-induced vapor bubble, *Lab Chip* **11**, 3816–3820 (2011).

59. L. Lin, X. Peng, Z. Mao, W. Li, M. N. Yogeesh, B. Bangalore Rajeeva, E. P. Perillo, A. K. Dunn, D. Akinwande, and Y. Zheng, Bubble-pen lithography, *Nano Lett.* **16**(1), 701–708 (2016).

60. B. Bangalore Rajeeva, L. Lin, E. P. Perillo, X. Peng, W. Yu, A. K. Dunn, and Y. Zheng, High-resolution bubble printing of quantum dots, *ACS Appl. Mater. Interfaces* **9**, 16725–16733 (2017).

61. F. Karim, E. S. Vasquez, Y. Sun, and C. Zhao, Optothermal microbubble assisted manufacturing of nanogap-rich structures for active chemical sensing, *Nanoscale* **11**(43), 20589–20597 (2019).

62. X. Shi, D. V. Verschueren, and C. Dekker, Active delivery of single DNA molecules into a plasmonic nanopore for label-free optical sensing, *Nano Lett.* **18**(12), 8003–8010 (2018).

63. J. W. Strutt (Lord Rayleigh), On the pressure developed in a liquid during the collapse of a spherical cavity, *Phil. Mag.* **34**(200), 94–98 (1917).

64. M. S. Plesset, The dynamics of cavitation bubbles, *J. Appl. Mech.* **16**(3), 228–231 (1949).

65. A. Ashkin, J. M. Dziedzic, J. E. Bjorkholm, and S. Chu, Observation of a single-beam gradient force optical trap for dielectric particles. *Opt. Lett.* **11**, 288–290 (1986).

66. P. Kumari, J. A. Dharmadhikari, A. K. Dharmadhikari, H. Basu, S. Sharma, and D. Mathur, Optical trapping in an absorbing medium: From optical tweezing to thermal tweezing, *Opt. Express* **20**(4), 4645–4652 (2012).

67. J. A. Sarabia-Alonso, J. G. Ortega-Mendoza, J. C. Ramirez-San-Juan, P. Zaca Moran, J. Ramirez-Ramirez, A. Padilla-Vivanco, F. M. Munoz-Perez, and R. Ramos-Garcia, Optothermal generation, trapping, and manipulation of microbubbles, *Opt. Express* **28**(12), 17672–17682 (2020).

68. J. A. Sarabia-Alonso, J. G. Ortega-Mendoza, S. Mansurova, F. M. Munoz-Perez, and R. Ramos-Garcia, 3D trapping of microbubbles by the Marangoni force, *Opt. Lett.* **46**(23), 5786–5789 (2021).

69. S. Rybalko, N. Magome, and K. Yoshikawa, Forward and backward laser-guided motion of an oil droplet, *Phys. Rev. E* **70**(4), 046301 (2004).

70. J. Z. Chen, S. M. Troian, A. A. Darhuber, and S. Wagner, Effect of contact angle hysteresis on thermocapillary droplet actuation, *J. Appl. Phys.* **97**, 014906 (2005).

71. S. Y. Park, M. A. Teitell, and E. P. Chiou, Single-sided continuous opto-electrowetting (SCOEW) for droplet manipulation with light patterns, *Lab Chip* **10**(13), 1655–1661 (2010).

72. M. Muto, M. Yamamoto, and M. Motosuke, A noncontact picoliter droplet handling by photothermal control of interfacial flow, *Anal. Sci.* **32**, 49–55 (2016).

73. P. Capobianchi, M. Lappa, and S. M. Oliveira, Walls and domain shape effects on the thermal Marangoni migration of three-dimensional droplets, *Phys. Fluids* **29**(11), 112102 (2017).

74. N.-A. Goy, N. Bruni, A. Griot, J.-P. Delville, and U. Delabre, Thermal Marangoni trapping driven by laser absorption in evaporating droplets for particle deposition, *Soft Matter* **18**(41), 7949–7958 (2022).

75. K. Grzeskiewicz, M. Belej, S. Bartkiewicz, and A. Miniewicz, Optically steered crystal growth of para nitroaniline with Marangoni effect (in Polish), *Postepy Fizyki* **72**(2), 9–15 (2021).

76. W. Singer, H. Rubinsztein-Dunlop, and U. Gibson, Manipulation and growth of birefringent protein crystals in optical tweezers, *Opt. Express* **12**(26), 6440–6445 (2004).

77. Y. Hosokawa, H. Adachi, M. Yoshimura, Y. Mori, T. Sasaki, and H. Masuhara, Femtosecond laser-induced crystallization of 4-(Dimethylamino)-N-methyl-4-stilbazolium Tosylate, *Cryst. Growth Des.* **5**(3), 861–863 (2005).

78. T. Sugiyama and H. Masuhara Laser-induced crystallization and crystal growth, *Chem. Asian J.* **6**, 2878–2889 (2011).

79. T.-H. Liu, K.-I. Yuyama, T. Hiramatsu, N. Yamamoto, E. Chatani, H. Miyasaka, T. Sugiyama, and H. Masuhara, Femtosecond-laser-enhanced amyloid fibril formation of insulin, *Langmuir* **33**, 8311–8318 (2017).

80. S. M. Block, Making light work with optical tweezers, *Nature* **360**(6403), 493–495 (1992).

81. Y. Sun, J. Jiang, G. Zhang, N. Yuan, H. Zhang, B. Song, and B. Dong, Visible light-driven micromotor with incident-angle-controlled motion and dynamic collective behavior, *Langmuir* **37**, 180–187 (2021).

82. J. Liu and S. Li, Capillarity-driven migration of small objects: A critical review, *Eur. Phys. J. E: Soft Matter Biol. Phys.* **42**, 1 (2019).

83. D. Crowdy, Collective viscous propulsion of a two-dimensional flotilla of Marangoni boats, *Phys. Rev. Fluids* **5**, 124004 (2020).

84. H. Wu, J. Luo, X. Huang, L. Wang, Z. Guo, J. Liang, S. Zhang, H. Xue, and J. Gao, Superhydrophobic, mechanically durable coatings for controllable light and magnetism driven actuators, *J. Colloid Interf. Sci.* **603**, 282–290 (2021).

85. D. E. Lucchetta, F. Simoni, N. Sheremet, V. Reshetnyak, and R. Castagna, Shape-driven optofluidic rotational actuation, *Eur. Phys. J. Plus* **136**, 445 (2021).

86. T. Zhang, H. Chang, Y. Wu, P. Xiao, N. Yi, Y. Lu, Y. Ma, Y. Huang, K. Zhao, and X.-Q. Yan, Macroscopic and direct light propulsion of bulk graphene material, *Nat. Photoni.* **9**(7), 471–476 (2015).

87. C. Maggi, F. Saglimbeni, M. Dipalo, F. De Angelis, and R. Di Leonardo, Micromotors with asymmetric shape that efficiently convert light into work by thermocapillary effects, *Nat. Commun.* **6**(7855), 1–15 (2015).

88. W. Wang, B. Han, Y. Zhang, Q. Li, Y. Zhang, D. Han, and H. Sun, Laser-induced graphene tapes as origami and stick-on labels for photothermal manipulation via Marangoni effect, *Adv. Funct. Mater.* **31**(1), 200617 (2021).

89. Y. Wang, Y. Dong, F. Ji, J. Zhu, P. Ma, H. Su, P. Chen, X. Feng, W. Du, and B.-F. Liu, Patterning candle soot for light-driven actuator via Marangoni effect, *Sens. Actuators B: Chem.* **347**, 130613 (2021).

90. G. Da Costa and J. Calatroni, Transient deformation of liquid surfaces by laser-induced thermocapillarity, *Appl. Opt.* **18**(2), 233–235 (1979).

91. G. Da Costa and R. Escalona, Time evolution of the caustics of a laser heated liquid film, *Appl. Opt.* **29**(7), 1023–1033 (1990).

92. B. A. Bezuglyi, N. A. Ivanova, and A. Y. Zueva, Laser–induced thermocapillary deformation of a thin liquid layer, *J. Appl. Mech. Tech. Phys.* **42**(3), 493–496 (2001).

93. G. Verma, H. Chesneau, H. Chraïbi, U. Delabre, R. Wunenburger, and J.-P. Delville, Contactless thin-film rheology unveiled by laser-induced nanoscale interface dynamics, *Soft Matter* **16**, 7904 (2020).

94. Y. Miao, Z. Qiu, Y. Jiang, X. Zhang, L. Han, Z. Wang, and N. Wang, Measuring multiscale capillary curvature using laser beam self-interference, *Opt. Commun.* **497**(15), 127149 (2021).

95. M. Belej, K. Grzeskiewicz, and A. Miniewicz, Laser light-induced deformation of free surface of oil due to thermocapillary Marangoni phenomenon: Experiment and computational fluid dynamics simulations, *Phys. Fluids* **34**, 082104 (2022).

96. A. Ambrosio, L. Marrucci, F. Borbone, A. Roviello, and P. Maddalena, Light-induced spiral mass transport in azo-polymer films under vortex-beam illumination, *Nature Commun.* **3**, 989 (2012).

97. J. Choi, W. Jo, and S. Lee, Flexible and robust superomniphobic surfaces created by localized photofluidization of azopolymer pillars, *ACS Nano* **11**, 7821–8 (2017).

98. I. Kitamura, K. Oishi, M. Hara, S. Nagano, and T. Seki, Photoinitiated Marangoni flow morphing in a liquid crystalline polymer film directed by super-inkjet printing patterns, *Sci. Rep.* **9**, 2556 (2019).

99. R. Elashnikov, P. Fitl, V. Svorcik, and O. Lyutakov, Patterning of ultrathin polymethylmethacrylate films by in-situ photodirecting of the Marangoni flow, *Appl. Surf. Sci.* **394**, 562–568 (2017).

100. M. Tenjimbayashia and K. Manabe, A review on control of droplet motion based on wettability modulation: Principles, design strategies, recent progress, and applications, *Sci. Technol. Adv. Mater.* **23**(1), 473–497, (2022).

101. L. Zhang, Y. Yuan, X. Qiu, T. Zhang, Q. Chen, and X. Huang, Marangoni effect-driven motion of miniature robots and generation of electricity on water, *Langmuir* **33**(44), 12609–12615 (2017).

102. M. Frenkel, A. Vilk, I. Legchenkova, S. Shoval, and E. Bormashenko, Mini-generator of electrical power exploiting the Marangoni flow inspired self-propulsion, *ACS Omega* **4**(12), 15265–15268 (2019).

103. P. Lovass, M. Branicki, R. Tóth, A. Braun, K. Suzuno, D. Ueyama, and I. Lagzi, Maze solving using temperature-induced Marangoni flow, *RSC Adv.* **5**, 48563 (2015).

104. F. Temprano-Coleto, F. J. Peaudecerf, J. R. Landel, F. Gibou, and P. Luzzatto-Fegiz, Soap opera in the maze: Geometry matters in Marangoni flows, *Phys. Rev. Fluids* **3**, 100507 (2018).

105. R. L. Yang, Y. J. Zhu, D. D. Qin, and Z. C. Xiong, Light-operated dual-mode propulsion at the liquid/air interface using flexible, superhydrophobic, and thermally stable photothermal paper, *ACS Appl. Mater. Interf.* **12**(1), 1339–1347 (2020).

106. L. Lin, M. Wang, X. Peng, E. N. Lissek, Z. Mao, L. Scarabelli, E. Adkins, S. Coskun, H. E. Unalan, B. A. Korgel, L. M. Liz-Marzán, E.-L. Florin, and Y. Zheng, Opto-thermoelectric nanotweezers, *Nat. Photon.* **12**(4), 195–201 (2018).

107. J. Li, L. Lin, Y. Inoue, and Y. Zheng, Opto-thermophoretic tweezers and assembly, *J. Micro Nano-Manuf.* **6**, 040801 (2018).

108. B. Kwak and J. Bae, Skimming and steering of a non-tethered miniature robot on the water surface using Marangoni propulsion. In *Proc. of the IEEE/RSJ International Conference on Intelligent Robots and Systems (IROS)*, Vancouver, Canada, pp. 3217–3222 (2017).

109. K. Dietrich, N. Jaensson, I. Buttinoni, G. Volpe, and L. Isa, Microscale Marangoni surfers, *Phys. Rev. Lett.* **125**(9), 098001 (2020).

110. H. Jin, A. Marmur, O. Ikkala, and R. H. A. Ras, Vapour-driven Marangoni propulsion: Continuous, prolonged and tunable motion, *Chem. Sci.* **3**(8), 2526 (2012).

111. D. Pan, D. Wu, P. Li, S. Ji, X. Nie, S. Fan, G. Chen, C. Zhang, C. Xin, B. Xu, S. Zhu, Z. Cai, Y. Hu, J. Li, and J. Chu, Transparent light-driven hydrogel actuator based on photothermal Marangoni effect and buoyancy flow for three-dimensional motion, *Adv. Funct. Mater.* **31**(14), 2009386 (2021).

112. X. Wang, L. Dai, N. Jiao, S. Tung, and L. Liu, Superhydrophobic photothermal graphene composites and their functional applications in microrobots swimming at the air/water interface, *Chem. Eng. J.* **422**, 129394 (2021).

113. D. M. Zhang, Y. Y. Sun, M. T. Li, H. Zhang, B. Song, and B. Dong, A Phototactic Liquid Micromotor, *J. Mater. Chem. C* **6**, 12234–12239 (2018.)
114. M. Cheng, G. Zhu, L. Li, S. Zhang, D. Zhang, A. J. C. Kuehne, and F. Shi, Parallel and precise macroscopic supramolecular assembly through prolonged Marangoni motion, *Angew. Chem. Int. Ed.* **57**, 14106–14110 (2018).
115. C. Farzeena and S. N. Varanakkottu, Patterning of metallic nanoparticles over solid surfaces from sessile droplets by thermoplasmonically controlled liquid flow, *Langmuir* **38**(6), 2003–2013 (2022).
116. D. E. Lucchetta, F. Simoni, L. Nucara, and R. Castagna, Controlled-motion of floating macro-objects induced by light, *AIP Advances* **5**, 077147 (2015).
117. H. Kim, S. Sundaram, J. H. Kang, N. Tanjeem, T. Emrick, and R. C. Hayward, Coupled oscillation and spinning of photothermal particles in Marangoni optical traps, *Proc. Natl. Acad. Sci. USA* **118**, e2024581118 (2021).
118. S. Nagelberg, J. F. Totz, M. Mittasch, V. Sresht, L. Zeininger, T. M. Swager, M. Kreysing, and M. Kolle, Actuation of Janus emulsion droplets via optothermally induced Marangoni forces, *Phys. Rev. Lett.* **127**(14), 144503 (2021).
119. C. Lv, S. N. Varanakkottu, T. Baier, and S. Hardt, Controlling the trajectories of nano/micro particles using light-actuated Marangoni flow, *Nano Lett.* **18**(11), 6924–6930 (2018).
120. Z. Mao, G. Shimamoto, and S. Maeda, Conical frustum gel driven by the Marangoni effect for a motor without a stator, *Colloids Surf. A* **608**, 125561 (2021).

https://doi.org/10.1142/9789811280603_0003

Chapter 3

Molecular Alignment Patterning Enabled by Novel Photopolymerization with Structured Light and its Optical Applications

Sayuri Hashimoto[*,†], Kyohei Hisano[*,†], Miho Aizawa[*,†,‡],
and Atsushi Shishido[*,†,§,¶]

[*]Laboratory for Chemistry and Life Science,
Institute of Innovative Research, Tokyo Institute of Technology,
4259 Nagatsuta, Midori-ku, Yokohama 226-8503, Japan
[†]Department of Chemical Science and Engineering,
Tokyo Institute of Technology, 2-12-1 Ookayama, Meguro-ku,
Tokyo 152-8552, Japan
[‡]PRESTO, JST, 4-1-8 Honcho, Kawaguchi 332-0012, Japan
[§]Living Systems Materialogy (LiSM) Research Group,
International Research Frontiers Initiative (IRFI),
Tokyo Institute of Technology, 4259 Nagatsuta,
Midori-ku, Yokohama 226-8503, Japan
[¶]ashishid@res.titech.ac.jp

Precise patterning of two-dimensional (2D) alignment of liquid crystals (LCs) has attracted great attention with a view to enhancing their optical and mechanical functionalities. However, current photoalignment methods still require doping of photo-responsive dyes and polarized light or multi-step process to induce molecular alignment. Our group has developed a novel photoalignment method based on the concept of scanning wave photopolymerization (SWaP). This method can generate 2D alignment patterning of LCs over large areas in a single step. The key to alignment is molecular diffusion induced by spatiotemporal photopolymerization, with the great advantages of requiring no polarized light, photo-responsive dyes, or specific surface treatment. Furthermore, various types of polymerization reactions can be applied to SWaP. We believe that SWaP has great potential to realize advanced optical devices and mechanical soft actuators based on 2D alignment patterning.

1. Introduction

Alignment control of liquid crystals (LCs) is of interest for a wide variety of applications to improve material functionalities, such as photonics, optoelectronics, and soft robotics.[1-4] While designing molecular structures of LCs, it can be noted that they exhibit new functions and properties due to their one-dimensional (1D), two-dimensional (2D), and three-dimensional (3D) nanostructures.[2,5] As photonic devices, a 1D-aligned LC film can act as a half-wave plate as a converter of linearly polarized light, where retardation ($R = \Delta n \cdot d$; d is the thickness of the film) exactly matches $\lambda/2$ (λ: wavelength of incident light). In the same manner, more complex 2D alignment patterns of LCs make one accomplish versatile geometrical phase modulations, producing various optical devices (e.g., q-plate lenses, optical vortex waveplate, and beam steering).[6-12] Moreover, since LCs have the potential to change their optical anisotropy depending on the degree of molecular alignment, they are expected to be applied to photonic tunable devices.

As a simple and robust method to align LC molecules, mechanical processes such as stretching films and rubbing the surface of substrates are practically used. Mechanical processes are fundamentally versatile in terms of target molecules. Still, they have some drawbacks, for instance, contamination of dust, generation of static electricity, and difficulty of fine alignment control with complex patterns. As a good alternative, photo-based alignment methods (photoalignment) have been developed in the modern era. When irradiated with the linearly polarized light, dye molecules are aligned along the polarization direction through photochemical reactions, such as photoisomerization, photocrosslinking, or photodegradation.[7,13-18] Azobenzene molecules can be reversibly aligned between homeotropic (out-of-plane) and homogeneous (in-plane) states by using trans-cis photoisomerization by irradiation with linearly polarized light over the surface layer.[19,20] The photoalignment method has the advantage of forming the complex 2D alignment pattern simply by controlling the polarization direction of the incident light. However, even with advanced photoalignment techniques, irradiating photo-responsive dyes with polarized light is still required. The doped dye might cause the degradation of materials' optical and mechanical functions. Two dye-free approaches have been explored extensively. One is a two-step alignment method; using a very thin dye-containing photoalignment layer to coat over on LC surface, the LCs are aligned and fixed by a subsequent polymerization.[3,21-23] In this alignment method, the complex molecular alignment pattern was achieved on a free surface by coating an azobenzene polymer with an ink-jet printing system.[22,24] The other utilizes surface topography. LCs can be aligned over a surface topography template fabricated by lithography, nanoimprinting, and ink-jet printing.[25-28] However, these methods should have a multi-step process to prepare these surfaces. Further, the surface topography makes it difficult to fabricate thin LC films because of their surface roughness derived from the templates.

Our group has conducted a series of research and proposed a novel photoalignment system based on a scanning wave photopolymerization (SWaP), which uses spatiotemporal scanning ultra-violet (UV) light. SWaP triggers mass flow in the film as the polymerization reaction proceeds, resulting in the alignment of LC molecules depending on the spatiotemporal optical pattern of the incident light.[29,30] This approach can generate arbitrary 2D alignment patterns over large areas without the necessity of any added dyes, polarized light, and subsequent or pre-processing steps (Fig. 1(a)). Note that the optical pattern determines the resultant molecular alignment pattern, and the size of the pattern is, in principle, restricted only by the light diffraction limit. For example, we successfully fabricated an array (500×300) of radial molecular alignment patterns that are 10^{-4} size ($\sim 27.4\,\mu m$) compared to conventional photoalignment methods (Fig.1(b)). Further, SWaP does not need specific surface treatment, allowing us to develop flexible bending films over a large area (Fig. 1(c) and 1(d)).

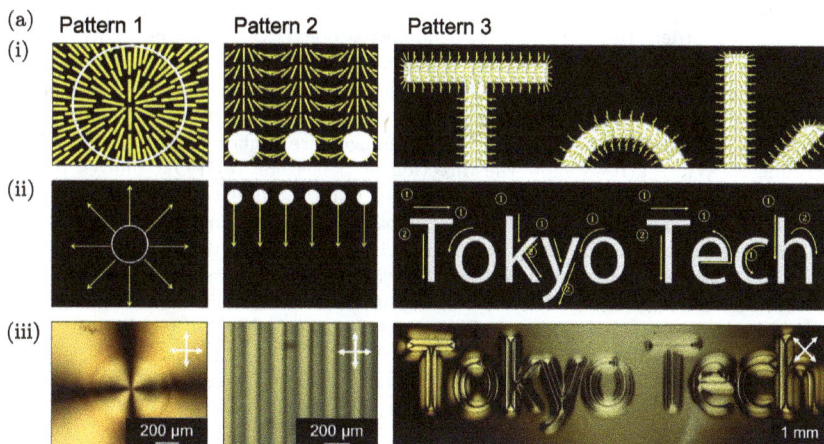

Figure 1.   (*Continued*)

(b)

(c)   (d)

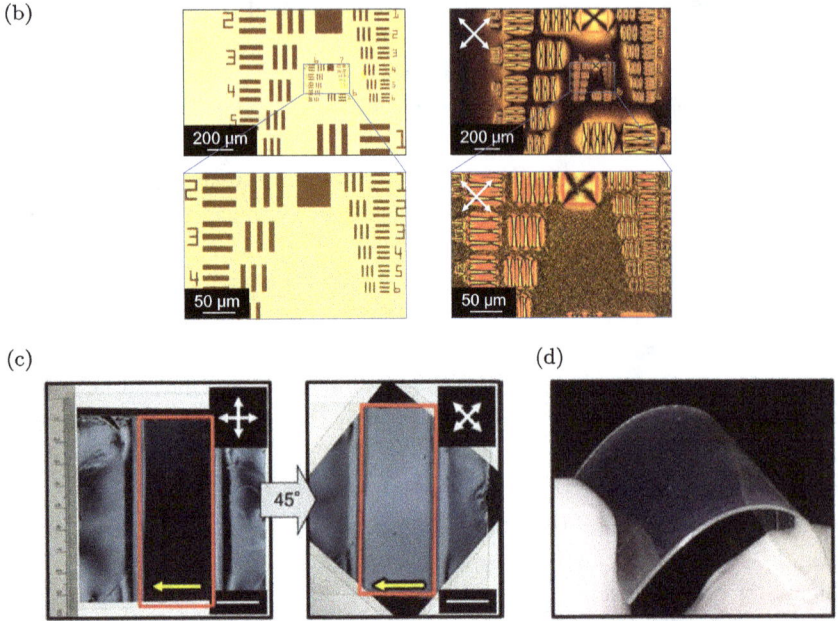

Figure 1.   (a) Arbitrary 2D alignment patterns developed by SWaP. (i) Schematic illustrations of desired patterns of alignment of mesogenic side-chain unit (yellow rod). (ii) Irradiated patterns: (1) toroid shape expanding at $27.4\,\mu$m/s with a width of $247\,\mu$m and a diameter of $493\,\mu$m in the initial state; (2) periodic dots scanned in 1D at $27.4\,\mu$m/s with a diameter of $137\,\mu$m and a center-to-center distance of $206\,\mu$m; and (3) the words "Tokyo Tech" where 178-μm-wide light was scanned at 13.7 mm/s and an exposure time of 1 min. White and black areas represent irradiated and non-irradiated regions, respectively. The yellow arrows show the scanning direction, and the numbers denote the scanning order. (iii) Polarized optical microscope (POM) images under crossed polarizers. White arrows show the direction of polarizers. (b) High-resolution patterning of molecular alignment by SWaP. Photograph of a test target used as a photomask having the highest resolution of $\sim$2-μm line space (left), and a POM image of the resultant polymer film (right). The bottom images are higher magnifications of the top images. White arrows show the direction of polarizers. (c) POM images rotated by $45°$ from one another of a molecularly aligned film over an extraordinarily large area of 5 cm $\times$ 9 cm on a grass substrate. The photopolymerized sample was composed of OXT-M and CPI-210S. The scale bar is 2 cm; white arrows show the alignment direction of the polarizers; yellow arrows show the light scanning direction; red rectangles represent the light scanned region. (d) Flexible bending of a molecularly aligned film over a COP. Reproduced with permission from Ref. 30. Copyright 2017, The Authors licensed under CC BY 4.0.

In this chapter, we focus on the photoalignment system, SWaP. First, Sec. 2 introduces the concept of SWaP, and Sec. 3 describes its mechanism. Section 4 focuses on the versatility of SWaP for aligning LCs in polymer films. Finally, Sec. 5 presents current research trends in optical functions.

2. A Novel Concept of Photoalignment Method Enabled by Molecular Diffusion

Molecular diffusion in photochemical reaction systems has been widely explored, especially in the field of photolithography. In photolithography, the photochemical reaction enables the creation of precise and fine nanoscale resolution. One of the critical challenges is the inscription of a clear line edge without roughness; thus, many studies have been conducted over the past few decades to suppress molecular diffusion.[6] By contrast, Broer *et al.*[31-33] revealed that photopolymerization of monomer mixtures with different reactivity could generate molecular diffusion and control the spatial patterning of concentrations between the irradiated and non-irradiated regions. Inspired by this work, Shishido *et al.* serendipitously found that photopolymerization-induced diffusion has the ability to align molecules even without any alignment layers, polarized incident light, and photo-responsive dyes.

The first demonstration of the concept of SWaP was carried out in 2017. Figure 2 shows examples of molecules in which SWaP can control 2D alignment. The detailed results and mechanism are described in Secs. 2 and 3, respectively. Figure 3 shows a representative result of SWaP. Here, the sample mixture composed of an acrylate monomer of cyanobiphenyl derivative (A6CB), a methacrylate crosslinker (HDDMA), and a photoinitiator (Irgacure 651) was used.[30] The sample was filled into a handmade glass cell (2 cm × 2 cm, thickness: 2–3 μm) and then irradiated with the scanned UV slit light (λ = 365 nm; light intensity, 1.2 mW/cm^2; scanning rate, 20 μm/s; slit width, 250 μm) to initiate free-radical photopolymerization. The photopolymerization was conducted at 100°C, where the resultant polymer film showed an LC phase, and the film was cooled below its glass transition temperature in liquid nitrogen.

Polarized optical microscopy (POM) observation of the resultant polymer film under crossed polarizers exhibited a clear contrast at every 45° rotation. At the same time, the image became completely dark when the polarization direction and the light scanning direction were parallel or perpendicular (Fig. 3(a)). Detailed POM observations with a

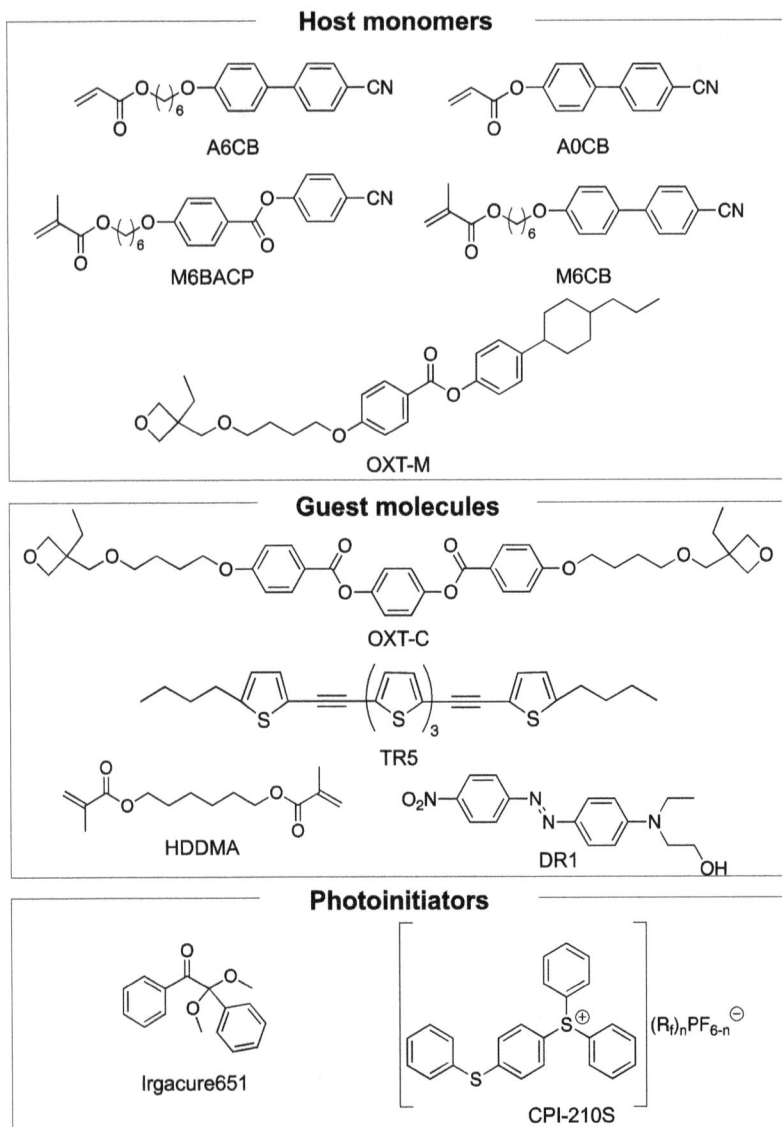

Figure 2. Representative examples of chemical structures of host monomers, guest molecules and photoinitiators that can be used for SWaP.

(a)   (b)   (c)

(d)

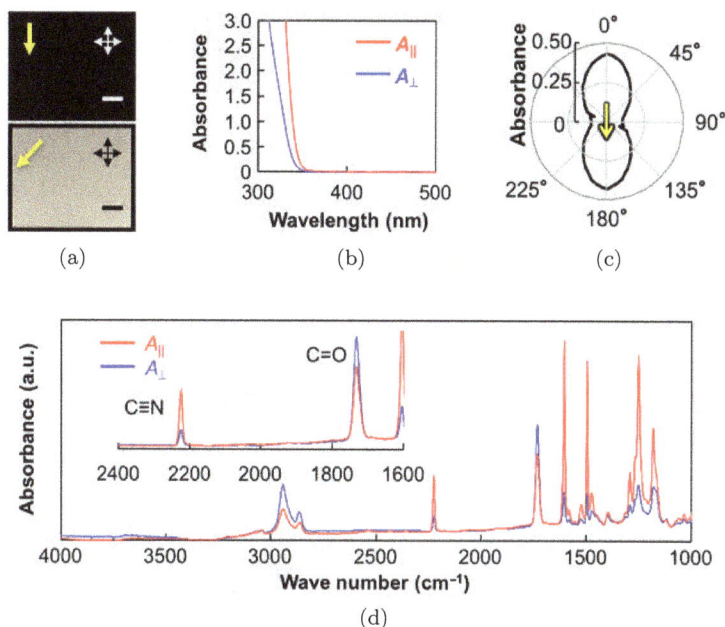

Figure 3. (a)–(d) Characterization of molecular alignment directed by SWaP. The photopolymerized sample was composed of A6CB, HDDMA and Irgacure 651. (a) POM images rotated by 45° from one another under crossed polarizers. Scale bars, 100 μm. Yellow arrows indicate the light scanning direction; black and white arrows depict the direction of the polarizers. (b) Polarized UV–vis absorption spectra, (c) a polar plot of UV–vis absorbance, and (d) polarized IR absorption spectra of the resultant film photopolymerized with light through a 250-μm slit scanned in 1D at 20 μm/s. Reproduced with permission from Ref. 30. Copyright 2017, The Authors licensed under CC BY 4.0.

tint plate with retardation of 137 nm revealed that the slow axis of uniform optical anisotropy was along the light scanning direction and almost the same as the range of typical birefringence levels of nematic LCs.[34] The optical anisotropy was investigated more quantitatively. Figure 3(b) shows polarized UV–visible (UV–vis) absorption spectra of the film, and Fig. 3(c) shows a polar plot of the absorbance in the wavelength region of the cyanobiphenyl mesogenic unit. The absorbance exhibited the highest value along the scanning direction. The order parameter (S) which is one of the evaluation methods for the in-plane molecular orientation was evaluated by the equation $S = (A_{\parallel} - A_{\perp})/(A_{\parallel} + 2A_{\perp})$, where $A_{\parallel}$ and $A_{\perp}$ are defined as the absorbance in which the direction of polarized incident light was parallel or perpendicular to the light scanning direction, respectively. The S value of the resultant film was 0.52. POM observation showed that the birefringence was greater than 0.1.[48] This means that the resulting polymer film possesses a uniform optical anisotropy arising from the unidirectional molecular alignment of mesogenic units. The value of S and birefringence is sufficient for industrial applications compared with similar chemical systems via conventional photoalignment methods.[7,35-37]

Further investigations were conducted to confirm the direction of the polymer main chain using polarized infrared (IR) absorption spectra of the 4-μm-thick freestanding film. Very interestingly, carbonyl moieties were aligned throughout the film in addition to cyanobiphenyl moieties (Fig. 3(d)). Moreover, we found the alignment direction of carbonyl moieties to be perpendicular to that of the cyanobiphenyl moieties. This result suggests that the side chains are aligned alongside their polymer main chain under a shear-flow field.[38-40] Consequently, the alignment direction of the carbonyl groups becomes perpendicular to the polymer main chain.[37,41] We have concluded that SWaP can align the polymer main chains along the light scanning direction.

3. The Detailed Mechanism of Molecular Alignment by SWaP

The mechanism of 2D molecular alignment achieved by SWaP is discussed in the following (Fig. 4(a)). Initially, the alignment of monomers is random. When UV slit light is scanned, monomers are photopolymerized only in the UV-irradiated area. A spatial gradient of chemical potential is generated around the boundary of irradiated and non-irradiated regions.[29,32,42-45] Based on the chemical potential gradient, molecular diffusion and molecular flow are induced. The 1D scanned light at a constant rate creates a stationary nonequilibrium state in which a thermodynamic mass flow is maintained along the light scanning direction. This flow can impose shear stress on monomers and polymers to align the polymer main chains and the side chain mesogens along the flow direction.[38-40] Finally, the whole cell is irradiated by UV light to complete photopolymerization and immobilize the resultant alignment. A stable macroscopic 2D molecularly aligned polymer film is obtained. In contrast, when spatiotemporal uniform UV light irradiates the LC samples, the initial random alignment of monomers is immobilized. Their isotropy is stabilized, leading to a fixed polydomain structure.

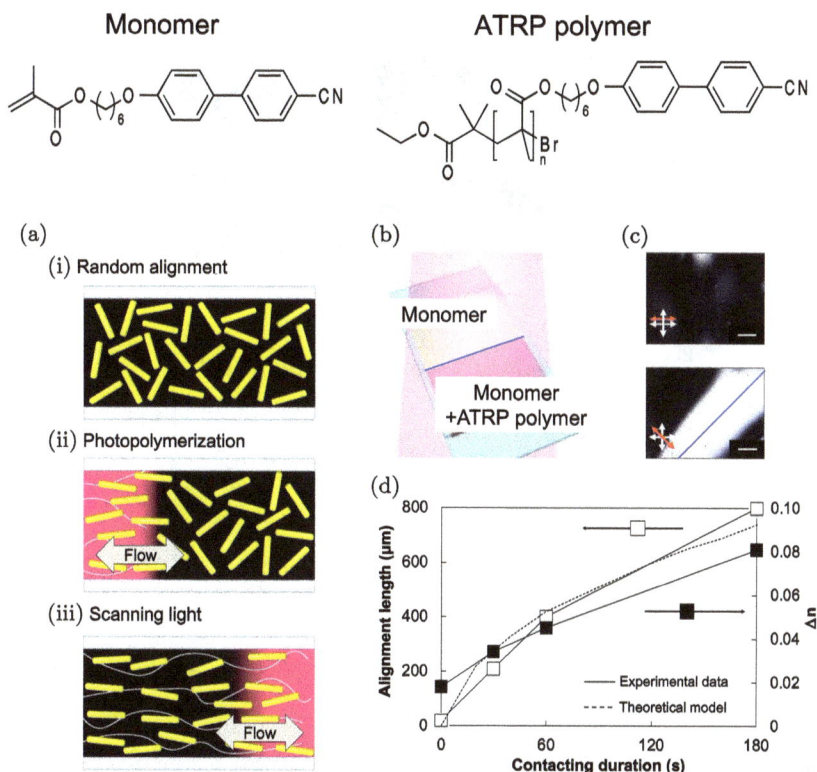

Figure 4. Evidence that molecular diffusion drives alignment. (a) Schematic illustrations of the generation of molecular alignment during SWaP where a mass flow triggered by photopolymerization aligns both the mesogenic side-chain units (yellow rod) and the polymer main chains (white curves) over large areas by scanning light. (b) A schematic illustration of photopolymerization at the boundary (blue line) of two mixtures injected into a glass cell from separate sides. The chemical structures of monomer and ATRP polymer are shown as above. (c) POM images under crossed polarizers offset 0° (top) and −45° (bottom) from vertical of resultant films photopolymerized 180 s after contacting the two mixtures. Red arrows show the injection direction; white arrows show the direction of polarizers. Scale bars, 200 µm. (d) Alignment length and birefringence emerged as a function of flow duration compared with a theoretical model. Reproduced with permission from Ref. 30. Copyright 2017, The Authors licensed under CC BY 4.0.

To investigate the molecular diffusion and molecular flow, Hisano *et al.*[29] fabricated molecularly aligned LC films by UV irradiation using a photomask with a striped pattern and observed periodic peak/valley structures in the surface of the LC film. Hashimoto *et al.*[46] explored the complex relationship between the surface relief structures and molecular alignment.

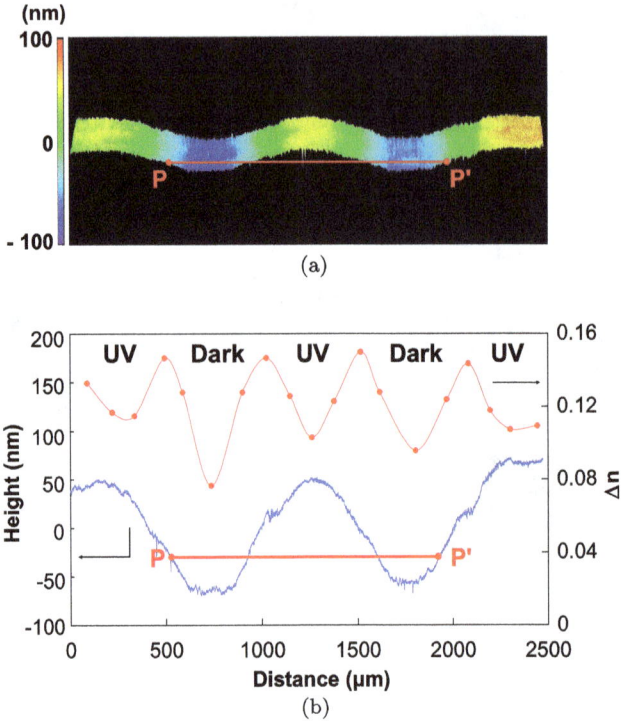

Figure 5.   (a) Laser confocal microscope image and (b) the birefringence and surface profile of the film. The film was formed by striped-pattern UV irradiation with the grating periods of 986 μm. Purple regions show the irradiated areas. Reprinted with permission from Ref. 46. Copyright 2019, The Authors licensed under CC BY 4.0.

The LC polymer film fabricated by striped-pattern UV irradiation formed periodic peak/valley structures with a maximum height of 120 nm. The period of these structures was the same as the grating slit width of the irradiated light patterns (Fig. 5). As shown in Fig. 5(b), the birefringence showed the highest value at the boundaries between the irradiated and non-irradiated areas. In contrast, smaller values were obtained in the centers of the irradiated and non-irradiated areas compared to the boundary areas. This result indicated the molecular alignment is disordered in the center areas, which reflects that the monomers diffuse into irradiated areas.

As a proof-of-concept experiment of molecular diffusion-triggered alignment, we designed the test photopolymerization system to exclude other possible factors, e.g., polymerization-induced heat and volume shrinkage.[30]

Two different sample mixtures were prepared: one contained a monomer of M6CB and photoinitiator of Irgacure 651; the other had the mixture of M6CB, Irgacure 651, and a polymer synthesized by atom transfer radical polymerization (ATRP) of M6CB [M_n (number-average molecular weight) = 8000]. These sample mixtures were injected into a glass cell from each side toward its center by capillary forces, allowing molecular diffusion to occur at the contacted boundary between them without photopolymerization (Fig. 4(b)). After the contacting duration of 0, 30, 60, and 180 s, flood UV irradiations of the glass cell were performed to immobilize any molecular alignment induced by diffusion at the boundary. Figure 4(c) shows POM images of resultant films photopolymerized 180 s after contacting the two mixtures. As a result, unidirectional molecular alignment perpendicular to the boundary was induced where the alignment direction was the same as the diffusion direction. Increasing the contacting duration increased the alignment length and the value of birefringence to 800 μm and 0.08, respectively (Fig. 4(d)).

Moreover, comparing the observed experimental data with the developed theoretical model of alignment length based on the assumption of Fick's law of diffusion, the theoretical analysis agrees very well with the experimental data. These results quantitatively supported the mechanism of molecular alignment by the SWaP system. Further, Ueda *et al.*[47] demonstrated and visualized the directional molecular diffusion during patterned photopolymerization by monitoring the position of doped luminescent quantum dots (QDs) of CdSe/ZnS core/shell type with red-emission (Fig. 6). The sample was composed of the monomer (A6CB) doped with the QDs at appropriate concentrations. Figure 6(a) shows the result of real-time monitoring of QD diffusion under UV irradiation using a photomask; numbers 1–6 denote the individual QDs tracked. Figure 6(b) shows the traces constructed from 600 consecutive fluorescence images taken every 0.1 s over 1 min. The motion of QD1, which was located in the masked region, reflects the random Brownian motion throughout the measured time interval. On the other hand, QDs 2–6, which were located in the polymerized region, exhibited translational diffusion to the masked region as time passed. This evidence of a directed molecular diffusion and mass flow field between polymerized and masked regions indicates that photopolymerization with structured light induces unidirectional mass flow. This flow would be the key to aligning molecules even without polarized light and dyes.

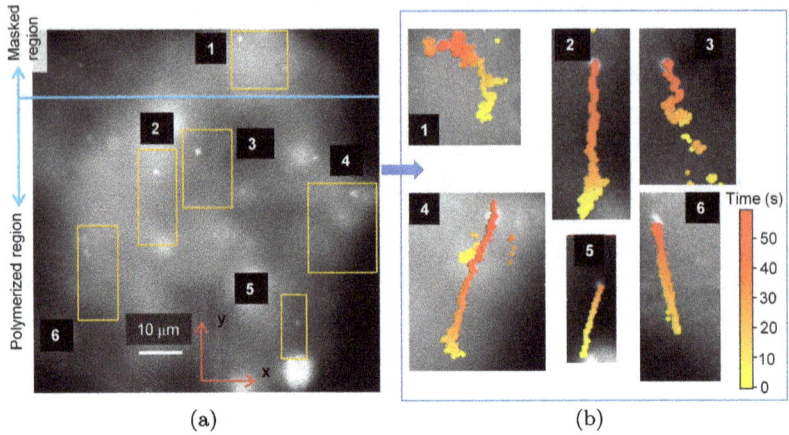

Figure 6. Fluorescence monitoring of QD diffusion near the photomask. (a) Fluorescence image of the whole Area 1 near the photomask taken after completion of the photopolymerization reaction. The numbers 1–6 denote individual QDs, and the yellow rectangle areas are magnified in (b); the red arrows define the directions x (parallel to the photomask) and y (perpendicular to it). (b) Diffusion traces of QDs 1–6 monitored over 60 s at 0.1 s steps. The positions in each step are determined from 2D Gaussian fitting of the fluorescence images. The blue horizontal line shows the edge of the photomask. Reprinted with permission from Ref. 47. Copyright 2021, The Authors licensed under CC BY 4.0.

4. Versatility of Polymerization System in SWaP for Uniform Molecular and Mesoscopic Alignment

As described above, SWaP controls the molecular alignment direction using the shear stress arising from photopolymerization-induced diffusion. Thus, SWaP has the significant advantages of (i) the extremely few restrictions for selecting molecular structures and polymerization systems and (ii) diffusion-directed alignment, which might not be affected by the surface condition of substrates. In fact, the versatility of SWaP is demonstrated in terms of the chemical structure of monomers, crosslinkers, photoinitiators, and dye dopants, polymerization method, substrate, and surface treatment.[48–53]

Ishizu et al.[48] investigated the effect of surface treatment on molecular alignment behavior induced by SWaP. A6CB monomer in a rubbing-treated glass cell was conducted by SWaP, where UV light was 1D scanned along or across the rubbing direction. The resultant films showed unidirectional molecular alignment parallel to the light scanning direction, independent of the rubbing direction. The S value of these films was found to be ~0.5, and both films had a birefringence of 0.11, indicating that surface anchoring from the substrates has no effect on the generation or the degree of molecular alignment achieved by SWaP. Furthermore, to compare the alignment ability of SWaP with the rubbing treatment, Ishiyama *et al.*[49] synthesized soluble polymers of M6BACP, which has a mesogenic unit of a phenyl benzoate derivative. Although the drop-cast of the soluble polymer solution over a rubbing treated substrate did not result in a uniform molecular alignment, SWaP of M6BACP led to induce uniform molecular alignment along the light scanning direction. This result indicates that SWaP is a novel pathway to align highly viscous polymers even without liquid crystallinity in the monomer state.

Our group demonstrated alignment control of dopant compounds and higher-order structures in addition to polymerized molecules. Aizawa *et al.*[50] utilized SWaP to align doped azobenzene molecules, one of the most common photo-responsive molecules. The sample mixture of A6CB, HDDMA, Irgacure 651, and doped DR1 at a concentration of 0.5 mol% was photopolymerized with a scanning light. The result of polarized UV–vis absorption spectra revealed that the cyanobiphenyl moieties and azobenzene molecules were well aligned parallel to the light scanning direction in the film. The S value of the cyanobiphenyl moieties and DR1 were 0.37 and 0.55, respectively, which revealed that SWaP does not require linearly polarized light to align azobenzene moieties. In general photoalignment systems, irradiation with polarized light is essential to induce the alignment of azobenzene. SWaP enables the alignment of doped azobenzene moieties with unpolarized light and by obtaining high-resolution alignment patterns.

Ishizu *et al.*[51] reported alignment control of smectic layer structures in liquid-crystalline polymers. Two types of monomers were prepared: A6CB monomer, which shows a nematic phase when polymerized, and A0CB monomer, which has a smectic phase without an alkyl spacer. Polarized UV–vis spectroscopy and polarized Fourier transform infrared (FTIR) spectroscopy of the films revealed that both side-chain mesogens and polymer main chains of the polymerized A6CB film were aligned parallel to the light scanning direction. By contrast, A0CB film had the parallel polymer main chain alignment but the mesogenic alignment perpendicular to the scanning direction. The grazing-incidence small-angle X-ray scattering (GI–SAXS) data analyzed the nanostructures of the polymerized A6CB and A0CB film (Fig. 7(a)). X-ray diffraction was not

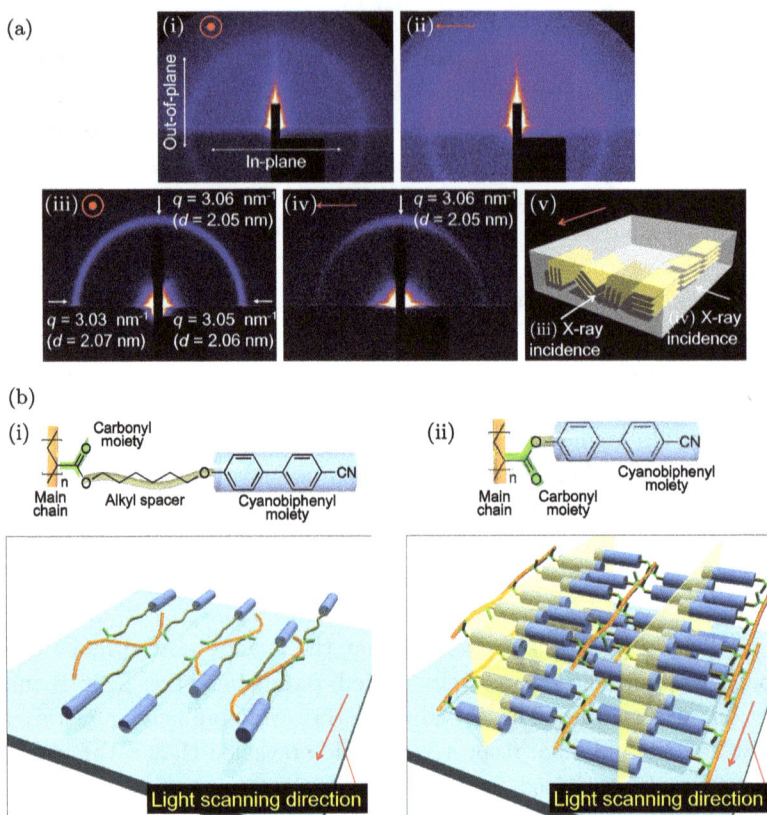

Figure 7.   (a) 2D GI-SAXS diffractograms for the polymerized A6CB film using parallel (i) or perpendicular (ii) X-ray incidence to the scanning direction, respectively, and for the polymerized A0CB film using parallel (iii) or perpendicular (iv) X-ray incidence, respectively. (v) Possible 3D nanostructures made of smectic layers (represented by yellow layers) in the polymerized A0CB film. Red circles and arrows represent the light scanning directions. (b) 3D schematics of the possible alignment structure of the films obtained by photopolymerization with the 1D scanned slit light of (i) A6CB and (ii) A0CB mixtures. Yellow planes indicate smectic layers. Reproduced with permission from Ref. 51. Copyright 2022, American Chemical Society.

observed for the polymerized A6CB film ((i) and (ii) in Fig. 7(a)), meaning that there was no periodic nanostructure due to the mesogen alignment in the film.

In contrast, the polymerized A0CB film showed a uniform X-ray diffraction pattern in both in-plane and out-of-plane directions when the incident X-ray beam entered parallel to the scanning direction ((iii) in Fig. 7(a)). In addition, a perpendicular incidence of the X-ray beam to the scanning direction resulted in only out-of-plane diffraction ((iv) in Fig. 7(a)). These data revealed that the polymerized A0CB film possessed a 3D anisotropic nanostructure of the smectic layers. These layers were aligned along the scanning direction but allowed to be rotated around the axis along the scanning direction ((ii) in Fig. 7(b)). Therefore, SWaP has a possibility toward the development of aligning not only LC mesogens but also mesoscopic anisotropic structures.

SWaP can be applied to cationic polymerization as well as radical polymerization. Hisano *et al.*[30] explored a SWaP system using cationic polymerization. SWaP of a cationic photoinitiator and cationic photopolymerizable compounds was conducted under ambient conditions over a glass substrate. Figure 1(b) shows POM images that indicate the successful production of 2D molecularly aligned polymer films by SWaP over a large area of 5 cm × 9 cm without a glass cell or a nitrogen atmosphere. In addition, SWaP achieved aligning molecules over a commercially available cyclo-olefin polymer (COP) film, which may apply SWaP to developing flexible devices (Fig. 1(d)).

One of the fascinating applications using LC polymers in the modern era is advanced mechanical materials such as soft actuators for soft robotics and haptics. For realizing them, LC-elastomer (LCE) and LC-polymer-network (LCN) films have attracted significant attention. In particular, the control of crosslinker density plays an essential role. Kobayashi *et al.*[52,53] succeeded in fabricating 2D-aligned LCN films by SWaP. The effect of crosslinker concentration on molecular alignment behavior was investigated in SWaP. The sample mixtures were prepared with the monomer of OXT-M and the crosslinker of OXT-C at the molar ratio of 100:0, 80:20, 50:50, 20:80, and 0:100, referred to as OXT_{C0}, OXT_{C20}, OXT_{C50}, OXT_{C80}, and OXT_{C100}, respectively. Scanning UV light to photopolymerize those samples induced in-plane alignment along the light scanning direction on the films of OXT_{C0}, OXT_{C20}, OXT_{C50}, and OXT_{C80}. The birefringence and the order parameter S showed maximum values of ∼0.08 and ∼0.28, respectively.

On the other hand, OXT_{C100} film only with the crosslinker showed slight perpendicular alignment to the scanning direction because OXT-C is an LC crosslinker with two functional groups. These results revealed that the main chain affects the direction of molecular alignment, and consequently, the mesogens in the OXT_{C100} film are aligned perpendicular to the light scanning direction. Moreover, the thermal stability of molecular alignment in the LCN films was enhanced by an increase in the amount of crosslinker concentration. SWaP enables us to fabricate the molecularly aligned LCN films with high thermal stability.

5. Two-dimensional Alignment Patterning and its Potential Applications to Optics

Representative applications of 2D-aligned materials include optical devices that can realize birefringence modulation and polarized conversion. In developing such advanced optical devices, making molecules controlled with precise patterning or array pattern is imperative. SWaP has the advantage of forming 2D alignment patterns only by irradiating complex light patterns, making it promising for next-generated optical applications. Our group demonstrated various applications with desired molecular alignment patterns. Hisano et al.[30] successfully induced arbitrary alignment patterns with fine control, such as radial patterns, cycloids, and the words developed by spatiotemporal scanning of UV light (Fig. 1(a)). Optical applications were reported utilizing the 2D-aligned LC polymer films fabricated by SWaP. Aizawa et al.[54] demonstrated that the LC polymer film acts as a highly efficient q-plate. The film showed radial molecular alignment over a large area induced by irradiation of periodically hexagonal array (Fig. 8(a)). The linearly polarized beam incident on the film was converted into a vector beam when the retardation of the patterned film was adjusted to the half-waveplate condition (Figs. 8(b)–8(d)).

Additionally, by sending a beam, more than 50 donut-shaped vector beams were generated over large areas in a single step (Fig. 8(e)). Hisano et al.[55] and Nakamura et al.[56] fabricated cycloidal diffractive waveplates (CDWs) that can convert an incident beam into right- and left-circularly polarized beams. The rod-shaped pattern during 1D scanning of light for photopolymerization generated cycloidal molecular alignment. SWaP can be a platform for inscribing 2D molecular alignment patterns quickly, efficiently, and inexpensively over large areas. Q-plates, CDWs, and various types of DWs with optical functionalities, such as optical vortex converting, and beam steering, will be industrially fabricated by SWaP.

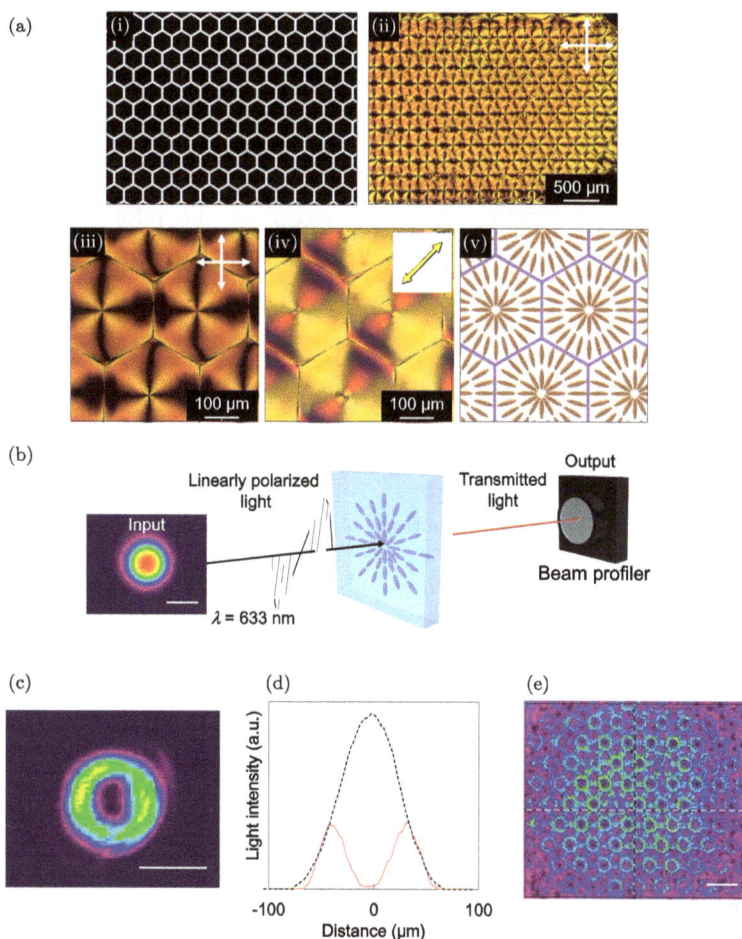

Figure 8. (a) Hexagonal array made by patterned photopolymerization. (i) Optical pattern of a hexagonal lattice. White lines represent the irradiated regions. (ii)–(v) POM images of the resultant polymer film under crossed polarizers without a tint plate (ii), (iii) and with a tint plate having a retardation of 137 nm (iv), respectively. (v) Schematic illustration of the resultant molecular alignment pattern generated from the photoirradiation with the optical pattern. Blue lines define the irradiated region, and orange ellipses indicate the long-axis direction of molecules. (b) Schematic illustration of the optical setup for evaluation of optical properties where a linearly polarized He–Ne laser beam is propagated. The inset shows the 2D intensity profile of an incident He–Ne laser beam with a Gaussian distribution. (c) Beam profile and (d) 1D intensity profiles of a He–Ne laser beam that propagated through the resultant polymer film of (iii) with radial alignment. In 1D profiles, red and black broken lines depict the 1D profile of the transmitted and of incident laser beams, respectively. The scale bar = 100 μm. (e) Beam profile of the transmitted light through the polymer film of (ii) with array patterns. Scale bar = 500 μm. Reprinted with permission from Ref. 54. Copyright 2019, The Optical Society.

6. Conclusion

This chapter demonstrated the novel, direct photoalignment system that allows for 2D alignment patterns over large areas without surface treatment, electrodes, polarized light sources, or photo-responsive dyes. The critical mechanism of SWaP is molecular diffusion based on chemical potential gradient by spatial selective photopolymerization. Therefore, various LC molecules can uniformly align in one direction or complex patterns, enabling the development of advanced optical devices. SWaP provides a new and powerful pathway to the straightforward design of highly functional organic materials and devices utilizing arbitrary, delicate molecular alignment patterns.

References

1. R. Lakes, Materials with structural hierarchy, *Nature* **361**(6412), 511–515 (1993).
2. J. Uchida, B. Soberats, M. Gupta and T. Kato, Advanced functional liquid crystals, *Adv. Mater.* **34**, 2109063 (2022).
3. T. J. White and D. J. Broer, Programmable and adaptive mechanics with liquid crystal polymer networks and elastomers, *Nat. Mater.* **14**(11), 1087–1098 (2015).
4. M. O'Neill and S. M. Kelly, Liquid crystals for charge transport, luminescence, and photonics, *Adv. Mater.* **15**(14), 1135–1146 (2003).
5. T. Kato, J. Uchida, T. Ichikawa, and T. Sakamoto, Functional liquid crystals towards the next generation of materials, *Angew. Chem. Int. Ed.* **57**(16), 4355–4371 (2018).
6. A. Shishido, Rewritable holograms based on azobenzene-containing liquid-crystalline polymers, *Polym. J.* **42**(7), 525–533 (2010).
7. V. Chigrinov, V. Kozenkov, and H. S. Kwok, *Photoalignment of Liquid Crystalline Materials: Physics and Applications*, John Wiley & Sons, England (2008).
8. S. R. Nersisyan, N. V. Tabiryan, D. M. Steeves, and B. R. Kimball, The promise of diffractive waveplates, *Opt. Photon. News* **21**(3), 40–45 (2010).
9. J. Kim, Y. Li, M. N. Miskiewicz, C. Oh, M. W. Kudenov, and M. J. Escuti, Fabrication of ideal geometric-phase holograms with arbitrary wavefronts, *Optica* **2**(11), 958–964 (2015).
10. S. V. Serak, D. E. Roberts, J.-Y. Hwang, S. R. Nersisyan, N. V. Tabiryan, T. J. Bunning, D. M. Steeves, and B. R. Kimball, Diffractive waveplate arrays, *J. Opt. Soc. Am. B* **34**(5), B56–B63 (2017).
11. K. Mehta, A. R. Peeketi, L. Liu, D. J. Broer, P. Onck, and R. K. Annabattula, Design and applications of light responsive liquid crystal polymer thin films, *Appl. Phys. Rev.* **7**(4), 041306 (2020).

12. G. P. Crawford, J. N. Eakin, M. D. Radcliffe, A. Callan-Jones, and R. A. Pelcovits, Liquid-crystal diffraction gratings using polarization holography alignment techniques, *J. Appl. Phys.* **98**(12), 123102 (2005).

13. O. Yaroshchuk and Y. Reznikov, Photoalignment of liquid crystals: Basics and current trends, *J. Mater. Chem.* **22**(2), 286–300 (2012).

14. T. Seki, S. Nagano, and M. Hara, Versatility of photoalignment techniques: From nematics to a wide range of functional materials, *Polymers* **54**(22), 6053–6072 (2013).

15. T. Seki, A wide array of photoinduced motions in molecular and macromolecular assemblies at interfaces, *Bull. Chem. Soc. Jpn.* **91**(7), 1026–1057 (2018).

16. A. Priimagi, C. J. Barrett, and A. Shishido, Recent twists in photoactuation and photoalignment control, *J. Mater. Chem. C* **2**(35), 7155–7162 (2014).

17. M. Schadt, K. Schmitt, and V. Chigrinov, Surface-induced parallel alignment of liquid crystals by linearly polymerized photopolymers, *Jpn. J. Appl. Phys.* **31**(7), 2155–2164 (1992).

18. N. Kawatsuki, Photoalignment and photoinduced molecular reorientation of photosensitive materials, *Chem. Lett.* **40**(6), 548–554 (2011).

19. K. Ichimura, Y. Suzuki, T. Seki, A. Hosoki, and K. Aoki, Reversible change in alignment mode of nematic liquid crystals regulated photochemically by "command surfaces" modified with an azobenzene monolayer, *Langmuir* **4**(5), 1214–1216 (1988).

20. K. Ichimura, Photoalignment of liquid-crystal systems, *Chem. Rev.* **100**(5), 1847–1873 (2000).

21. T. H. Ware, M. E. McConney, J. J. Wie, V. P. Tondiglia, and T. J. White, Voxelated liquid crystal elastomers, *Science* **347**(6225), 982–984 (2015).

22. K. Fukuhara, S. Nagano, M. Hara, and T. Seki, Free-surface molecular command systems for photoalignment of liquid crystalline materials, *Nat. Commun.* **5**(1), 1–8 (2014).

23. M. Schadt, H. Seiberle, and A. Schuster, Optical patterning of multi-domain liquid-crystal displays with wide viewing angles, *Nature* **381**(6579), 212–215 (1996).

24. T. Seki, Light-directed alignment, surface morphing and related processes: Recent trends, *J. Mater. Chem. C* **4**(34), 7895–7910 (2016).

25. V. K. Gupta and N. L. Abbott, Design of surfaces for patterned alignment of liquid crystals on planar and curved substrates, *Science* **276**(5318), 1533–1536 (1997).

26. Y. Xia, G. Cedillo-Servin, R. D. Kamien, and S. Yang, Guided folding of nematic liquid crystal elastomer sheets into 3D via patterned 1D microchannels, *Adv. Mater.* **28**(43), 9637–9643 (2016).

27. R. Lin and J. A. Rogers, Molecular-scale soft imprint lithography for alignment layers in liquid crystal devices, *Nano Lett.* **7**(6), 1613–1621 (2007).

28. M. Reznikov, A. Sharma, and T. Hegmann, Ink-jet printed nano particle alignment layers: Easy design and fabrication of patterned alignment layers for nematic liquid crystals, *Part. Part. Syst. Charact.* **31**(2), 257–265 (2014).

29. K. Hisano, Y. Kurata, M. Aizawa, M. Ishizu, T. Sasaki, and A. Shishido, Alignment layer-free molecular ordering induced by masked photopolymerization with non-polarized light, *Appl. Phys. Express* **9**(7), 072601 (2016).
30. K. Hisano, M. Aizawa, M. Ishizu, Y. Kurata, W. Nakano, N. Akamatsu, C. J. Barrett, and A. Shishido, Scanning wave photopolymerization enables dye-free alignment patterning of liquid crystals, *Sci. Adv.* **3**(11), e1701610 (2017).
31. C. F. Van Nostrum, R. J. Nolte, D. J. Broer, T. Fuhrman, and J. H. Wendorff, Photoinduced opposite diffusion of nematic and isotropic monomers during patterned photopolymerization, *Chem. Mater.* **10**(1), 135–145 (1998).
32. C. M. Leewis, A. M. de Jong, L. J. Van IJzendoorn, and D. J. Broer, Reaction-diffusion model for the preparation of polymer gratings by patterned ultraviolet illumination, *J. Appl. Phys.* **95**(8), 4125–4139 (2004).
33. C. Sánchez, B. J. De Gans, D. Kozodaev, A. Alexeev, M. J. Escuti, C. Van Heesch, B. Thijs, U. S. Schubert, C. W. M. Bastiaansen, and D. J. Broer, Photoembossing of Periodic Relief Structures Using Polymerization-induced Diffusion: a Combinatorial Study, *Adv. Mater.* **17**(21), 2567–2571 (2005).
34. D. Demus, J. W. Goodby, G. W. Gray, H. W. Spiess, and V. Vill, *Handbook of Liquid Crystals*, John Wiley & Sons, England (1998).
35. N. Kawatsuki, T. Washio, J. Kozuki, M. Kondo, T. Sasaki, and H. Ono, Photoinduced orientation of photoresponsive copolymers with N-benzylideneaniline and nonphotoreactive mesogenic side groups, *Polymers* **56**, 318–326 (2015).
36. T. Seki, New strategies and implications for the photoalignment of liquid crystalline polymers, *Polym. J.* **46**(11), 751–768 (2014).
37. Y. Wu, Y. Demachi, O. Tsutsumi, A. Kanazawa, T. Shiono, and T. Ikeda, Photoinduced alignment of polymer liquid crystals containing azobenzene moieties in the side chain. 2. Effect of spacer length of the azobenzene unit on alignment behavior, *Macromolecules* **31**(4), 1104–1108 (1998).
38. S. V. Fridrikh and E. M. Terentjev, Polydomain-monodomain transition in nematic elastomers, *Phys. Rev. E* **60**(2), 1847 (1999).
39. C. Pujolle-Robic and L. Noirez, Observation of shear-induced nematic-isotropic transition in side-chain liquid crystal polymers, *Nature* **409**(6817), 167–171 (2001).
40. A. Agrawal, A. C. Chipara, Y. Shamoo, P. K. Patra, B. J. Carey, P. M. Ajayan, W. G. Chapman, and R. Verduzco, Dynamic self-stiffening in liquid crystal elastomers, *Nat. Commun.* **4**(1), 1–6 (2013).
41. B. R. Nair, V. G. Gregoriou, and P. T. Hammond, FT-IR studies of side chain liquid crystalline thermoplastic elastomers, *Polymers* **41**(8), 2961–2970 (2000).
42. K. V. V. Krongauz, E. R. Schmelzer, and R. M. Yohannan, Kinetics of anisotropic photopolymerization in polymer matrix, *Polymers* **32**(9), 1654–1662 (1991).
43. D. J. Broer, J. Lub, and G. N. Mol, Wide-band reflective polarizers from cholesteric polymer networks with a pitch gradient, *Nature* **378**(6556), 467–469 (1995).

44. R. A. M. Hikmet and H. Kemperman, Electrically switchable mirrors and optical components made from liquid-crystal gels, *Nature* **392**(6675), 476–479 (1998).

45. R. L. Sutherland, V. P. Tondiglia, L. V. Natarajan, and T. J. Bunning, phenomenological model of anisotropic volume hologram formation in liquid-crystal-photopolymer mixtures, *J. Appl. Phys.* **96**(2), 951–965 (2004).

46. S. Hashimoto, M. Aizawa, N. Akamatsu, T. Sasaki, and A. Shishido, Simultaneous formation behaviour of surface structures and molecular alignment by patterned photopolymerisation, *Liq. Cryst.* **46**(13–14), 1995–2002 (2019).

47. K. Ueda, M. Aizawa, A. Shishido, and M. Vacha, Real-time molecular-level visualization of mass flow during patterned photopolymerization of liquid-crystalline monomers, *NPG Asia Mater.* **13**(1), 1–7 (2021).

48. M. Ishizu, M. Aizawa, N. Akamatsu, K. Hisano, S. Fujikawa, C. J. Barrett, and A. Shishido, Effect of surface treatment on molecular alignment behavior by scanning wave photopolymerization, *Appl. Phys. Express* **12**(4), 041004 (2019).

49. T. Ishiyama, Y. Kobayashi, H. Nakamura, M. Aizawa, K. Hisano, S. Kubo, and A. Shishido, Solubility and molecular alignment behavior of liquid-crystalline polymers by scanning wave photopolymerization, *Proc. SPIE* **12207**, 82–89 (2022).

50. M. Aizawa, K. Hisano, M. Ishizu, N. Akamatsu, C. J. Barrett, and A. Shishido, Unpolarized light-induced alignment of azobenzene by scanning wave photopolymerization, *Polym. J.* **50**(8), 753–759 (2018).

51. M. Ishizu, K. Hisano, M. Aizawa, C. J. Barrett, and A. Shishido, Alignment control of smectic layer structures in liquid-crystalline polymers by photopolymerization with scanned slit light, *ACS Appl. Mater. Interfaces* **14**, 48143–48149 (2022).

52. Y. Kobayashi, R. Taguchi, N. Akamatsu, and A. Shishido, Effect of the concentration gradient on molecular alignment by scanning wave photopolymerization, *J. Photopolym. Sci. Technol.* **33**(3), 291–294 (2020).

53. Y. Kobayashi, K. Hisano, M. Aizawa, M. Ishizu, N. Akamatsu, and A. Shishido, Liquid crystal polymer networks directed by scanning wave photopolymerization of oxetane monomer and crosslinker, *Mol. Cryst. Liq. Cryst.* **713**(1), 37–45 (2021).

54. M. Aizawa, M. Ota, K. Hisano, N. Akamatsu, T. Sasaki, C. J. Barrett, and A. Shishido, Direct fabrication of a Q-plate array by scanning wave photopolymerization, *J. Opt. Soc. Am. B* **36**(5), D47–D51 (2019).

55. K. Hisano, M. Ota, M. Aizawa, N. Akamatsu, C. J. Barrett, and A. Shishido, Single-step creation of polarization gratings by scanning wave photopolymerization with unpolarized light, *J. Opt. Soc. Am. B* **36**(5), D112–D118 (2019).

56. H. Nakamura, Y. Kobayashi, M. Ota, M. Aizawa, S. Kubo, and A. Shishido, Fabrication of diffractive waveplates by scanning wave photopolymerization with digital light processor, *J. Photopolym. Sci. Technol.* **34**(3), 225–230 (2021).

© 2024 World Scientific Publishing Company
https://doi.org/10.1142/9789811280603_0004

Chapter 4

Nonlinear Optical Propagation in Heliconical Cholesteric Liquid Crystals

Ashot H. Gevorgyan* and Francesco Simoni[†,‡]

*Institute of High Technologies and Advanced Materials,
Far Eastern Federal University, 10 Ajax Bay, Russky Island,
690922 Vladivostok, Russia
[†] Università Politecnica delle Marche, 60100 Ancona, Italy
and Institute of Applied Sciences and Intelligent Systems CNR,
80072 Pozzuoli, Italy
[‡] f.simoni@photomat.it

We show that Heliconical Cholesteric Liquid Crystals have peculiar nonlinear optical properties that make them represent a new paradigm for nonlinear optics of liquid crystals. The key point is the presence of the bend deformation of the conical structure, which allows strong coupling with the optical field and gives rise to optical reorientation of the molecular director. The consequence is the light-induced redshift of the Bragg resonance that has been recently demonstrated in pump–probe experiments. Here we discuss how this effect leads to strong nonlinear optical behavior originated by the amplitude change on a light beam powerful enough to induce optical reorientation in this material.

Using a multilayer approach, it is found that by increasing the light intensity after the initial redshift of the photonic bandgap, self-oscillations take place, while a further increase leads to chaotic behavior. By performing this calculation for the whole spectrum of visible wavelengths around the Bragg resonance, it is possible to show that the mentioned regimes occur at different intensities for different wavelengths, and the photonic bandgap disappears at high intensities. A simple model explains the onset of the oscillatory regime.

Then, Heliconical Cholesteric Liquid Crystals represent quite unique systems where the strong optical nonlinearity is due to light-induced modulation of the Bragg resonance, and it leads to different regimes for optical transmission that can be driven and controlled by static electric field and light intensity.

1. Introduction

Cholesteric liquid crystals (CLC) are well-known materials investigated since 1888 when the Austrian botanist and chemist Friedrich Reinitzer[1] discovered the first liquid crystal by studying cholesteryl benzoate. They are characterized by a helical structure around a specific axis (the helical axis): the average molecular direction given by the unit vector **n** (the director) lies on a plane perpendicular to the helical axis and turns around in a regular way, therefore, a 2π director rotation defines the pitch p of the structure along the axis. Light propagation through CLC has been deeply investigated by many scientists and their achievements are well referenced in different books on liquid crystal properties.[2-5] It has been shown that light has two main propagation modes inside the material circularly polarized in opposite directions. The one with the same handedness of the CLC helix experiences strong reflection if the wavelength falls in a restricted range of values due to the Bragg diffraction originated by the periodic structure that corresponds to that of one-dimensional (1D) photonic crystals. The peak of the photonic bandgap (PBG) is at $\lambda_B = n_{av}p$, where $n_{av} = (n_{\parallel} + n_{\perp})/2$ is the average refractive index of the material, with $n_{\parallel}$ refers to the polarization parallel to the director and $n_{\perp}$, the polarization perpendicular to it. The approximate width of the PBG is $\Delta\lambda = \Delta n \cdot p$, with $\Delta n = n_{\parallel} - n_{\perp}$ being the refractive index anisotropy. The circularly polarized mode with opposite-handedness passes through the medium with minor changes, therefore, a strong rotatory power is also associated with the Bragg reflection. Of course, these peculiar optical properties have been studied together with the effects of the magnetic field and low and high frequency [optical] electric field applied to the CLC structure driven by the possible applications for displays and photonic devices. Among these studies, it is remarkable that a conical configuration of CLC was theoretically expected to occur when a low-frequency electric field E_S (denominated *static* in the following) is applied along the direction of the helix axis, under the condition of the bend elastic constant K_3 lower than the twist elastic constant K_2, that is $K_3 < K_2$.[6,7] It was necessary to wait more than four decades to discover the existence of such a configuration, when a team headed by Oleg Lavrentovich succeeded in fulfilling this condition realizing a cholesteric mixture including a dimeric liquid crystal with two rigid rod-like units connected by a flexible chain with an odd number of links. This configuration is called Oblique Heliconical Cholesteric (ChOH). Then in ChOH, the molecular director $\mathbf{n} \equiv (\sin\theta\cos\phi, \sin\theta\sin\phi, \cos\theta)$ rotates in space on a conical

(a) CLC

(b) ChOH

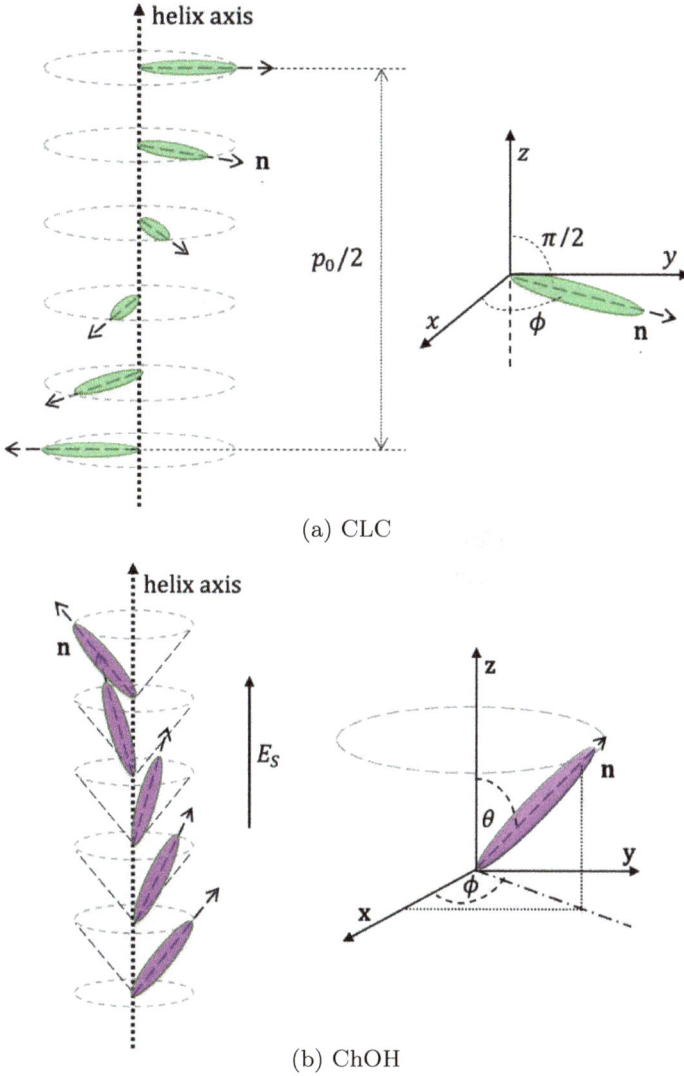

Figure 1. Spatial arrangement of the molecular director for (a) conventional cholesteric (CLC) and (b) heliconical cholesteric liquid crystal (ChOH).

surface, forming an angle θ with the helix axis and the azimuthal angle is periodically modulated, $\phi(z) = \left(\frac{2\pi}{p}\right) z = qz$, where p is the helix pitch.

A sketch of the conventional CLC and ChOH configurations is given in Fig. 1.

With the usual procedure of minimization of the free energy density suitable to find the stability conditions, it is easily found that if $\kappa = K_3/K_2 < 1$ the ChOH configuration is stable when the applied static field fulfills the following condition:

$$E_{NC^*} < E_S < E_{NC}, \tag{1}$$

where

$$E_{NC} = \left(\frac{2\pi}{p_0}\right) \frac{K_2}{\sqrt{\Delta\epsilon_S K_3/4\pi}} \tag{2}$$

is the critical field necessary to unwind the helix, over which the material has a nematic order, and

$$E_{NC^*} \approx \frac{\kappa[2 + \sqrt{2(1-\kappa)}]}{1+\kappa} \tag{3}$$

refers to the lower critical field, below which value the stable configuration corresponds to a conventional CLC with pitch p_0 and axis rotated by $\pi/2$ with respect to the field direction. Here $\Delta\epsilon_S$ is the anisotropy of the dielectric permittivity at the frequency of the field E_S.

The relationship between the conical angle θ and the applied field is found to be

$$\sin^2 \theta = \frac{\kappa}{1-\kappa} \left(\frac{E_{NC}}{E_S} - 1\right) \tag{4}$$

and the one with the helix pitch is

$$\sin^2 \theta(1-\kappa) + \kappa = \frac{p}{p_0} \tag{5}$$

Combining Equations (4) and (5), we get

$$\frac{p}{p_0} = \kappa \frac{E_{NC}}{E_S} \tag{6}$$

These are the basic equations describing the behavior of the heliconical cholesteric liquid crystal submitted to a static electric field: increasing the field, the conical angle becomes smaller and smaller until unwinding of the helix is achieved at $E_S = E_{NC}$ and a nematic order is established. In a similar way, the pitch becomes shorter and shorter before the disruption of the helical order. In this way it is expected and has been experimentally demonstrated that the Bragg resonance can be tuned by a moderate voltage applied to the ChOH sample over the whole visible range.[8-10]

As a matter of fact, the conical arrangement of these materials provides a new phenomenology with respect to CLC for what concerns coupling between the optical field of a light beam traveling through this structure and the molecular director **n**. In fact, due to the bend deformation present in ChOH, an optical torque arises in a way similar to the one that originates Giant Optical Nonlinearity (GON) in nematic liquid crystals. The consequence is optical reorientation leading to light-induced tuning of the Bragg resonance.[11] This phenomenon could not be observed previously in CLC because in this case coupling between optical field and molecular director involves only the twist deformation being much less effective in inducing optical reorientation.[12,13] We should remark that the previously reported effect of light-induced modifications and shift of the Bragg resonance observed in pure CLC is not due to optical reorientation but to photochemical transformation of the molecules undergoing trans-cis photoisomerization.[14-16]

In this chapter, we start by briefly recalling the electromagnetic approach used to describe the optical propagation through a helical structure showing that what applies to CLC can be easily extended to ChOH by substituting the permittivity $\epsilon_\parallel$ with an effective value dependent on θ that takes into account the conical configuration. Then we will discuss the optical reorientation in ChOH to show that it leads to an increase in the conical angle and pitch with a consequent redshift of the Bragg wavelength λ_B. In the following sections, we will review the results obtained by considering the nonlinear effects arising on a light beam traveling along the helix axis when it induces optical reorientation. The approximation of constant intensity through the ChOH sample will be considered first to show the expectations of analytical solutions, while the achievements obtained by a careful calculation exploiting a multilayer approach will be finally presented.

2. Electromagnetic Treatment of Light Propagation

We consider light propagation along the **z** axis parallel to the helix axis that is perpendicular to the sample boundaries (i.e., normal incidence). In this condition, the wave equations for electric and magnetic fields are decoupled, and for the electric field components E_x and E_y, we can write the following matrix equation:

$$\frac{d^2}{dz^2}\mathbf{J} + \frac{\omega^2}{c^2}\mathbf{M} \cdot \mathbf{J} = 0 \tag{7}$$

where

$$\mathbf{J} = \begin{bmatrix} E_x \\ E_y \end{bmatrix} \qquad \mathbf{M} = \begin{bmatrix} M_{11} & M_{12} \\ M_{21} & M_{22} \end{bmatrix} \tag{8}$$

with ω being the light frequency (times 2π) and c, the speed of light in vacuum.

While the components of the dielectric matrix are[17,18]

$$M_{11} = \epsilon_{xx} - \frac{\epsilon_{xz}^2}{\epsilon_{zz}} = \frac{\epsilon_\perp}{\epsilon_\perp + \Delta\epsilon \cos^2\theta}$$
$$\times (\epsilon_\perp + \Delta\epsilon \cos^2\theta + \Delta\epsilon \sin^2\theta \cos^2\phi), \tag{9a}$$

$$M_{12} = M_{21} = \epsilon_{xy} - \frac{\epsilon_{xz}\epsilon_{yz}}{\epsilon_{zz}} = \frac{\epsilon_\perp}{\epsilon_\perp + \Delta\epsilon \cos^2\theta}$$
$$\times (\Delta\epsilon \sin^2\theta \sin\phi \cos\phi), \tag{9b}$$

$$M_{22} = \epsilon_{yy} - \frac{\epsilon_{yz}^2}{\epsilon_{zz}} = \frac{\epsilon_\perp}{\epsilon_\perp + \Delta\epsilon \cos^2\theta}$$
$$\times (\epsilon_\perp + \Delta\epsilon \cos^2\theta + \Delta\epsilon \sin^2\theta \sin^2\phi), \tag{9c}$$

where ϵ_{ij} are the elements of the dielectric tensor; $\epsilon_\|$, $\epsilon_\perp$ are the ones parallel and perpendicular to the axis of a generic uniaxial system, with $\Delta\epsilon = \epsilon_\| - \epsilon_\perp$ being the anisotropy at optical frequencies. It is possible to write these matrix elements in a more useful way in order to point out that a simple substitution of parameters allows using the same analytical expressions for CLC and ChOH.

In fact, after a simple manipulation, we can write

$$M_{11} = \frac{1}{2}\left(\frac{\epsilon_\perp \epsilon_\|}{\epsilon_\perp + \Delta\epsilon \cos^2\theta} + \epsilon_\perp\right) + \frac{1}{2}\left(\frac{\epsilon_\perp \epsilon_\|}{\epsilon_\perp + \Delta\epsilon \cos^2\theta} - \epsilon_\perp\right)\cos 2\phi, \tag{10a}$$

$$M_{12} = M_{21} = \frac{1}{2}\left(\frac{\epsilon_\perp \epsilon_\|}{\epsilon_\perp + \Delta\epsilon \cos^2\theta} - \epsilon_\perp\right)\sin 2\phi, \tag{10b}$$

$$M_{22} = \frac{1}{2}\left(\frac{\epsilon_\perp \epsilon_\|}{\epsilon_\perp + \Delta\epsilon \cos^2\theta} + \epsilon_\perp\right) - \frac{1}{2}\left(\frac{\epsilon_\perp \epsilon_\|}{\epsilon_\perp + \Delta\epsilon \cos^2\theta} - \epsilon_\perp\right)\cos 2\phi.$$

$$(10c)$$

We can recognize that by setting $\theta = \pi/2$, we get the usual dielectric permittivity tensor of CLC, as follows:

$$\mathbf{M}^{CLC} = \begin{bmatrix} \epsilon_{av} + \dfrac{\Delta\epsilon}{2}\cos(2qz) & \dfrac{\Delta\epsilon}{2}\sin(2qz) \\[2mm] \dfrac{\Delta\epsilon}{2}\sin(2qz) & \epsilon_{av} + \dfrac{\Delta\epsilon}{2}\cos(2qz) \end{bmatrix}, \qquad (11)$$

where $\epsilon_{av} = \frac{\epsilon_\| + \epsilon_\perp}{2}$ is the average value of the dielectric response.

By looking at Eqs. (10), we can define

$$\tilde{\epsilon} = \frac{1}{2}\left(\frac{\epsilon_\perp \epsilon_\|}{\epsilon_\perp + \Delta\epsilon \cos^2\theta} + \epsilon_\perp\right), \qquad (12a)$$

$$\Delta\epsilon_{eff} = \left(\frac{\epsilon_\perp \epsilon_\|}{\epsilon_\perp + \Delta\epsilon \cos^2\theta} - \epsilon_\perp\right), \qquad (12b)$$

to have a similar matrix for ChOH

$$\mathbf{M}^{ChOH} = \begin{bmatrix} \tilde{\epsilon} + \dfrac{\Delta\epsilon_{eff}}{2}\cos(2qz) & \dfrac{\Delta\epsilon_{eff}}{2}\sin(2qz) \\[2mm] \dfrac{\Delta\epsilon_{eff}}{2}\sin(2qz) & \tilde{\epsilon} + \dfrac{\Delta\epsilon_{eff}}{2}\cos(2qz) \end{bmatrix}. \qquad (13)$$

In other words, for ChOH we can use the same analytical results obtained for cholesteric liquid crystals by the following simple substitution:

$$\epsilon_\| \rightarrow \frac{\epsilon_\perp \epsilon_\|}{\epsilon_\perp + \Delta\epsilon \cos^2\theta}. \qquad (14)$$

This can also be shown by transformation of matrix $\mathbf{M}^{ChOH}$ into the local frame applying a rotation $\phi = 2qz$ to get the dielectric tensor in the local frame, as follows:

$$\mathbf{M}^{ChOH}_{local} = \begin{bmatrix} \dfrac{\epsilon_\perp \epsilon_\|}{\epsilon_\perp + \Delta\epsilon \cos^2\theta} & 0 \\[2mm] 0 & \epsilon_\perp \end{bmatrix}. \qquad (15)$$

The consequence is that we can use the analytical results already obtained for light propagation in CLC also for ChOH provided that the substitution indicated in Eq. (14) is made in the mathematical expressions.

Therefore, in ChOH under normal incidence we also have two main propagation modes circularly polarized in opposite directions, left (L) and right (R), as follows:

$$n_L = \tilde{\epsilon}^{1/2} - \frac{\Delta\epsilon_{eff}^2}{32\tilde{\epsilon}^{3/2}\left(\dfrac{\lambda}{\lambda_B}\right)\left(1 - \dfrac{\lambda}{\lambda_B}\right)}, \tag{16a}$$

$$n_R = \tilde{\epsilon}^{1/2} + \frac{\Delta\epsilon_{eff}^2}{32\tilde{\epsilon}^{3/2}\left(\dfrac{\lambda}{\lambda_B}\right)\left(1 + \dfrac{\lambda}{\lambda_B}\right)}, \tag{16b}$$

considering a left-handed helix, with λ as the light wavelength in vacuum, and the Bragg wavelength is now written as

$$\lambda_B = \tilde{\epsilon}^{1/2}p. \tag{17}$$

The mode with the same handedness of the helix (left-handed in this case) will suffer Bragg diffraction, being strongly reflected in a band around λ_B :

$$\Delta\lambda \approx \Delta n_{eff}\,p, \tag{18}$$

where

$$\Delta n_{eff} = n_\parallel - \frac{n_\perp n_\parallel}{\sqrt{n_\perp^2 \sin^2\theta + n_\parallel^2 \cos^2\theta}}. \tag{19}$$

The big difference with CLC is the dependence on the tilt angle θ in all these equations. Anyway, the stronger one is included in the pitch p as determined by Eq. (5) that allows large tunability of the Bragg resonance according to Eq. (6). This dependence is the reason why ChOH becomes interesting from the point of view of nonlinear optics. In fact, it allows optical reorientation to take place, which changes the optical propagation parameters for the wave in a significant way if the wavelength is close to the PBG. This leads to a very original nonlinear behavior driven by the light-induced shift of the Bragg resonance.

3. Optical Reorientation in ChOH

It is easy to take into account the interaction between the optical field and the molecular director starting from Meier's theory[7] including the contribution of the optical field $\mathbf{E}_{OPT}$ in the free energy density, as follows:

$$f = \frac{1}{2}K_1(\nabla \cdot \mathbf{n})^2 + \frac{1}{2}K_2(\mathbf{n} \cdot \nabla \times \mathbf{n} - q_0)^2 + \frac{1}{2}K_3(\mathbf{n} \times \nabla \times \mathbf{n})^2$$

$$- \frac{1}{8\pi}\Delta\epsilon_S(\mathbf{n} \cdot \mathbf{E}_S)^2 - \frac{1}{16\pi}\Delta\epsilon(\mathbf{n} \cdot \mathbf{E}_{OPT})^2, \tag{20}$$

where K_1 is the splay elastic constant, $q_0 = 2\pi/p_0$. We recall that $\Delta\epsilon_S = \epsilon_\parallel^S - \epsilon_\perp^S$ is the dielectric anisotropy at low frequency and $\Delta\epsilon = \epsilon_\parallel - \epsilon_\perp$ is the one at optical frequency.

The low-frequency field is applied along the helix axis direction: $\mathbf{E} = E_S\hat{\mathbf{z}}$. The optical field in the medium is a superposition of the right-handed and left-handed circularly polarized modes: $\mathbf{E}_R = E_{0R}(\hat{\mathbf{x}} + i\hat{\mathbf{y}})$ and $\mathbf{E}_L = E_{0L}(\hat{\mathbf{x}} - i\hat{\mathbf{y}})$, respectively, therefore the term describing the interaction with the optical field becomes

$$(\mathbf{n} \cdot \mathbf{E}_{OPT})^2 = (\mathbf{n} \cdot \mathbf{E}_{OPT})(\mathbf{n} \cdot \mathbf{E}_{OPT})^*$$

$$= \sin^2\theta(E_{0R}^2 + E_{0L}^2 + 2E_{0R}E_{0L}\cos 2\phi)$$

$$\approx \sin^2\theta(E_{0R}^2 + E_{0L}^2) \equiv \sin^2\theta A^2, \tag{21}$$

if averaged over a director rotation of π (corresponding to $p/2$), then

$$f_{OPT} = -\frac{1}{16\pi}\Delta\epsilon A^2 \sin^2\theta, \tag{22}$$

defining A as an effective amplitude of the optical field.

The equilibrium condition derived from the Euler–Lagrange equations gives $\partial f/\partial\theta = 0$, i.e.,

$$\frac{\partial f}{\partial \theta} = 2K_2(\sin^2\theta)\left(\frac{\partial\phi}{\partial z}\right)^2 - 2K_2q_0\frac{\partial\phi}{\partial z} + K_3(\cos^2\theta - \sin^2\theta)\left(\frac{\partial\phi}{\partial z}\right)^2$$

$$+ \frac{1}{4\pi}\Delta\epsilon_S E_S^2 - \frac{1}{16\pi}\Delta\epsilon A^2 = 0. \tag{23}$$

It is straightforward to highlight that this equation leads to the same results obtained in the absence of optical field provided that instead of the static electric field an effective field E_{eff} is considered, as follows:

$$E_{eff} = \sqrt{E_S^2 - \frac{1}{2}\frac{\Delta\epsilon}{\Delta\epsilon_S}A^2}. \tag{23}$$

In other words, as expected by the orientation of the fields, the optical torque acts in the opposite direction with respect to the static field torque, therefore, its action leads to a weakening of the effect of the static field

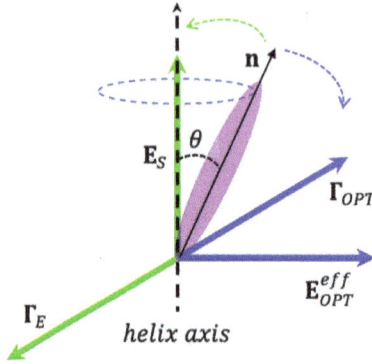

Figure 2.   Orientation of the optical torque Γ_{OPT} and the static field torque Γ_E acting on the molecular director **n** toward the opposite direction. $\mathbf{E}_s$ is the static field and $\mathbf{E}_{OPT}^{eff}$ represents the component of the optical field effective at a given time.

necessary to stabilize the conical structure causing an increase in the conical angle and an increase of the helix pitch. This is easily shown by using Eqs. (4)–(6) using E_{eff} in place of E_S.

The competing actions of the applied static field and of the optical field are shown in Fig. 2.

Using the intensity I rather than the field amplitude, we can write

$$A^2 \approx \frac{8\pi}{cn_m} I \tag{24}$$

(the sign $\approx$ means that for light wavelength close to the Bragg wavelength λ_B, one should take into account the resonant behavior of the mode circularly polarized with the same handedness of the helix. In Eq. (24), we have defined $\boldsymbol{n_m} = \tilde{\epsilon}^{1/2}$ the average refractive index of the conical structure. Then the effective field becomes:

$$E_{eff}(I) = \sqrt{E_S^2 - \frac{4\pi\Delta\epsilon}{cn_m\Delta\epsilon_S} I}. \tag{25}$$

In this way, taking into account Eqs. (23 and 24) from Eq. (4) we have:

$$\sin^2\theta(I) = \frac{\kappa}{1-\kappa}\left(\frac{E_{NC}}{E_{eff}(I)} - 1\right) \tag{26}$$

or:

$$\theta(I) = \arcsin\sqrt{\frac{\kappa}{1-\kappa}\left(\frac{E_{NC}}{E_{eff}(I)} - 1\right)}. \tag{27}$$

This is the key equation of optical reorientation in ChOH where the dependence of the tilt angle θ on the light intensity is highlighted. In fact, all the basic relationships for this material will be dependent on the intensity I, because they include the dependence on the conical angle θ. As already mentioned, this effect is not present in conventional CLC.

In order to plot the function $\theta(I)$, we use the parameters of the material studied in Ref. 10:

$$\Delta\epsilon_S = 4.79; \quad \epsilon_{\parallel} = 2.79; \quad \epsilon_{\perp} = 2.19; \quad P_0 = 1.4\,\mu\text{m}; \quad \kappa = \frac{K_3}{K_2} = 0.10;$$

$$E_{NC} = 4.88\frac{V}{\mu m}; \quad E_{NC^*} = 1.53\frac{V}{\mu m}.$$

Even if E_{N^*C} does not appear in Eq. (26), nevertheless it determines the minimum value of E_{eff} suitable for the conical configuration, i.e., the maximum value of the intensity, over which the system transforms into a conventional CLC with the helix axis rotated by 90° with respect to the ChOH configuration. The result of this calculation is reported in Fig. 3 for three different values of the applied static field.

It is clear from the figure that the useful range for optical reorientation is narrow for fields slightly over $\boldsymbol{E}_{NC^*}$ because at zero intensity the tilt angle is close to its maximum value, while this range is much broader for higher fields corresponding to a smaller value of θ. On the other hand, by increasing the static field, a higher intensity will be necessary to reach the critical field $E_{eff} = E_{NC^*}$.

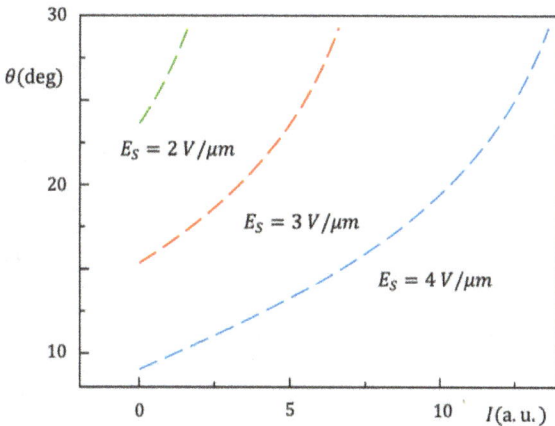

Figure 3.   Dependence of the conical angle θ versus light intensity I for different values of the applied static field.

The most relevant effect of the reorientation is the shift of the Bragg wavelength λ_B due to the consequent change of the helix pitch. From Eq. (6), we have

$$p(I) = p_0 \frac{\kappa \, E_{NC}}{\sqrt{E_S^2 - \frac{4\pi\Delta\epsilon}{cn_m\Delta\epsilon_S}I}}. \tag{28}$$

Then the explicit dependence of λ_B on the intensity is

$$\lambda_B = \tilde{\epsilon}^{1/2}p = \sqrt{\frac{1}{2}\left(\frac{\epsilon_\perp\epsilon_\parallel}{\epsilon_\perp + \Delta\epsilon\cos^2\theta(I)} + \epsilon_\perp\right)}p_0\frac{\kappa \, E_{NC}}{\sqrt{E_S^2 - \frac{4\pi\Delta\epsilon}{cn_m\Delta\epsilon_S}I}}. \tag{29}$$

As observed from Fig. 3, with our parameters in the best case we have an angle change from about $10°$ to about $30°$, it means that the average refractive index represented by the first square root term changes by about 1%, while the pitch may have a substantial increase by 50% or more. For this reason, we can concentrate only on $p(I)$ and write the following:

$$\lambda_B \cong n_m p_0 \frac{\kappa \, E_{NC}}{\sqrt{E_S^2 - \frac{4\pi\Delta\epsilon}{cn_m\Delta\epsilon_S}I}}. \tag{30}$$

This expression points out that the Bragg wavelength moves to longer wavelengths by increasing the light intensity I with a consequent redshift of the stop band.

4. Nonlinear Light Propagation at Constant Intensity

We can use these results and the ones reported in the former section to modify the analytical solutions in order to get an exact analytical solution for a ChOH in the case of propagation of light along the helix axis, following the approach of A. Gevorgyan[19] that allows calculating the reflectivity and transmittivity of a layer of finite thickness for an anisotropic material with helical arrangement of the dielectric and magnetic permeability. In the rotating coordinate frame, one gets the following values for wave vectors of the propagating modes:

$$k_{1,2,3,4} = \pm\frac{2\pi}{\lambda}\sqrt{\tilde{\epsilon}\left[1 + \left(\frac{\lambda}{p\sqrt{\tilde{\epsilon}}}\right)^2 \pm \gamma\right]} \tag{31}$$

where

$$\gamma = \sqrt{4\left(\frac{\lambda}{p\sqrt{\bar{\epsilon}}}\right)^2 + \delta^2} \quad \text{and} \quad \delta = \frac{\Delta\epsilon_{eff}}{2\bar{\epsilon}} \tag{32}$$

In the rotating coordinate frame (x', y', z'), the z' axis coincides with the z axis of the laboratory frame, and the x' and y' axes rotate around z axis in such a way that for each value of z the dielectric tensor of ChOH is given by the matrix $\mathbf{M}_{local}^{ChOH}$ given by Eq. (15).

On both sides of the ChOH layer, the boundaries correspond to isotropic half-spaces with the same refractive indices equal to n_{av} to minimize the influence of dielectric boundaries. The boundary conditions, consisting of the continuity of the tangential components of the electric and magnetic fields, give a system of eight linear equations with eight unknowns. Thus, solving this boundary-value problem in the same way as in[19] for ordinary CLCs, we obtain the following expressions for the reflection $R_{1,2}$ and transmission $T_{1,2}$ coefficients:

$$R_{1,2} = \left|\frac{u\delta s_{1,2}}{a_{1,2}}\right|^2 \quad \text{and} \quad T_{1,2} = \left|\frac{1}{a_{1,2}}\right|^2 \tag{33}$$

with

$$u = \frac{\pi d\sqrt{\bar{\epsilon}}}{\lambda}; \quad s_{1,2} = \frac{\sin(k_{1,2}d)}{k_{1,2}d} \tag{34}$$

$$\text{and} \quad a_{1,2} = \cos(k_{1,2}d) \mp iu(\gamma \pm 2)s_{1,2}$$

and d is the ChOH layer thickness. These expressions correspond to the case when $n_{av} = \sqrt{\frac{\varepsilon_\| + \varepsilon_\perp}{2}}$ and when the incident light polarizations coincide with two eigenpolarizations (EPs). In the most general case, very complex expressions are found (see, for instance, the expressions for reflection and transmission coefficients for usual CLC layer in Ref. 19). The EPs correspond to the two polarizations, which do not change when light travels through the system. They coincide with the polarizations of the eigen modes. Under certain conditions, e.g., in the case of $n_{av} = \sqrt{\frac{\varepsilon_\| + \varepsilon_\perp}{2}}$, the EPs of the ChOH layer under the normal incidence of light practically coincide with orthogonal circular polarizations, although in the general case they can significantly differ from circular ones.

With the previously used material parameters and setting $E_S = 2.03\,V/\mu m$, we have calculated the spectrum of reflectivity for different values of light intensity: $I_0 = 0$, corresponding to the reflectivity of the structure without any optical reorientation; $I_1 = 4.33 \cdot 10^6\,W/cm^2$ and

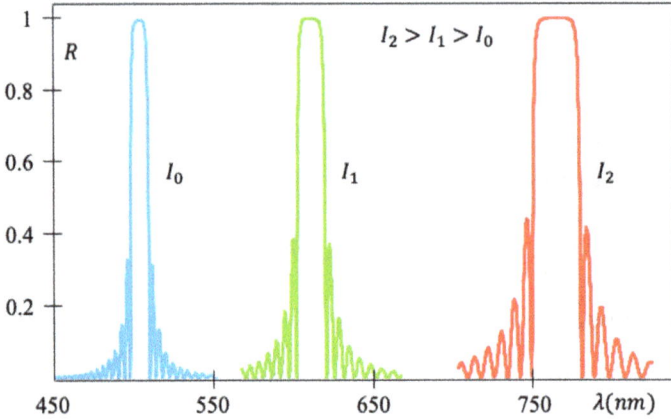

Figure 4.   Reflectivity spectrum for increasing values of the light intensity I at $E_S = 2.03\,V/\mu m$. Calculations have been performed using the material parameters reported in the text; $I_0 = 0\,\frac{W}{cm^2}$; $I_1 = 4.33\cdot 10^6\,\frac{W}{cm^2}$; $I_2 = 7.66\cdot 10^6\,\frac{W}{cm^2}$.

$I_2 = 7.66\cdot 10^6\,W/cm^2$ corresponding to light-induced increase of the conical angle. Figure 4 shows the results of this calculation, pointing out that the whole range of the visible spectrum can be spanned as far as the maximum intensity allows fulfilling the condition $E_{eff} < E_{NC^*}$.

This effect has been experimentally demonstrated in a pump–probe experiment where a redshift of the Bragg resonance larger than 120 nm has been observed within the visible range using light intensities typical of optical reorientation in nematic liquid crystals.[11]

Another remarkable effect due to the redshift of the Bragg resonance concerns the refractive index of the resonant mode given by Eq. (16a). In fact, we see from Eqs. (16) that when λ approaches λ_B, we have a fast increase of the refractive index for the mode with the same handedness of the helix, while no significant change occurs for the other mode. Therefore, when the Bragg resonance is shifted, this propagation mode sees a big change of the index affecting its propagation. This is actually linked to the high reflectivity in this range of wavelength, producing the anomalous behavior of the real part of refractive index in the presence of an absorption band.

The described light-induced shift of the Bragg reflection gives rise to a remarkable nonlinear optical response on the transmission of the light beam. In fact, by considering a light beam with wavelength λ initially close to the Bragg resonance, the occurrence of optical reorientation may tune

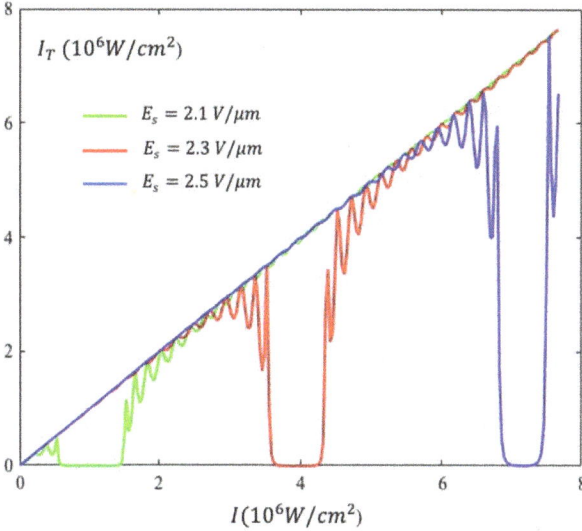

Figure 5.   The transmitted intensity I_T vs the impinging intensity I of a ChOH sample for a circularly polarized beam with the same handedness of the helix and $\lambda = 514\,nm$. Material parameters are the ones reported in the text. The sample thickness is $d = 20\,\mu m$. The transmitted intensity is plotted for different values of the applied static field E_s.

the PBG through a range of values crossing λ, thus a stop band for a definite range of the impinging intensity appears. This effect has been highlighted in Ref. 20 where the transmitted intensity $I_T = T \cdot I$ vs I has been reported. By performing the same calculation that uses a standard expression of the reflectivity of a CLC sample adapted to the ChOH case,[20] we report in Fig. 5 the transmitted intensity I_T vs I for different values of the applied static field E_s,[20] pointing out, as expected, that E_s can be used as a parameter for tuning the stop band for the incoming light. It is clear that for a definite value of E_s the location of the stop band is dependent on the value of λ_0.

We must underline that the present result has been analytically calculated by taking the intensity I constant through the sample. This is certainly a good approximation in a pump–probe configuration, where the pump beam (at wavelength far from the resonance) induces the reorientation and one measures the transmission of a weak monochromatic probe beam, not affecting the director orientation.

We may also expect this effect to be self-induced, namely, considering a single beam traveling through the sample with intensity high enough to act with an effective optical torque on the director, if its wavelength is close to the initial location of the Bragg resonance.

However, in this case the actual situation may be more complex because the intensity of the traveling beam is affected by the change of transmittivity induced by the beam itself. Therefore, after an initial shift of the PBG as described above, we expect that the consequent change of light intensity decreases or increases the possibility of affecting the director orientation, thus making the beam transmittivity less predictable. Hence, it is necessary to use a more careful approach to study the propagation of a light beam under these conditions to be able to take into account the propagation with nonlinear change of the intensity through the ChOH sample.

5. Multilayer Approach

The limit of the previously considered approximation of constant intensity through the medium is highlighted when investigating the light localization peculiarities in a homogeneous ChOH layer.

The calculation is performed in the same way as for ordinary CLCs in Ref. 21. Figure 6 shows the z dependence of the normalized intensity $|E_{in}(z)|^2$ defined as

$$|E_{in}(z)|^2 = \frac{I(z)}{I_i} \tag{35}$$

with I_i being the intensity of the incident light.

The figure reports the plot of $|E_{in}(z)|^2$ vs z for different wavelengths of the Bragg resonance centered at $\lambda_1 = \lambda_B = 502.4\,\text{nm}$, using the same material parameters as in the previous figures; $\lambda_2 = 497.25\,\text{nm}$, $\lambda_3 = 496.5\,\text{nm}$, $\lambda_4 = 495.35\,\text{nm}$, $\lambda_5 = 494.3\,\text{nm}$, and $\lambda_6 = 493\,\text{nm}$. A comparison of these results with those for ordinary CLCs shows that weak oscillations appear in these dependences even with the minimal influence of dielectric boundaries.

It is clear from these curves that, near the PBG, we have an inhomogeneous distribution of the optical field inside the liquid crystal layer, and, for different wavelengths, the nature of the inhomogeneity is different.

At low intensities of the incident light, these inhomogeneities in the volume distribution have negligible effect on the orientation of the ChOH molecules. However, we expect that the situation changes significantly when

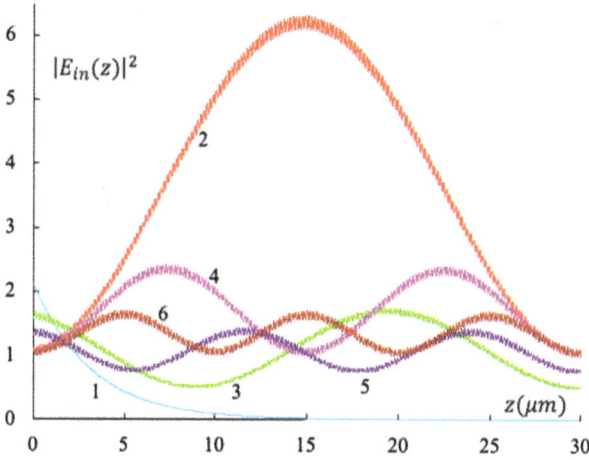

Figure 6. The dependences $|E_{in}(z)|^2$ for the wavelengths $\lambda_B \equiv \lambda_1 = 502.04\,\text{nm}$ (inside the PBG in its center; curve 1), $\lambda_2 = 497.25\,\text{nm}$ (on the first shortwavelength minimum of the diffraction reflection; curve 2), $\lambda_3 = 496.5\,\text{nm}$ (on the first shortwavelength maximum; curve 3), $\lambda_4 = 495.35\,\text{nm}$ (on the second shortwavelength minimum; curve 4), $\lambda_5 = 494.3\,\text{nm}$ (on the second shortwavelength maximum; curve 5), and $\lambda_6 = 493.0\,\text{nm}$ (on the third shortwavelength minimuum; curve 6).

optical reorientation takes place. Thus, it is necessary to adopt an approach that takes into account the field distribution inside the sample affecting the conical angle of the molecular director and the pitch of the helix as well. This is done in the following way.

The amplitudes of the electric field of the incident $\mathbf{E}_i$, reflected $\mathbf{E}_r$ and transmitted $\mathbf{E}_t$ plane waves are written as

$$\mathbf{E}_{i,r,t} = E_{i,r,t}^p \mathbf{u}^p + E_{i,r,t}^s \mathbf{u}^s = \begin{pmatrix} E_{i,r,t}^p \\ E_{i,r,t}^s \end{pmatrix}, \tag{36}$$

where $\mathbf{u}^p$ and $\mathbf{u}^s$ are the unit vector for the p- and s-polarizations, respectively, and $E_{i,r,t}^p$ and $E_{i,r,t}^s$ are the corresponding amplitudes for the incident, reflected, and transmitted waves.

Then we divide the ChOH layer of thickness d into a large number of thin sublayers of thicknesses $d_1, d_2, d_3, \ldots, d_L$. If the maximal thickness is small enough, we can assume that the optical characteristic parameters are constant in each anisotropic sublayer. In this way the problem of calculating the reflection and transmission of the whole sample is reduced to the solution of the following system of matrix difference equations

(Ambartsumian's modified layer addition method[22]):

$$\hat{R}_m = \hat{r}_m + \tilde{\hat{t}}_m \hat{R}_{m-1} \left(\hat{I} - \tilde{\hat{r}}_m \hat{R}_{m-1} \right)^{-1} \hat{t}_m.$$
$$\hat{T}_m = \hat{T}_{m-1} \left(\hat{I} - \tilde{\hat{r}}_m \hat{R}_{m-1} \right)^{-1} \hat{t}_m,$$
(37)

with initial conditions $\hat{R}_0 = \hat{0}$ and $\hat{T}_0 = \hat{I}$.[22] Here, $\hat{R}_m$, $\hat{T}_m$, $\hat{R}_{m-1}$, and $\hat{T}_{m-1}$ are the reflectance and transmittance matrices of the system with m and $(m-1)$ sublayers, respectively; $\hat{r}_m$, $\hat{t}_m$ are the reflectance and transmittance matrices of the m-th sublayer, $\hat{0}$ is the zero matrix; $\hat{I}$ is the unit matrix, and the respective matrices for the reverse light propagation are denoted by tilde. Thus, the problem is reduced to finding the matrices of reflectance and transmittance for an anisotropic layer. The solution of this problem is well known (see, for example,[22]).

The electric field at the boundary of any two adjacent i and j sublayers is determined as follows. The layer addition method is written as

$$\hat{R}_{i+j} = \hat{R}_i + \tilde{\hat{T}}_i \hat{S} \hat{T}_i$$
$$\hat{T}_{i+j} = \hat{T}_j \hat{P} \hat{T}_i.$$
(38)

The matrices $\hat{S}$ and $\hat{P}$ describe the resulting waves that arise in the boundary between the i and j sublayers. Then, the amplitude of the light electric field on this sewing plane, propagating in the forward direction, is

$$\mathbf{E}_\rightarrow = \hat{P}\hat{T}_i \mathbf{E}_i,$$
(39)

and correspondingly for the light propagating in the backward direction is

$$\mathbf{E}_\leftarrow = \hat{S}\hat{T}_i \mathbf{E}_i.$$
(40)

Therefore, the total wave field arising in the sewing plane between these two sublayers has the following form:

$$\mathbf{E}_{\text{total}} = (\hat{P} + \hat{S})\hat{T}_i \mathbf{E}_i.$$
(41)

Taking into account Eq. (36), we have

$$\hat{S} = \hat{R}_{m-1}(\hat{I} - \tilde{\hat{r}}_m \hat{R}_{m-1})^{-1}$$
$$\hat{P} = (\hat{I} - \tilde{\hat{r}}_m \hat{R}_{m-1})^{-1}$$
(42)

We write the light intensity of the total wave excited in the ChOH layer as $I(z) = |E_{\text{in}}(z)|^2 I_i$, where I_i is the intensity of incident light. In the linear limit, we will take $I_i = I_0 = 1$, and in general $I_i = NI_0$, that is we assume that the amount of increase of the intensity of the incident light is uniformly distributed in each plane z in the ChOH layer. As shown above, the function $I(z)$ in a ChOH layer has complex behavior and it can be either much higher or much lower than the intensity of the incident light. Then, we investigate how the intensity $I(z)$ affects the structure of the ChOH layer and how this effect can be taken into account in light propagation. To this aim, we organize our calculations as follows. In the first step $(j = 1)$, we take

$$E_{\textit{eff}} = \sqrt{E_S^2 - \frac{\Delta\epsilon}{\Delta\epsilon_S}IN} = \text{const}; \quad p = \text{const}; \quad s = \sin^2\theta = \text{const} \quad (43)$$

With these parameters we calculate the reflection and transmission coefficients as well as the light localization in the ChOH layer, that is the $I(z,\lambda) = |E_{\text{in}}(z,\lambda)|^2 I_i$. In the second step $(j = 2)$, we take $E_{\textit{eff}}(z,\lambda) = \sqrt{E_S^2 - \frac{\Delta\epsilon}{\Delta\epsilon_S}|E_{\text{in}}(z,\lambda)|^2 N}$ and take into account how it affects the helix pitch and angle θ. Therefore, we have $p(z,\lambda,j) = \kappa\frac{E_{NC}p_0}{E_{\textit{eff}}(z,\lambda,j)}$ and $s(z,\lambda,j) = \sin^2\theta = \frac{\kappa}{1-\kappa}\left(\frac{E_{NC}}{E_{\textit{eff}}(z,\lambda,j)} - 1\right)$. Now, we calculate the reflection and transmission coefficients as well as light localization in the ChOH layer, that is the $I(z,\lambda,j) = |E_{\text{in}}(z,\lambda,j)|^2 N$ for these new parameters, and so on with $j = 3, 4, 5, \ldots$.

The material parameters chosen for the calculation are the ones reported in Ref. 23: $\epsilon_\parallel = 2.79$, $\epsilon_\perp = 2.19$, $\Delta\epsilon = 4.79$, and $\kappa = 0.1$; the unperturbated helix pitch is $p_0 = 1,400\,\text{nm}$; the layer thickness is $d = 20\,\mu\text{m}$ with critical fields $E_{NC} = 4.88\,\text{V}/\mu\text{m}$ and $E_{N*C} = 1.53\,\text{V}/\mu\text{m}$; and $n_{av} = \sqrt{(\epsilon_\parallel + \epsilon_\perp)/2}$ the refractive index of the ChOH layer. The static field applied along the helix direction is $E_S = 2.03\,\text{V}/\mu\text{m}$. In this calculation, all the sublayers have the same thickness: $d_1 = d_2 = d_3 = \cdots = d_L = 8\,\text{nm}$. Details of the obtained results are reported in Ref. 24.

We underline that the step index j is related to the elapsed time from the excitation. In fact, each time j increases by one, the new field distribution is considered, which leads to new values for the conical angle θ and the pitch p. Then j can be considered as the elapsed time in units of the response time τ_{on} required by the material to adapt its parameters to the new field values:

$$j = \frac{t}{\tau_{\text{on}}}. \quad (44)$$

Therefore, any dependence on the index j appearing for physical quantities calculated in this way should be considered as a time dependence.

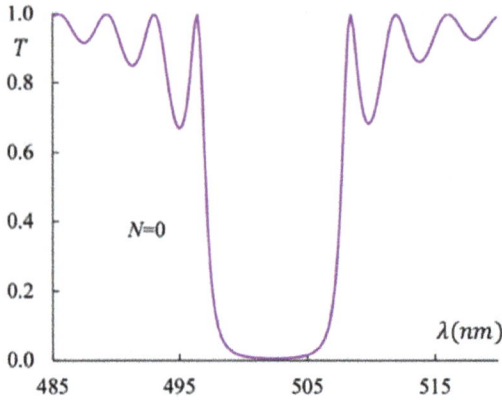

Figure 7. The transmission spectrum of a ChOH sample at $I = 0$ ($N = 0$). (Adapted from Fig. 2 of Ref. 24).

It is important to remark that the spectrum calculated at zero intensity has the well-known shape and is obviously not dependent on the step index j, being constant in time. The spectrum at $N = 0$ is reported in Fig. 7.

Significant results of the calculations performed for increasing values of light intensity are shown in Fig. 8 where the transmission spectra of the ChOH layer are reported vs increasing values of the step index j.

In the figure, data for $N = 1$ to $N = 5,000$ are displayed, with N being the normalized intensity as described above.

At $N = 1$, as shown in Ref. 23, the change of j has a negligible effect on the spectrum that is approximately symmetric with respect to the center of the PBG and is constant in time. For $N = 100$ at the red side edge of the Bragg resonance we observe alternate yellow and red colors, meaning regular oscillations of transmittivity form about 0.6 to about 0.8.

The increase of the intensity at $N = 300$ shows a higher modulation of the transmittivity (alternate green and red colors) occurring for a wider range of wavelengths. The oscillations start also on the blue edge of the Bragg resonance with a weaker modulation and a narrower range of wavelengths.

A remarkable narrowing of the PGB is clear at $N = 500$ together with irregular variable transmission on the red side edge of the resonance. Finally, at much higher intensity ($N = 5,000$) we observe irregular transmission and no more PBG.

The main features that can be observed in these figures are: the strong asymmetric behavior with respect to the center of the PBG; the transitions

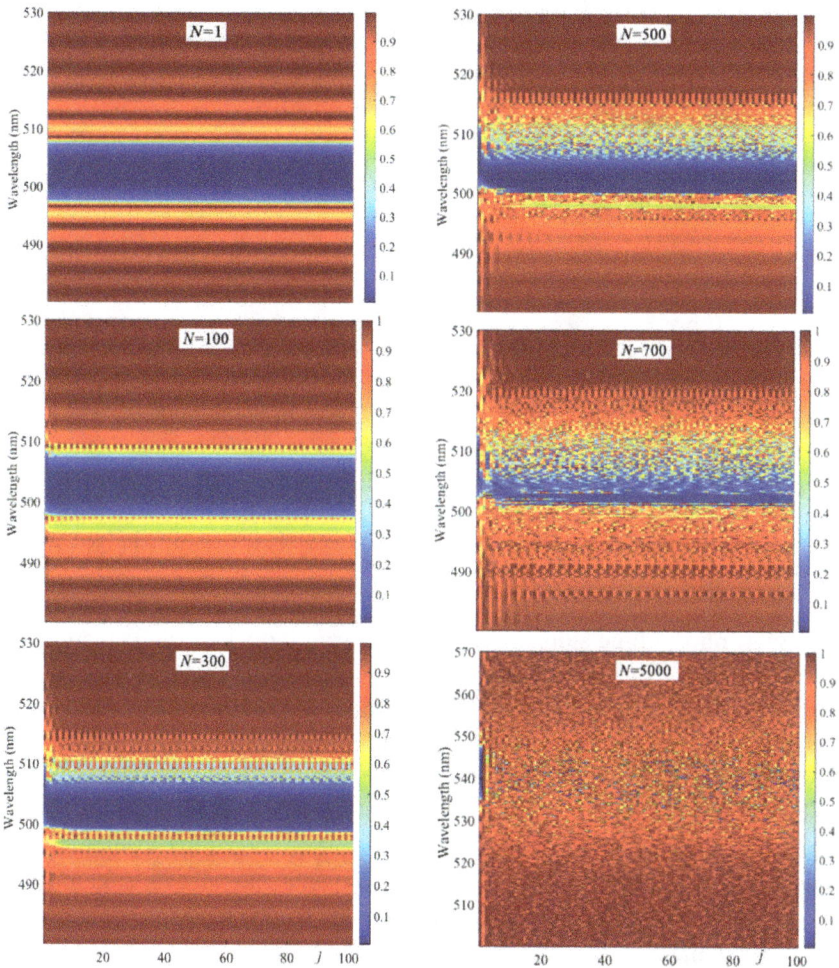

Figure 8. The transmission spectra vs the step index j for increasing values of the normalized intensity N. Colors give the normalized value of the transmittivity. The index j is proportional to the elapsed time. Adapted from Fig. 2 of Ref. 24.

from a stationary to an oscillatory and to a chaotic behavior at each wavelength by increasing the light intensity; the final disappearing of the PBG at high intensities.

The asymmetric behavior has been highlighted in Ref. 24 by comparing the time evolution for wavelengths on symmetric sides of the Bragg resonance. It has been shown that by increasing the light intensity while on the red edge of the resonance the transmission starts showing an oscillatory

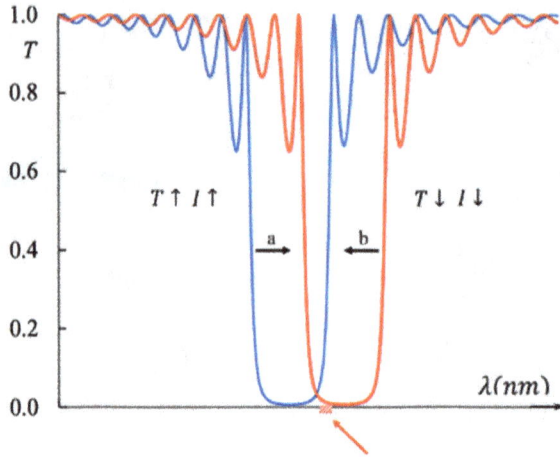

Figure 9.   The transmission spectrum at $j = 1$ undergoing alternate redshift (a) and blueshift (b) following the change of intensity of the light beam due to the consequent change in transmission for the range of wavelengths indicated by the red arrow. Reproduced from Fig. 5 of Ref. 24.

behavior, on the blue side it is still stationary; and when the transmission becomes chaotic on the red side, the blue side still has oscillatory behavior.

With the used parameters the self-induced oscillatory behavior occurs at intensities easily achievable by a c.w. laser with moderate power, in fact, $N = 500$ (just above the threshold intensity for $\lambda_2 = 510.4\,\text{nm}$) corresponds to $I = 1.9 \cdot 10^5\,\text{W/cm}^2$. In Fig. 9, a simple model is sketched that allows understanding the onset of the oscillations. Equation (30), demonstrated by experimental data reported in Refs. 11 and 25, tells us that optical reorientation leads to a redshift of the Bragg resonance.

By looking at Fig. 9, we notice that for a narrow range of wavelengths on the red side edge of the PGB (indicated by the red arrow) the redshift of the Bragg resonance leads to a decrease in transmittivity (red curve). This leads to a decrease in light intensity. According to Eq. (25), we get an increase of the effective static field E_{eff} leading to blueshift of the resonance with consequent increase in transmittivity and light intensity. In this way the process starts again (blue curve). This model accounts for the strong asymmetry mentioned above, consisting of an oscillatory behavior occurring at higher intensities for wavelengths on the blue side edge of the PBG, because the described process is related, in this case, to the secondary maxima of the resonant spectrum, while it is related to the peak resonance for the red side wavelengths. This explains why on the blue side edge of the

PBG the oscillatory behavior is observed at a higher intensity and shows a lower amplitude of modulation.

As shown in Fig. 8, the transitions to oscillatory and to chaotic behavior occur at higher and higher intensities as the considered wavelength is farther and farther from the Bragg wavelength λ_B. Therefore, since the value of λ_B is determined by the applied static field E_S, its value can be used to drive the time behavior of the transmittivity at a given wavelength.

In other words, the optical transmission at a fixed wavelength and light intensity can be stable, oscillatory, or chaotic in time depending on the applied electric field.

We show here as Fig. 10 the example given in Ref. 24 where the transmittivity vs j is reported at $\lambda_1 = 494.4$ nm for two values of the intensity and the applied static field. All the material parameters are the same as the former figures. We observe that with $N = 150$ f from $E_S = 2.03$ V/μm to $E_S = 2.09$ V/μm, the system goes from a stable transmittivity to an oscillatory behavior; while with $N = 650$, the same change of applied static field drives the system from oscillatory behavior to a chaotic one.

This example points out that the combination of two parameters, static electric field and light intensity, drives the transmission state of the system at fixed wavelength and may offer new opportunities for sensor applications.

As already remarked, the data reported in Figs. 8 and 10 show the actual time dependence of the transmitted intensity because the step index j represents the normalized elapsed time t. Taking into account Eq. (42), one is able to foresee the time range of the observed oscillations. In fact, according to the above-presented model, they are related to small changes in the tilt angle θ determining the bend distortion of the helical structure. We know that $\tau_{on} \leq \tau_{off}$ with τ_{off} being the relaxation time necessary for the system to come back to equilibrium after switching off the exciting optical field, in fact τ_{on} is expected to decrease as the intensity increases in a way similar to what happens for other reorientational effects.[26] Then, the order of magnitude of the response time can be estimated using the expression valid for a bend distortion in homeotropic nematics[26] as follows:

$$\tau_{off} = \frac{\gamma_1}{K_3} \left(\frac{d}{\pi} \right)^2$$

since also in this case optical reorientation involves a change in bend distortion. With typical values of the bend elastic constant K_3 and the viscosity γ_1 for a sample thickness of 20 μm we get $\tau_{off} \approx 40$ ms. However,

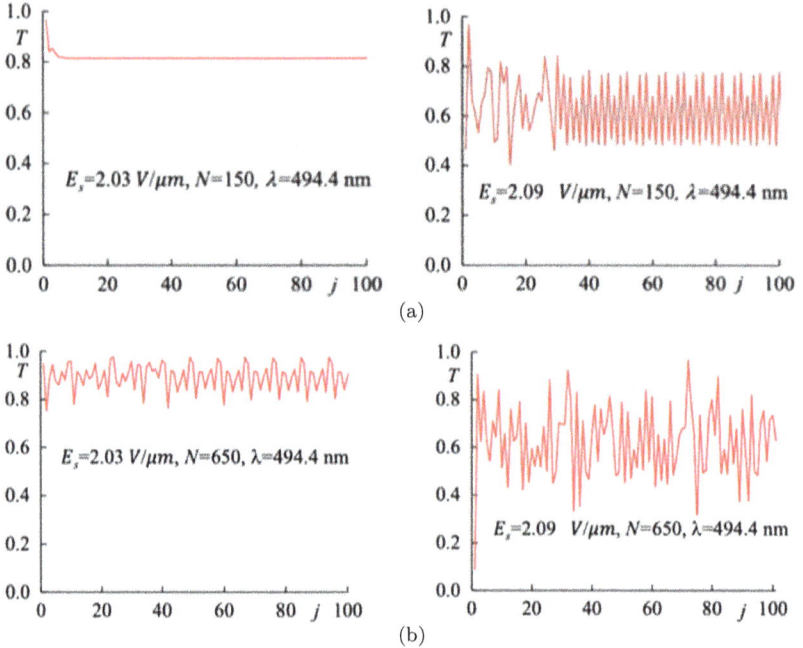

Figure 10. The transmittivity vs the index j at wavelengths $\lambda_1 = 494.4$ nm for two different values of the applied static field: $E_S = 2.03$ V/μm (curves on the left) and $E_S = 2.09$ V/μm (curves on the right): (a) $N = 150$; (b) $N = 650$. Reproduced from Fig. 6 of Ref. 24.

since ChOH exists only when $K_3 < K_2$, and its value in ChOH is 2–3 times lower than in nematics,[27] we may expect a longer response time. These arguments suggest that oscillations can occur in the range of ten/hundred milliseconds.

6. Conclusions

We summarize here the main concepts that make ChOH a very original material in the landscape of nonlinear optics of liquid crystals.

First of all, we recall that the configuration of oblique heliconical cholesterics has been foreseen over five decades ago in the theoretical work of De Gennes[6] and Meyer,[7] but only in 2014 was the experimental demonstration of its existence given[8] after synthesizing a chiral nematic having the bend elastic constant K_3 lower than the twist elastic constant K_2. This configuration is stable within a definite range of values of the applied static electric field E_S necessary for its appearance. In this range of the order of few $V/\mu m$, by varying the field it is possible to tune the Bragg reflection band over the whole visible spectrum, something that it is not possible to achieve with conventional CLC.

Based on this peculiar director arrangement ChOH represents a very new structure from the point of view of nonlinear optics of liquid crystals. In fact, the bend deformation, not present in planar CLC, gives a new opportunity to optical reorientation. This is the key point: in CLC the interaction with the optical field involves twist deformation and is much less effective requiring light intensity several orders of magnitude higher to induce any observable effect.[12,13] On the contrary, in ChOH the bend deformation allows strong coupling between the optical field and the molecular director. In this way the induced optical torque is quite efficient and can compete with the one induced by the static field for determining the value of the conical angle θ and the pitch p of the helix.

The important consequence is a light-induced redshift of the Bragg resonance, easily shown in the pump–probe experimental setup.[11,25] This observation suggests that a strong nonlinear effect can occur on a light beam powerful enough to induce optical reorientation while traveling through a ChOH sample. In fact, in this case it gives rise to a shift of the Bragg resonance that changes the amplitude and phase of the beam itself.

To date the nonlinear optical behavior originated by the amplitude change has been deeply analyzed[20,23,24] for the eigenmodes, which are the main solutions of the wave equation in the material. A multilayer approach allows a detailed calculation of the optical field distribution inside the ChOH sample in order to take into account the effect of the shift of the Bragg resonance during light propagation. The basic result is that by increasing the light intensity after the initial redshift of the PBG, self-oscillations take place, while a further increase leads to chaotic behavior. By performing this calculation for the whole spectrum of visible wavelengths around the Bragg resonance, it is possible to show that the mentioned regimes occur at different intensities for different wavelengths, lower for wavelengths close to the PBG, that disappears for pretty high intensities. A simple model explains the onset of the oscillatory regime as the result of the light-induced oscillatory shift of the Bragg resonance.

In conclusion, we have a quite unique system where the strong optical nonlinearity is due to the light-induced modulation of the Bragg resonance, and it leads to different regimes for optical transmission that can be driven and controlled by two parameters: static electric field E_S and light intensity I.

References

1. F. Reinitzer, Contribution to the knowledge of cholesterol, *Liq. Cryst.* **5**, 7–18 (1989) (translation from the original paper published in German in *Monatshefte für Chemie* **9**, 421–441 (1888)).
2. P. G. De Gennes, *Physics of Liquid Crystals*, Clarendon Press, Oxford (1975).
3. L. Blinov, *Structure and Properties of Liquid Crystals*, Springer, Dordrecht (2011).
4. V. A. Belyakov and V. E. Dmitrienko, *Optics of Chiral Liquid Crystals*, Harwood Academic, London (1989).
5. P. Yeh and C. Gu, *Optics of Liquid Crystal Displays*, Wiley, Hoboken (2010).
6. P. G. De Gennes, Calcul de la distorsion d'une structure cholesterique par un champ magnetique, *Solid State Commun.* **6**, 163–165 (1968).
7. R. B. Meier, Effects of electric and magnetic fields on the structure of cholesteric liquid crystals, *Appl. Phys. Lett.* **12**, 281–282 (1968).
8. J. Xiang, S. V. Shiyanovskii, C. T. Imrie, and O. D. Lavrentovich, Electrooptic response of chiral nematic liquid crystals with oblique heliconical director, *Phys. Rev. Lett.* **112**, 217801 (2014).
9. J. Xiang, Y. Li, Q. Li, D. A. Paterson, J. M. D. Storey, C. T. Imrie, and O. D. Lavrentovich, Electrically tunable selective reflection of light from ultraviolet to visible and infrared by heliconical cholesterics, *Adv. Mater.* **27**, 3014–3018 (2015).
10. M. Mrukiewicz, O. S. Iadlovska, G. Babakhanova, S. Siemianowski, S. V. Shiyanovskii, and O. D. Lavrentovich, Wide temperature range of an electrically tunable selective reflection of light by oblique helicoidal cholesteric, *Liq. Cryst.* **46**, 1544–1550 (2019).
11. G. Nava, F. Ciciulla, O. S. Iadlovska, O. D. Lavrentovich, F. Simoni, and L. Lucchetti, Pitch tuning induced by optical torque in heliconical cholesteric liquid crystals, *Phys. Rev. Res.* **1**, 033215-5 (2019).
12. B. Ya Zel'dovich and N. V. Tabiryan, Orientational effect of a light wave on a cholesteric mesophase, *Sov. Phys. JETP* **55**, 99–104 (1982).
13. E. Santamato, G. Abbate, P. Maddalena, and Y. R. Shen, Optically induced twist Freedericksz transitions in planar-aligned nematic liquid crystals, *Phys. Rev. A* **36**, 2389–392 (1987).
14. J. P. Vernon, A. D. Zhao, R. Vergara, H. Song, V. P. Tondiglia, T. J. White, N. V. Tabiryan, and T. J. Bunning, Photostimulated control of laser transmission through photoresponsive cholesteric liquid crystals, *Opt. Exp.* **21**, 1645–1655 (2013).
15. U. A. Hrozhyk, S. V. Serak, N. V. Tabiryan, T. J. White, and T. J. Bunning, Nonlinear optical properties of fast, photoswitchable cholesteric liquid crystal bandgaps, *Opt. Mat. Exp.* **1**, 943–952 (2011).
16. S. V. Serak, N. V. Tabiryan, and T. J. Bunning, Nonlinear transmission of photosensitive cholesteric liquid crystals due to spectral bandwidth autotuning or restoration, *J. Nonlin. Opt. Phys. Mat.* **16**, 471–483 (2007).
17. D. W. Berreman, Optics in stratified and anisotropic media: 4×4 matrix formulation, *J. Opt. Soc. Am.* **62**, 502–510 (1972).

18. D. W. Berreman, Twisted Smectic C Phase: Unique Optical Properties, *Mol. Cryst. Liq. Cryst.* **22**, 175–184 (1973).

19. A. H. Gevorgyan, Reflection and transmission of light for a layer with dielectric and magnetic helicities. I. Jones Matrices. Natural polarizations, *Opt. Spectroscopy* **89**, 631–638 (2000).

20. F. Simoni, Nonlinear optical propagation near the Bragg resonance in heliconical cholesteric liquid crystals, *Opt. Lett.* **45**, 6510–6513 (2020).

21. A. H. Gevorgyan, S. S. Golik, and T. A. Gevorgyan, On peculiarities in localization of light in cholesteric liquid crystals, *J. Exp. Theor. Phys.* **131**, 329–336 (2020).

22. A. H. Gevorgyan and M. Z. Harutyunyan, Chiral photonic crystal with an anisotropic defect layer, *Phys. Rev E* **76**, 031701 (2007).

23. A. H. Gevorgyan and F. Simoni, Onset of optical instabilities in the nonlinear optical transmission of heliconical cholesteric liquid crystals, *Photonics* **9**, 139–148 (2022).

24. A. H. Gevorgyan and F. Simoni, Light-induced self-oscillations and spoiling of the bragg resonance due to nonlinear optical propagation in heliconical cholesteric liquid crystals, *Photonics* **9**, 881–892 (2022).

25. G. Nava, R. Barboza, O. S. Iadlovska, O. D. Lavrentovich, F. Simoni, and L. Lucchetti, Optical control of light polarization in heliconical cholesteric liquid crystals, *Opt. Lett.* **47**, 2967–2970 (2022).

26. F. Simoni, *Nonlinear Optical Properties of Liquid Crystals and PDLC*, World Scientific, Singapore (1997).

27. O. S. Iadlovska, G. Babakhanova, G. H. Mehl, C. Welch, E. Cruickshank, G. J. Strachan, J. M. D. Storey, T. I. Corrie, S. V. Shiyanovskii, and O. D. Lavrentovich, Temperature dependence of bend elastic constant in oblique helicoidal cholesterics, *Phys. Rev. Res.* **2**, 013248-7 (2020).

Chapter 5

Liquid Crystals for Displays, Smart Windows, Tunable Metamaterials, Plasmonic Nanostructures, Micro-ring Resonators, and Ultrafast Laser Manipulations

I. C. Khoo* and T.-H. Lin

Electrical Engineering Department,
The Pennsylvania State University
University Park, PA 16802, USA
*ick1@psu.edu

This chapter gives a brief review of the physical and optical properties of liquid crystals for developing tunable/reconfigurable optical materials and nanostructures, and a discussion of some exemplary applications such as conventional display and smart window, and more specialized topics in metamaterials, plasmonic nanostructures, micro-ring resonators, optical limiter and chiral photonic crystals for polarizations, and manipulations of ultrafast pulsed lasers.

1. Introduction

Liquid crystals (LC) are novel optical materials formed by natural self-assembly of organic molecules into various ordered phases, cf. Fig. 1.[1,2] Among the myriad of ordered phases discovered to date, nematic and chiral nematic [often called cholesteric] liquid crystals, NLC and CLC for short, are among the most investigated and widely used for optical and photonic applications.

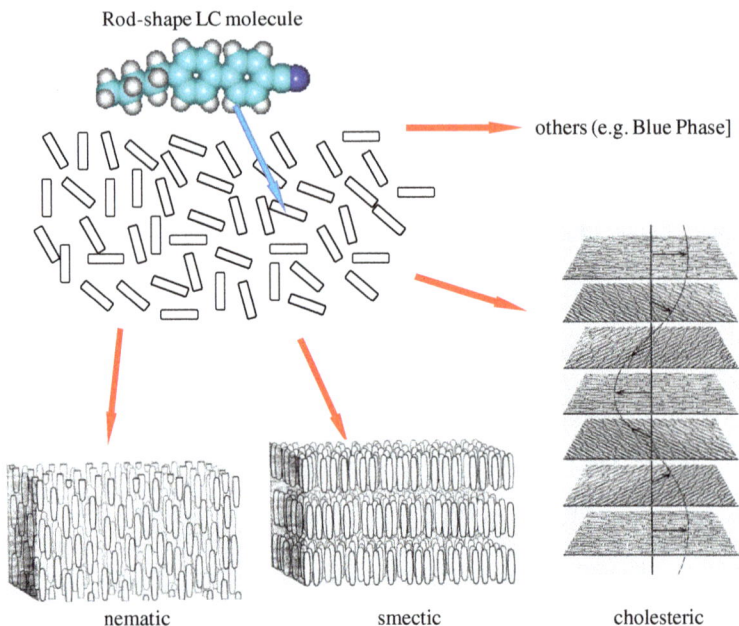

Figure 1. Self-assembly of rod-shaped organic molecules into various ordered phases exemplified by nematic, smectic, and cholesteric liquid crystals.

Liquid crystals exhibit linear and nonlinear optical responses arising from individual molecular origins as well as collective crystalline mechanisms. Individual molecular responses include fundamental electronic wave functions' perturbation by the optical electric field, field-induced individual molecular and nuclear reorientations, electrostriction, etc., with characteristic response times typically in the picoseconds–femtoseconds time scales. Collective crystalline responses depend on the specific ordered phase and include field-induced crystalline reorientation, lattice distortion, order parameter modifications, photorefractivity, flows, etc., with response times typically in the milliseconds–microseconds range, but can be much longer or shorter.

Liquid crystals possess many unique characteristics well suited for applications. Nematics, for example, are endowed with large index birefringence throughout the visible, near-, and mid-infrared, Terahertz, and longer wavelength regimes[1–7]; such large birefringence manifests as large circular birefringence in chiral nematics (cholesteric liquid crystals). Their fluid nature allows easy integration in/with various nano- and micro-structures

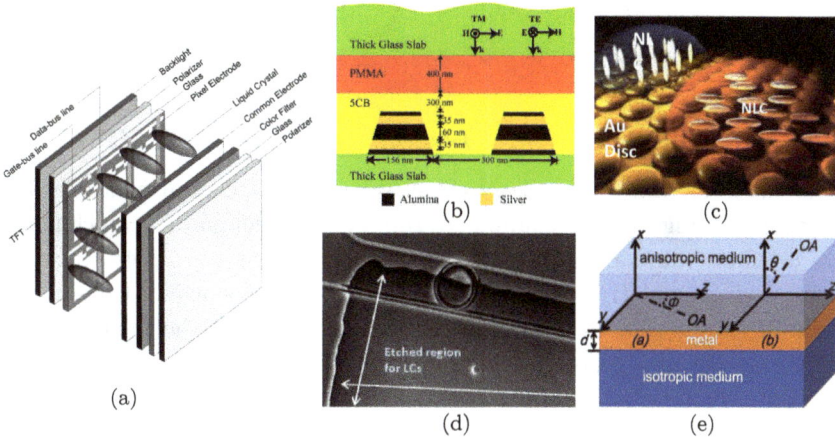

Figure 2.   (a) Typical LC pixel in ubiquitous display devices; (b)–(e) Tunable micro-
and nano-photonic structures incorporating liquid crystals. (b) Magnetic metamaterials;
(c) plasmonics nanostructure (nano-gold-disk array); (d) tunable micro-ring resonator;
(e) plasmonic waveguide.

and incorporation of photo-sensitive agents such as azo-dyes, fullerene,
and plasmonic nanoparticles, giving rise to a host of novel materials and structures such as tunable metamaterials, metasurfaces, plasmonic-
waveguides, -switches, and -quantum emitters, micro-ring resonators, and
other electromagnetic structures, cf. Fig. 2. In conjunction with wide-
ranging mechanisms and dynamical responses, these materials and structures provide a fertile ground for fundamental research as well as photonic
applications. This chapter is intended to present a concise summary and
a glimpse of some fundamentals and exemplary photonic applications/
potentials; details can be found in the quoted references in the text.

2. Optical Properties of Liquid Crystals — General Discussion

Optical properties of liquid crystals fall under two distinct types: (i) properties arising from molecular electronics responses and (ii) those associated
with specific ordered/crystalline phase with distinctive arrangement of
the LC molecules. Here we first discuss some fundamentals of molecular
electronic optical responses that apply to all liquid crystalline phases
(including the isotropic liquid phase). For (ii), we focus on the 1-D photonic

crystalline properties arising from the chiral arrangement of director axis
in cholesteric liquid crystals.

2.1. *Molecular electronic responses to optical fields*

The electronic and energy level structures of liquid crystal molecules are
complex, cf. Fig. 3, and require powerful quantum computation techniques
to calculate fundamental optical parameters such as dipole moment,
energy levels and transition probabilities/rates/frequencies.[8] Considering
only dipole interaction between the optical fields and the molecule, the
field-induced dipole moment of a molecule (*usually ground state*) is of the
form of a power series in $\boldsymbol{E}$:

$$\boldsymbol{d}(t) \sim \boldsymbol{\alpha} : \boldsymbol{E} + \boldsymbol{\beta} : \boldsymbol{EE} + \boldsymbol{\gamma} : \boldsymbol{EEE}. \tag{1}$$

Here the double dots signify tensorial operation between the polarizability
tensors $\boldsymbol{\alpha}$, $\boldsymbol{\beta}$, and $\boldsymbol{\gamma}$ with the vector fields Es.

Since $\boldsymbol{E}$ is a vector described by three Cartesian components ($\boldsymbol{E}_i$, $\boldsymbol{E}_j$,
$\boldsymbol{E}_k$), α, β, and γ are tensors of second, third, and fourth rank, respectively;
$\alpha = \{\alpha_{ij}\}, \beta = \{\beta_{ijk}\}$, and $\gamma = \{\gamma_{ijkh}\}$ are, respectively, linear, second-
order, and third-order molecular polarizabilities. The optical field-induced
electric polarization $\boldsymbol{P}$, defined as the dipole moment per unit volume,

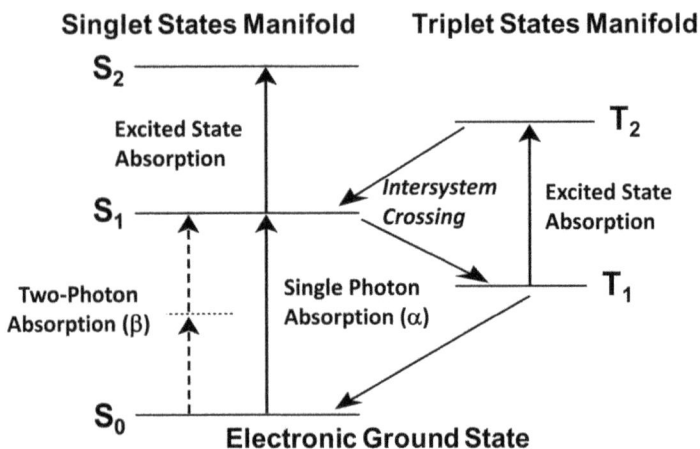

Singlet States Manifold   **Triplet States Manifold**

S_2

Excited State
Absorption

T_2

Intersystem
Crossing

Excited State
Absorption

S_1

Two-Photon
Absorption (β)

Single Photon
Absorption (α)

T_1

S_0 **Electronic Ground State**

Figure 3.   Molecular energy levels of typical organic molecules.

is therefore of the form

$$P = \varepsilon_o[\chi^{(1)} : E + \chi^{(2)} : EE + \chi^{(3)} : EEE]. \tag{2}$$

The first, second, and third terms on the right-hand side of (2) are, respectively, the linear, second-, and third-order nonlinear polarizations, with $\chi^{(1)}, \chi^{(2)}$, and $\chi^{(3)}$ the respective susceptibility tensors. The macroscopic susceptibility tensors $\chi^{(n)}$ are related to the respective microscopic polarizability tensors α, β, and γ by the molecular number density N weighted by the *local field correction factor* $L^{(n)}$;[1] e.g., $\chi^{(3)} = NL^{(3)}\gamma$.

It is very important to note here that to be physically meaningful to account for relaxations processes, all susceptibility tensors, and physical parameters such as μ, ε, and index of refraction, are complex entities with real and imaginary parts.

2.2. *Linear optical response and refractive indices*

Linear optical properties of liquid crystals are the single most important physical parameter in any discussions on photonics. In the present context, we focus our attention on the index of refraction, which in liquid crystalline media is anisotropic due to their molecular structures. In the crystalline coordinates, the refractive indices $n_i(i = 1, 2, 3)$ are related to their respective components of the permeability tensor μ and permittivity tensor ε by

$$n_i = (\mu_i \varepsilon_i / \mu_o \varepsilon_o)^{1/2}, \tag{3}$$

where μ_o and ε_o are, respectively, vacuum permeability and permittivity.

For a non-magnetic medium, $\mu = \mu_o$ and so $n_i = (\varepsilon_i/\varepsilon_o)^{1/2}$ where ε_i is the ith-component of the dielectric tensor $\varepsilon = (1 + \chi^{(1)})$. The real part of $\chi^{(1)}$ is involved in dispersion-related optical processes, while the imaginary part of $\chi^{(1)}$ is involved in linear (single-photon) absorption. Most liquid crystals are birefringent; in that case, the refractive indices are usually denoted by $n_e(\theta)$ and n_o for extraordinary and ordinary light waves, respectively. The value of $n_e(\theta)$ depends on the director axis orientation angle θ relative to the optical electric field and varies from a maximum value of $n_{//}$ for light polarized along the molecular axis to $n_\perp$ for light polarized perpendicular to the axis. Both n_e and n_o are functions of ambient temperature through their dependence on the order parameter S and, to a lesser degree, on the density ρ.[1]

From Eq. (3), one can see that the index of refraction n is a complex number with a real part [which is commonly called the refractive index] and an imaginary part associated with light absorption or amplification. The value of the real part of n can assume negative values or sub-unity (<1) or near-0 value, i.e., values less than that of vacuum. The field of study on materials that have such characteristics is now rapidly evolving, with the emergence of what are referred to as metamaterials or meta-surfaces and nanostructured plasmonic or dielectric structures. Naturally, if liquid crystal is present as an active constituent, the effective refractive indices and other physical and optical properties will become tunable or reconfigurable[9-22] by applied fields/stimuli via several molecular or crystalline mechanisms with response times ranging from slow (seconds–milliseconds) to fast (microseconds–nanoseconds) and ultrafast (picosecond–femtoseconds).[1,2,22-24] More on this is presented in Secs. 3.2–3.4.

2.3. *Nonlinear optical responses and intensity-dependent refractive index change*

Most liquid crystals, including the nematic and cholesteric liquid crystals under present consideration, exhibit centro-symmetry and thus $\chi^{(2)} = 0$. The symmetry can be 'broken' by applying a DC field (e.g., in electric field-induced second harmonic generation) or at an interface (e.g., in surface second harmonic generation).

On the other hand, all materials possess nonlinear third-order susceptibility $\chi^{(3)} = \mathrm{Rl}\,\chi^{(3)} + i\mathrm{Im}\,\chi^{(3)}$. Among the multitudes of terms present in the real part of the third-order nonlinear polarization $\boldsymbol{P}^{(3)} = \varepsilon_o \mathrm{Rl}\,\chi^{(3)}$: $\boldsymbol{EEE}$, one can identify a term of the form $\varepsilon_o \mathrm{Rl}\,\chi^{(3)}(\boldsymbol{EE}^*)\boldsymbol{E} = \Delta\varepsilon\boldsymbol{E}$ that corresponds to a laser-induced change in the material's dielectric constant, $\Delta\varepsilon$, that is proportional to the laser intensity I. This is equivalent to an intensity-dependent index change $\Delta n = n_2 I$. The nonlinear index coefficient n2 is often termed optical Kerr constant in analogy to AC field-induced Kerr effect in electro-optics crystals. The intrinsic off-resonant nonlinear index coefficient n_2 of transparent liquid crystals are about $10^{-14} - 10^{-13}\,\mathrm{cm}^2/\mathrm{W}$.[8,22-24]

It is important to note here that such nonlinearity that originates from the electronic origins with response time in the femtoseconds time scale are distinctly different from intensity-dependent index changes associated with

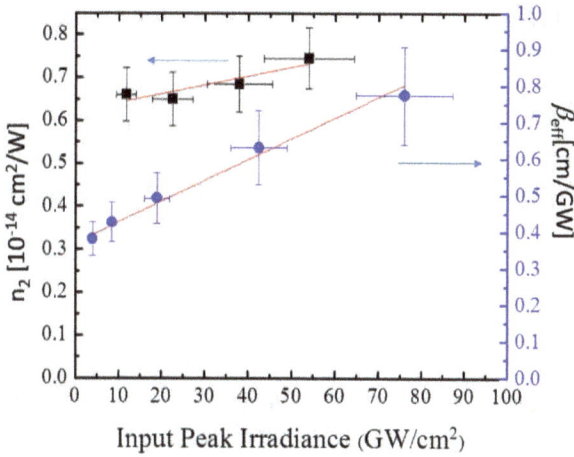

(a)

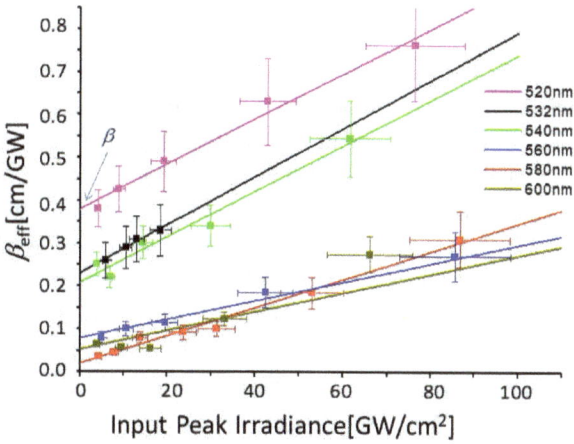

(b)

Figure 4.   Peak irradiance dependence of (a) Nonlinear index coefficient n_2 and effective two photon absorption coefficients β_{eff} of an organic liquid (L34) and (b) β_{eff} obtained in femtosecond Z-scan measurement;[8] the intrinsic value β is obtained by the (intercept) at zero intensity.

much slower processes in the liquid crystalline phase such as director axis reorientation, order parameter modifications, thermal and density changes, flows, lattice or helical pitch distortions, photorefractivity, etc.; those processes produce extraordinarily large optical nonlinearities with response times ranging from seconds through milliseconds to nanoseconds, and have been thoroughly investigated for nematic and cholesteric [including Blue-Phase] liquid crystals.[1,22–34]

Cholesteric liquid crystals that are 1-D photonic crystals; the electronic optical nonlinearities can experience enhancement near the photonic band-edges due to the increase in the optical density of states (slowdown of light's group velocity) by orders of magnitude to reach $10^{-11} - 10^{-10}$ cm^2/W.[35–37] The magnitude of Im $\chi^{(3)}$ of liquid crystals is also among the largest of all known optical materials.[8,38–41]

Since real transitions are involved and the fact that singlet-triplet inter-system crossing processes usually occur at much different time scales than instantaneous two- and multi-photon absorptions, the observed effective nonlinear absorption coefficient β_{eff} depends critically on the laser pulse duration for a given laser energy fluence [energy/area]. For example, the β_{eff} value of the transparent neat liquid L34(4-propyl 4'-butyl diphenyl acetylene) was observed to vary from ~1 cm/GW in the femtoseconds time scale to over 25 cm/GW in the nanosecond's regime.[8] If the molecules have strong excited-state absorption following the two-photon absorption from the ground state, β_{eff} also tends to exhibit an increasing-dependence with laser intensity.[38–41] A quantitative modeling[38] of the dynamical evolution of the level populations shows that *for the same material*, the observed β_{eff} *could vary by orders* of magnitude as the pulse duration or laser intensity is varied. Therefore, *characterizations of what is referred to as the nonlinear absorption ability of a nonlinear optical material can only be meaningful if the observed data are accompanied by the actual pulse duration, wavelength, and energy fluence used in the measurements; nonlinear absorption coefficients observed in one time scale (or intensity) do not hold in another time scale (or intensity) at all.*

Since the most excited-state absorption and inter-system crossing processes occur in the femtoseconds–nanoseconds time scale, nonlinear absorption-based applications such as passive optical limiting for eye/sensor protection against pulsed lasers work well in these time scales,[8,22] cf. Fig. 5. Against lasers of longer pulse duration (100's ns–μs and longer), however, such a highly intensity-dependent optical limiting action becomes less efficient because of the proportional decrease in laser intensity for a given laser pulse energy.

Therefore, it is highly desirable to have nonlinear optical materials such as liquid crystals that have *simultaneously* multiple mechanisms that respond effectively over a very wide temporal range spanning femtoseconds–microseconds to CW, cf. Fig. 6. In combination with specialized optical

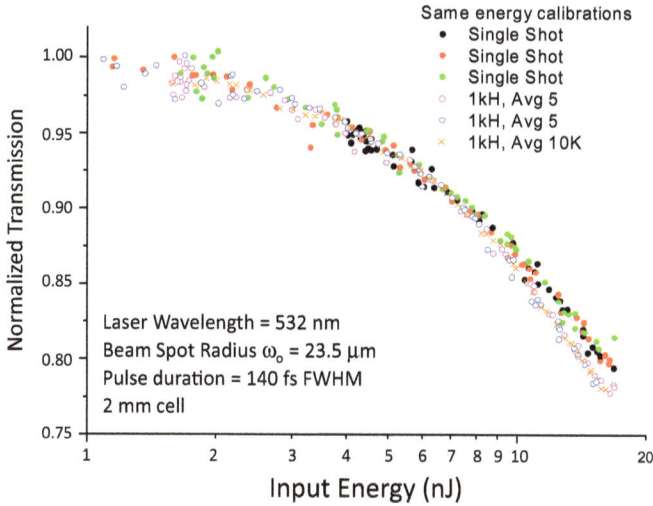

Figure 5. Optical limiting performance of a 2-mm L34 cell irradiated by a 140-fs laser under various pulse repetition rates. Transmission switching to lower normalized transmission value begins at an input laser threshold energy of ~1 nanoJoule.[22]

structures such as nonlinear liquid-cored fiber arrays and planar LC optics,[38–48], the resulting device is capable of passive limiting of agile frequency lasers over a wide-temporal range.

As 'summarized' in Fig. 6, besides modification of electronic wave function leading to an equivalent index change, optical fields induce birefringent crystalline axis reorientations, order parameter changes, photorefractive effects and excited dye-LC intermolecular torque ... etc.;[22–24] these mechanisms produce intensity-dependent index changes characterized by n_2 that are orders of magnitude larger than the electronic nonlinearity, cf. Fig. 6, and have been employed for optical switching and limiting, image processing, wave-mixing and optical phase conjugation, and tunable metamaterials, micro-ring resonators, and plasmonic nanostructures.[22–34]

2.4. *Electro-optical responses of liquid crystals — A brief introduction*

Electro-optical responses commonly employed in optical modulation devices, where the refractive indices are modified with an applied AC or DC electric field, fall into two classes. In Pockels cell effect, the induced index change is proportional to the applied field E and is therefore sensitive to the

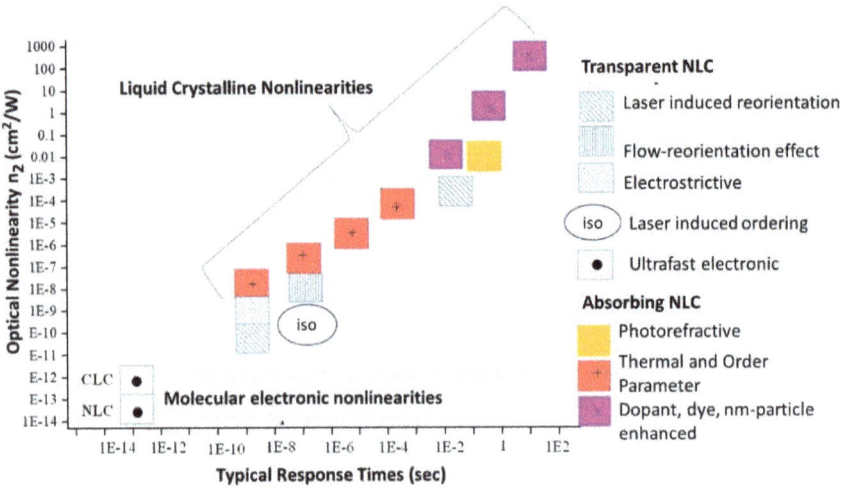

Figure 6. Major mechanism for optical nonlinearity n_2 and typical response times of nematic and cholesteric liquid crystals.

direction (either positive or negative) and magnitude of the applied field. In Kerr effect, the index change is proportional to the square modulus of the applied electric field E, i.e., $\Delta n \sim \kappa E^2$, where κ is the Kerr constant.

It is important to note here that the response of liquid crystals to external stimuli such as electric field is intricately related to the 'crystalline' make up of a particular ordered phase (nematic, cholesteric, Blue-phase, or heliconical cholesteric liquid crystals, ferroelectric, smectic, bend-core, etc.) as well as the presence of other 'agents', 'dopants', or ions.[2,49] Such applied field-induced index changes do not fall into simple well-defined classes such as Pockels or Kerr.

It is also clear that in the presence of a bias field [AC or DC], both linear and nonlinear optical responses of the liquid crystals will be substantially and often profoundly modified as the applied field changes all the physical properties of the liquid crystals such as the director axis configurations, lattice structures, space charge fields, shifts in photonic bandgaps, and a myriad of other characteristics.[1,2,22-24,49]

More on electro-optical effects employed in the current fast developing field of liquid-crystal displays (LCDs) and smart window technologies is presented in Sec. 3.1.

2.5. Optical properties of chiral nematic [cholesteric] liquid crystal — A 1D photonic crystal

Cholesteric liquid crystals (CLCs) are formed by introducing a chiral agent into a nematic liquid-crystal host. Upon cooling from the isotropic phase, the molecules self-assemble into an helical arrangement as depicted in Fig. 7. Owing to the birefringence of the nematic constituent, an incident light 'sees' a periodically varying refractive index between n_e and n_o.

Optically, therefore, CLC is a chiral Bragg grating or a 1D chiral photonic crystal and shows interesting properties such as photonic bandgap and strong dispersion near the band-edges, cf. Fig. 7. The long- and short-wavelength band-edges are located at $\lambda_L = Pn_e$ and $\lambda_S = Pn_o$, respectively, where $p = 2\Lambda$ is the pitch of the cholesteric helix and Λ is the index grating period. The photonic bandgap has a width of $\lambda_L - \lambda_S$ and is centered at $\lambda_0 = P(n_o + n_e)/2$. The rich variety of nematic liquid crystals with positive or negative dielectric anisotropy enable fabrication of CLCs with photonic bandgaps covering the entire spectral range from visible to near- and mid-infrared.

Conventional CLCs are made by molecular self-assembly; the fluidic nature and the limited extent of cell-boundary anchoring forces can ensure good planar alignment characterized by uniform standing helices for cell thickness up to at most ∼100-μm. In thicker samples, non-standing helices

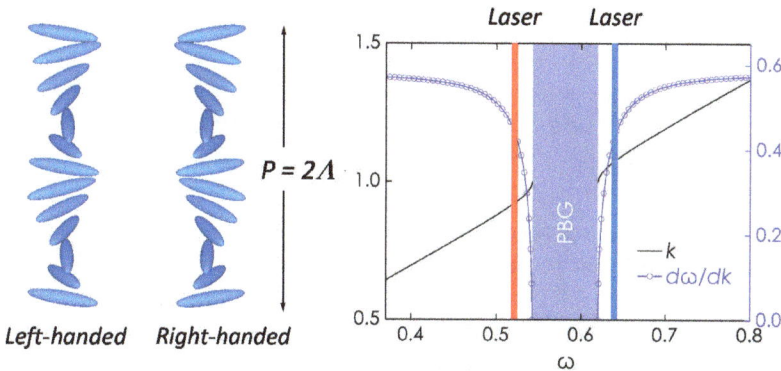

Figure 7. (Left) Schematic depiction of the helical arrangement of the director axis in a cholesteric liquid crystal. Both right- and left-handed twist are possible. (Right) Plot of the wave vector k (normalized to the cholesteric spiral wave vector $q = 2\pi/P$) and group velocity $d\omega/dk$ (normalized to c/q) in the vicinity of the CLC photonic bandgap, using an optical dielectric anisotropy $\Delta\varepsilon = 3$, and $n_e = 1.585$.

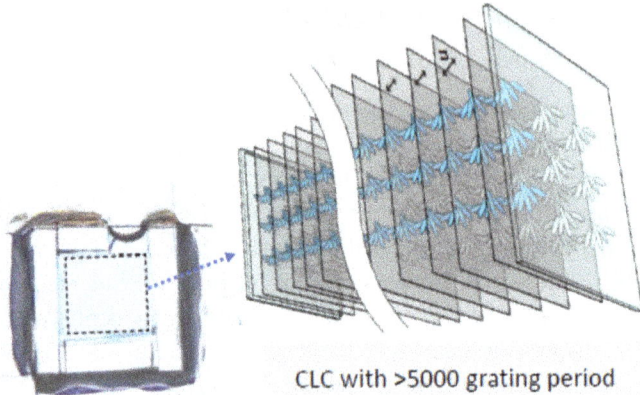

CLC with >5000 grating period

Figure 8.   Photograph of a fabricated extraordinary-thick CPC with extremely high grating periodic number; the clear aperture of ~cm^2 (larger if needed) allows free-space optical polarization rotation and pulse modulation applications with un-focused laser beams.

tend to appear in the bulk and gather to form line defects that take on the appearance of oily streaks. A recent detailed study[52] reported the development of a room-temperature Dual-Frequency Field Assembly [DFFA] technique that enables robust fabrication of stable well-aligned cholesteric liquid crystals to unprecedented thickness cf. Fig. 8. DFFA involves successive applications of a low- and a high-frequency electric field on a thick cell of CLC starting mixture. The low-frequency field creates conductive hydrodynamical instabilities that break up all the thick oily streaks present and render the mixture to a state with completely randomized orientation of the cholesteric helices. Then a high-frequency field (at a field strength below the hydrodynamic instability) is applied to align these randomly distributed helices into uniform standing helices.

Such a room-temperature DFFA technique enables fabrication of stable well-aligned CLC to record setting thicknesses with super-high grating period number N [thickness/Λ] of several 1000s and circumvents all the drawbacks and undesirable effects caused by large temperature change involved in conventional LC self-assembly (with or without field assist) techniques.

CLCs have found applications in various photonic and electro-optic applications[49–61,66,73–82] ranging from electrically or optically tunable elements/devices with selective transmission or reflection bands, through low-threshold lasing actions due to the increased density of states near the photonic band-edge, to energy-saving windows and, more recently, ultrafast (femtoseconds) pulse laser modulations and polarization switching,[62–82] to name a few; more on the latter is presented in Sec. 3.4.

3. Photonics Applications

3.1. *Displays and smart windows*

Most modern liquid-crystal display screens employ the mechanism of field-induced reorientation of the director axis of nematic liquid crystals sandwiched between two ITO-coated glass windows. In a 90° TN (twisted nematic) device, the liquid crystal is aligned (by an alignment layer over the ITO-glass surfaces) in directions perpendicular to each other so that the director axis is initially twisted by 90° from one end of the cell to the other. The polarization of incident light follows adiabatically the twisting of the director axis so that it is fully transmitted (bright state) if the LC is placed between two crossed polarizers. When an electric field is applied between two ITO substrates, the director axis is aligned perpendicular to the substrate; the polarization of the incident light does not rotate as it passes through the liquid-crystal layer and the exiting light will be blocked by the crossed polarizer (dark state). By controlling the voltage applied to the liquid crystal of each pixel, light can have different transmittance in the pixel matrix of a display panel for image display.[2,49] Such a display is often referred to as the transmissive type and involves backlighting to provide the incident light.

For many practical reasons, image display is increasingly performed in the reflective mode; the brightness of reflective displays comes from reflecting the ambient light. Absent a backlight module, it can achieve energy-saving or even non-energy-consuming operation. Reflective displays can be much lighter than transparent displays seen in the market today and are even easier to use in flexible/portable products such as electronic paper, electronic labels, and electronic signage.

Cholesteric liquid crystals have both reflective and bistable display capabilities. The operation principle of such a display is based on the Bragg reflection band arising from the spatially spiraling helices; Red, Green, or Blue (reflection) display is achieved by adjusting the concentration of the chiral dopant in the cholesteric liquid crystal which dictates the width of the pitch. The well-aligned planar texture can be switched to a randomly aligned focal conic texture by an electric field to produce the (non-reflective) dark mode. Another important feature is the bistable property.[65] Under the proper driving scheme of electric field, it can switch between planar texture (bright state) and Focal conic texture (dark state), and the state remains upon field removal. This bistable feature enables excellent low power consumption.[65-67]

Liquid-crystal display (LCD) technology is rapidly advancing, making it possible to produce displays with even better image quality and versatility. An example is the use of dynamic backlighting with quantum dots, which enhances the color gamut and contrast of LCDs.[68] This technology has made it possible for LCDs to achieve a level of performance that is comparable to that of light-emitting diode (LED) displays, while still being affordable. As technology continues to develop, applications of LCDs become more diverse, with a wide range of new and innovative products entering the market. This includes transparent displays,[69] which allow the display to blend seamlessly into the environment. Another growing application of LCD technology is in augmented reality (AR) and virtual reality (VR) devices, where head-mounted near-eye displays are becoming increasingly popular.[70] These displays offer an immersive experience and are essential for a wide range of AR and VR applications,[71,72] from gaming to education, training, and simulation.

In conclusion, LCD technology continues to evolve, with new developments being made all the time as exemplified by smart windows. Smart windows, which dynamically adjust light transmittance/reflection and other desirable optical properties for privacy protection and information display, have existed for quite some time — photochromic, thermochromic, or simply electric field-controlled color changes. Here we review smart windows made of liquid-crystal materials and classify them according to the mechanisms underlying their operating principles: *absorption, scattering,* and combination of *scattering and absorption,* cf. Fig. 9.

The absorption mode refers to keeping a low haze level of the material while controlling its transmittance or absorption rate. For example, in the case of building or car windows, the proportion of strong sunlight entering the interior in the daytime can be controlled to provide heat insulation without losing the background information or image clarity that the original glass window should provide. In places where the difference between light and dark is more significant, instantaneous adjustment of transmittance can help the eyes adapt to the environmental differences. There are various methods to achieve an absorption function in liquid crystal devices, by mixing dichroic dyes in the liquid crystal material or using an external polarizer.[2,49]

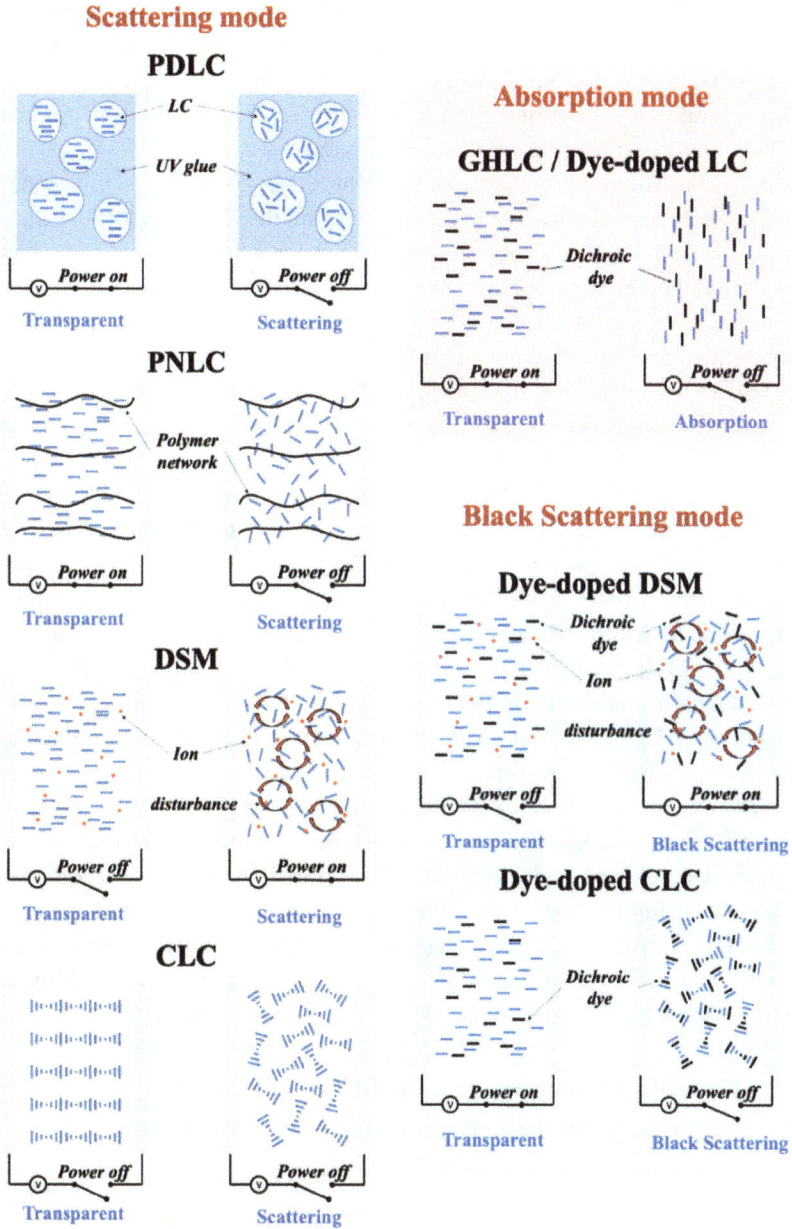

Figure 9.   Illustrations of liquid-crystal-based smart window technologies.

Switchable scattering (SS) mode is easier to achieve in materials with birefringence, such as liquid crystals. The purpose is to produce a high-haze effect in the material to shield or blur the background image. SS smart windows can provide added privacy protection, so they are commonly used in architectural glass features, display cases, car windows, meeting rooms, or bathroom glass partitions. By integrating projection function, SS windows can be used as screens for information display. Polymer-Dispersed Liquid Crystals (PDLCs) are the most used for such SS operations based on the field-controlled matching or mismatching of the refractive indices of the liquid crystal and the polymer.[73,74] Other variants of PDLC[75,76] have emerged, each with their merits and limitations.

For an initially clear operation, cholesteric liquid crystals (CLCs) are promising alternatives as they can be switched from a well-aligned clear state to a highly scattering focal-conic state. Moreover, CLC exhibits bi-stable property — it can maintain the transparent (planar) and scattering (focal conic) states without applying an electric field. Compared with the liquid-crystal smart window technology mentioned earlier, the CLC-based window has a slower switching response time, mainly because it takes a longer time to self-assemble and arrange into the planar or focal conic state. In addition, the scattering performance of cholesteric liquid crystal in the focal conic state is currently not as good as PDLC or its variants.

These mechanisms (absorption and scattering) can be combined in polymer-stabilized cholesteric texture (PSCT), which uses a polymer network to stabilize the planar or focal conic state of cholesteric liquid crystal.[77-79] Although the bi-stable property is no longer possible, the response time is reduced; also, there are more refractive index interfaces between the liquid crystal and the polymer network, resulting in better scattering performance. The Bragg reflection provided by the (clear) planar state can be retained for ultraviolet, visible, to infrared light, resulting in a smart window with color reflection or 'transflective' effect.[80-82]

3.2. *Tunable near-zero and negative index of refraction; metamaterials and metasurfaces*

Consider the complex refractive index given by

$$
\begin{aligned}
n_{+z} &= \sqrt{\frac{\mu\varepsilon}{\mu_0\varepsilon_0}} = \sqrt{(\varepsilon_r' + i\varepsilon_r'')(\mu_r' + i\mu_r'')} \\
&= \sqrt{(\varepsilon_r'\mu_r' - \varepsilon_r''\mu_r'') + i(\varepsilon_r'\mu_r'' + \varepsilon_r''\mu_r')} \\
&= n' + in'',
\end{aligned}
\tag{4}
$$

where n_{+z} indicates that we *choose the sign of the square root for which the imaginary part n'' is positive*, i.e., we follow the convention where power flow is in the $+z$ direction, so a positive imaginary part n'' corresponds to an exponential attenuation of the fields propagating through a lossy medium.[16]

From Eq. (4), a sufficient condition to achieve negative index (Re$\{n_{+z}\} = n'$ is negative) is to have $\varepsilon_r' < 0$ and $\mu_r' < 0$, given that both the imaginary permittivity and the imaginary permeability are positive. However, even if the real part of the permittivity or the permeability is positive, n' can still assume negative value if $\varepsilon_r'\mu_r'' + \varepsilon_r''\mu_r' < 0$, i.e., if we have $\varepsilon_r' < -\frac{\varepsilon_r''\mu_r'}{\mu_r''}$.

Large and negative permittivity is commonly found in plasmonic materials in the vicinity of a plasmonic resonance. With the introduction of liquid crystals in the fabrication of metamaterials, the effective ε_r' can be modulated by a wide range of stimuli to develop metamaterials with tunable/reconfigurable index of refraction and other optical properties. Another ingredient for tunable metamaterial is the permittivity $\mu = \mu_r' + i\mu_r''$. Xiao et al.[11] first succeeded in fabricating the hard-to-get tunable permeability μ in the optical frequency domain, cf. Fig. 10. In this study, tuning of the magnetic structure was enabled by changing the ambient temperature of the structure, which changes the dielectric constant of the NLC (5CB) and therefore the magnetic structure. To date, various tunable and reconfigurable metamaterial structures have been developed

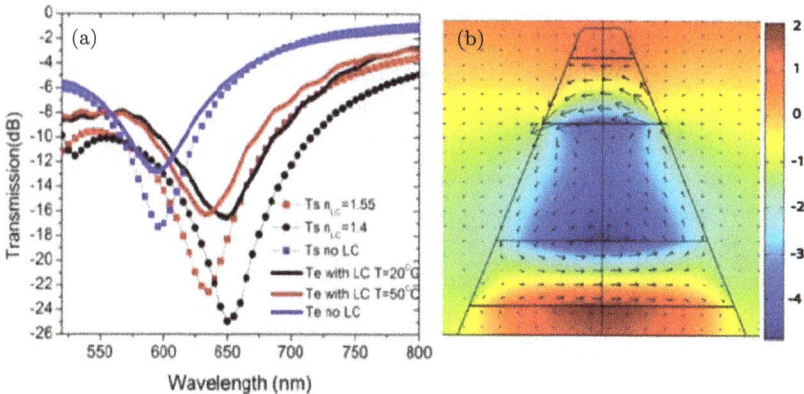

Figure 10. (Left) Transmission measurements of a thermally tunable magnetic response in a metamaterial (11). Solid lines show the experimental data, and dashed lines connecting colored dots are simulated results without LCs (blue lines), with LCs at 20°C (black lines), and at 50°C (red lines). (Right) Simulated electric displacement and magnetic field strength distributions at the magnetic resonance wavelength.

for applications in spectral regimes spanning the optical, Terahertz, to the RF, and micro-wave.

The approach of using nanoparticles dispersed uniformly in aligned bulk nematic liquid crystal to achieve metamaterials with *tunable* negative-, near-zero and sub-unity index material was first reported in Ref. 9. From the Mie scattering coefficients, the effective dielectric constant and permeability of the $NLC(host) + nanoparticles$ system are calculated using the Maxwell–Garnet mixing rule[9,83–85]:

$$\varepsilon_r^{\text{eff}} = \varepsilon_{\text{host}} \left(\frac{k_{\text{host}}^3 + j4\pi N a_1}{k_{\text{host}}^3 - j2\pi N a_1} \right); \quad \mu_r^{\text{eff}} = \frac{k_{\text{host}}^3 + j4\pi N b_1}{k_{\text{host}}^3 - j2\pi N b_1}. \tag{5}$$

Here k_{host} is the optical wave vector in the NLC host, $\varepsilon_{\text{host}}$ is the LC permittivity, a_1 and b_1 are the MIE scattering coefficients,[83] and N is the volume density ($N = 3f/4\pi r^3$; r = radius of nm-spheres) of the nanoparticulates, and f is the filling fraction of the nanoparticles in the nematic host. The Maxwell–Garnet mixing rule is valid for small filling fraction ($f \ll 1$) and long wavelength limit (optical wavelength $\gg$ particle size). Using this approach, a variety of nanospheres have been studied, including (a) solid-Drude spheres, (b) Drude-coated silica spheres, and (c) Drude-polaritonic core-shell spheres. In general, core-shell configuration gives a wider tuning range and a greater variation in the real refractive index than solid spheres. Solid gold nanospheres, for example, give rise to effective refractive indices that are greater than unity, while gold-covered silica nanospheres may exhibit sub-unity real refractive index.[16]

Most (including some very recent) studies of LC containing suspended nanoparticles, or nanostructures infiltrated with LC, neglect surface anchoring effects such as non-uniform alignment in non-planar surfaces and immobile LC molecules around the particles and alignment-treated surfaces. As shown in Refs. 86, 87 such surface anchoring effects, even in well-aligned LC metamaterials, give rise to significant deviation from results obtained under the assumption of uniform reorientation by an external field. Another serious drawback in the development of metamaterials is the large optical loss [coming from imaginary part of the index n'']. To reduce loss, one possibility is incorporation of gain materials such as dyes;[16,88,89] however, to date, large optical losses still pose almost unsurmountable challenges to developing *negative-index* metamaterials. Most current studies are focused on metamaterials with sub-unity (but positive) or near-zero index values where the optical losses are significantly reduced to acceptable levels for practical applications.

In recent years, such μ and ε engineering of metamaterials have evolved into the development of all-dielectric metasurfaces where the unit cell is further sub-divided into smaller units that impart patterned phase shift on the impinging optical (and longer wavelength electromagnetic) fields. By virtue of their dielectric constituents and sub-wavelength thick planar nature, these metasurfaces suffer little optical losses and show remarkable capabilities such as high-resolution imaging, sensing, and non-planar reflection. Naturally, incorporation of LC in these structures has enabled development of tunable, reconfigurable, and nonlinear optical devices.[90-93]

3.3. *Tunable plasmonic nanostructures, micro-ring resonators, enhanced nonlinearity, and spontaneous emission*

Two kinds of resonant interactions occur between an optical field and the electrons of a plasmonic material: (a) Surface Plasmon Resonance (SPR) — a resonant interaction due to the collective oscillations of the electrons near the surface of the metal; (b) Localized Surface Plasmon Resonances [LSPR) — associated with light interaction with isolated metal nanoparticles. There are two ways of combining the tunability of liquid crystals with SPR or LSPR: (i) dissolving or suspending nanosize plasmonic particles in bulk NLC cell and (ii) incorporating LC in plasmonic nanostructures. The most used LC in both cases are nematic liquid crystals (NLC).

If the light intensity is sufficiently high to generate index changes in the liquid crystals and/or the plasmonic particles, these induced changes in turn will act on the impinging field to affect its phase and amplitude, giving rise to enhanced nonlinear optical responses of the LC–plasmonic composite.

Since electronic optical nonlinearities are relatively insensitive to the mesophase order, we shall consider here the case of a small volume fraction f ($f \ll 1$) of nanospheres suspended in an *isotropic(liquid)-phase* liquid crystal such as L34; this simplifies the consideration as the system is now completely isotropic (polarization independent). The effective optical dielectric constant, ε_{com}, calculated using the Maxwell–Garnet mixing rule[94] becomes

$$\varepsilon_{\text{com}} = \frac{1 + 2f\gamma}{1 - f\gamma}\varepsilon_{\text{host}}, \quad \gamma = \frac{\varepsilon_{Au} - \varepsilon_{\text{host}}}{\varepsilon_{Au} + 2\varepsilon_{\text{host}}}. \tag{6}$$

Here $\varepsilon_{\mathrm{au}}$ is the complex dielectric constant of the Au nanospheres described by a Drude model and $\varepsilon_{\mathrm{host}}$ is the dielectric constant of L34. The real part of refractive index of L34 is 1.61, and the imaginary part is derived from the linear absorbance using the relationship[94,95]: $n''_{\mathrm{host}} = \frac{\lambda}{4\pi}\alpha$ [α: absorbance in cm^{-1}]. One can then calculate the nonlinear susceptibility using the formalism developed in[94-96] to yield the enhancement factor of the third-order susceptibility of the composite, as follows:

$$g^{(3)} \equiv \frac{\chi^{(3)}_{\mathrm{com}}}{\chi^{(3)}_{\mathrm{host}}} = \frac{1}{5}\left|\frac{\varepsilon_{\mathrm{com}} + 2\varepsilon_{\mathrm{host}}}{3\varepsilon_{\mathrm{host}}}\right|^2 \left(\frac{\varepsilon_{\mathrm{com}} + 2\varepsilon_{\mathrm{host}}}{3\varepsilon_{\mathrm{host}}}\right)^2 [8f|\gamma|^2\gamma^2 + 6f|\gamma|^2\gamma$$

$$+ 2f\gamma^3 + 18f(|\gamma|^2 + \gamma^2) + 5(1 - f)]. \tag{7}$$

By expressing $\chi^{(3)}$ in terms of the nonlinear refractive index: $n_2 = \frac{3}{4\varepsilon_0 c|n_0|^2}\left(1 - i\frac{n''_0}{n'_0}\right)\chi^{(3)}$, the nonlinear refractive index enhancement factor $g^{(2)}$ becomes

$$g^{(2)} \equiv \frac{n_{2,\mathrm{com}}}{n_{2,\mathrm{host}}} = \frac{|n_{0,\mathrm{host}}|^2}{|n_{0,\mathrm{com}}|^2} \frac{1 - i(n''_{0,\mathrm{com}}/n'_{0,\mathrm{com}})}{1 - i(n''_{0,\mathrm{host}}/n'_{0,\mathrm{host}})} g^{(3)} \tag{8}$$

Theoretical estimate of the enhancement is in good agreement with measurements[97,98] for the nonlinear absorption coefficient of an Au-nm spheres doped L34 cell with picoseconds pulsed laser, cf. Fig. 11(a). With a doping concentration of $\sim 0.5\%$, the effective nonlinear absorption coefficient β_{eff} value is enhanced by a factor of 2.5. Recalling that in the picoseconds time scale β_{eff} may have contributions from the excited states as well as the ground state, the observed enhancement factor *may not* reflect accurately the contribution from the plasmonic particles alone. The calculation was thus repeated using instead the *intrinsic two-photon absorption coefficient* β obtained with femtosecond lasers, which is an order of magnitude smaller than β_{eff} obtained with picosecond lasers [8, 22], cf. Fig. 4.

The calculation for the L34+Au composite for various laser wavelengths (500 nm–600 nm) of the femtosecond laser used are shown in Fig. 11(b). Notice that throughout the entire wavelength regime, the enhancement factor obtained for the intrinsic coefficient β is around 2–3, similar to the β_{eff} and confirming that the observed enhancement in both time scales is due to the presence of the plasmonic nanoparticles. Although the 100-μm thick L34 contains only 0.5% Au particles and incurs a transmission loss of $\sim 25\%$, it is noteworthy that its nonlinear absorption coefficient can be enhanced by over 250%.

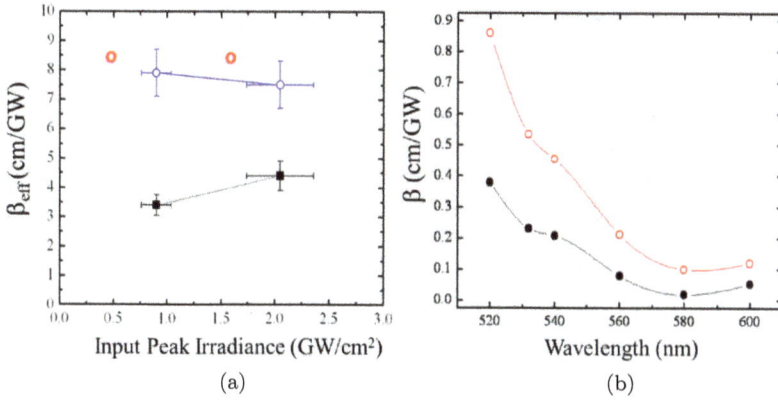

Figure 11. (a) Observed effective nonlinear absorption coefficient β_{eff} [equivalent to n_2''] with 83 ps laser at $\lambda = 460$ nm: (Blue circles) L34 + Au nm-spheres (4-nm in diameter); (Lower black squares) Observed effective nonlinear absorption coefficient β_{eff} of L34; (Red circles) Calculated values for L34 + nm Au particles with a filling fraction of 0.5%. (b) Wavelength dependence of the intrinsic two-photon absorption coefficient β: (Lower dark squares) experimental measured values for L34 deduced, (Upper red circles) Calculated values for L34 + Au nm-spheres.

Owing to their fluid nature and compatibility with almost all opto-electronic materials, NLC can be easily incorporated in nanostructures such as photonic-crystals inverse opal,[99] frequency selective surfaces (FSS),[12–14] micro-channel waveguides or photonic crystal fiber,[100,101] micro-ring resonators,[102,103] tunable transmissive devices, and dual-band perfect absorbers,[104–106] cf. Figs. 12–14, and many other tunable plasmonic metamaterials and structures mentioned previously. Plasmonic waveguides cladded with liquid crystal metamaterials with a quantum emitter embedded in the 'hot spot' have also been proposed for simultaneous enhancement and modulation of the spontaneous emission rate (Purcell factor) of the emitter.[107,108]

3.4. *Cholesteric liquid crystals for polarization rotation of complex laser vector fields and ultrafast (ps-fs) laser pulse modulation*

Cholesteric liquid crystals of thicknesses in microns- to tens of microns range provide adequate scattering, haze, absorption, and transmission/reflection modulation for conventional applications such as display and smart windows to function properly. While these applications do involve photonic

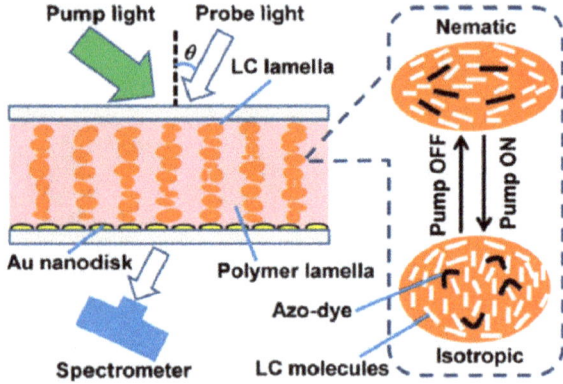

Figure 12. Schematic representation of the experimental setup for all-optical modulation of localized surface plasmon coupling in an azo-dye-doped holographic PDLC + nano-Au disk.[104]

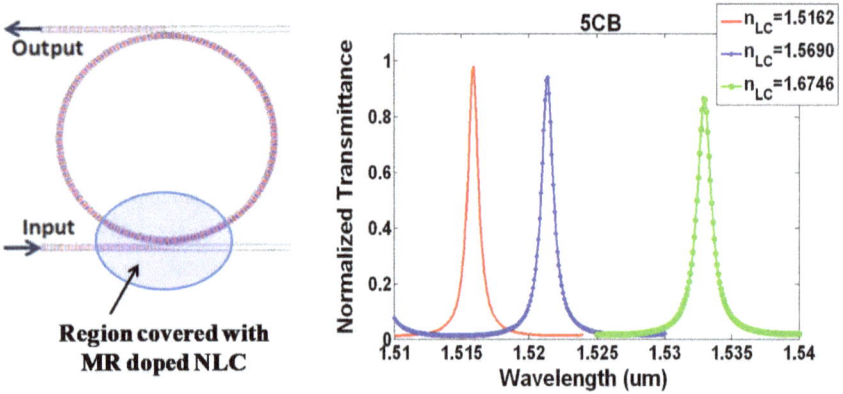

Figure 13. (Left) Schematic representation of a ring resonator with an etched region C for liquid crystals [Methyl-Red (MR)-doped 5CB] overlay and simulation result for TE modes of the ring resonator structure. (Right) Theoretical prediction of resonances as a function of changing 5CB liquid crystal cladding index with light of various polarizations: red (linear), blue (circular polarization, director axis not reoriented), green (orthogonal linear).[102]

crystal properties such as the photonic bandgap for selective and tunable reflection bands, they generally do not utilize another unique advantage of chiral photonic crystals, namely, polarization rotation mediated by the circular birefringence $\Delta n_c = n_- - n_+$.

Consider the case of a linearly polarized light traversing a right-handed CLC as shown in Fig. 15. Inside the CLC, it breaks up equally into two

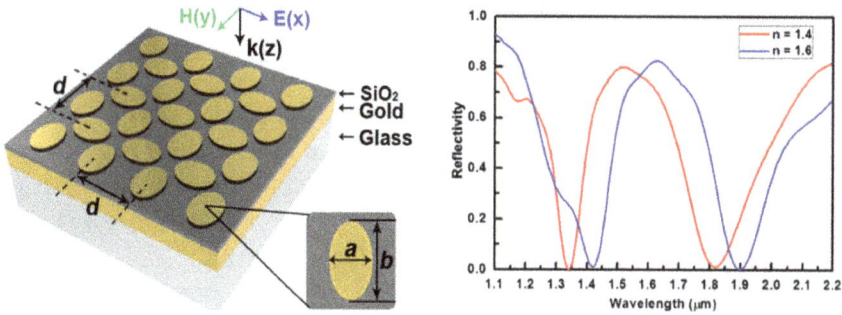

Figure 14. (Left) Schematic representation of an asymmetric nanogold disc structure; (Right) Tunable perfect absorption bands of the nanogold disc structure with a nematic liquid-crystal overlayer.[106]

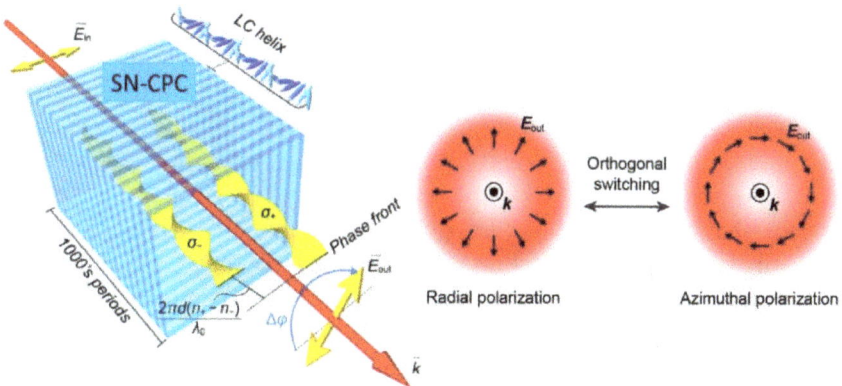

Figure 15. Cholesteric liquid crystal as a 1D chiral photonic crystal for polarization manipulations of linear as well as complex laser vector fields.

circular polarizations: left (σ_-) and right (σ_+) circular polarizations with respective refractive indices n_- and n_+. Due to the circular birefringence $\Delta n_c = n_- - n_+$, a phase difference (Γ) is built up between the σ_- and the right σ_+ after propagation through the CLC. The phase difference or retardation results in a rotation of the polarization vector by an angle $\Delta\varphi = \Gamma/2 = \Delta n_c \pi L/\lambda_0$, where L is the propagation length in CLC and λ_0 is the optical wavelength in vacuum. Importantly, such polarization rotation is independent of the input vector orientation, and it can be applied to laser vector beam of complex (azimuthal or radial) polarizations depicted in Fig. 15. In the case of orthogonal polarization switching [$\Delta\varphi = 90°$ or

$\pi/2$ radian] between azimuthal and radial, the required phase difference caused by the circular birefringence $\Gamma = \pi$.

Based on the theoretical formalism in[109] where the CLC is divided into $N/2$ lamellae with each lamella comprising a complete helicoidal twist of LC director, i.e., the lamella thickness is given by the pitch $p = 2\Lambda$, the polarization rotation capability of a CLC as a function of the number of grating period N [thickness/Λ] has been explicitly calculated in Ref. 51. For large N, the optical rotation power (ORP) is given by

$$\text{ORP} = \frac{\pi\Lambda(n_e - n_o)^2}{2\lambda_0^2} + \frac{\pi(\lambda_{\text{mid}} - \lambda_0)}{2\Lambda\lambda_0}\left[1 - \sqrt{1 - \frac{Q^2}{\varepsilon^2}}\right]. \qquad (9)$$

Here $Q \approx [\pi(n_e - n_o)/n_{\text{avg}}]\sin(\varepsilon)/\varepsilon$ is the reflection coefficient per lamella, and $\varepsilon = \delta - 2\pi = 2\pi(\lambda_{\text{mid}} - \lambda_0)/\lambda_0$ describes the phase difference (per lamella) between the lights at the operating wavelength (λ_0) and mid-gap wavelength (λ_{mid}), and δ is the accumulated optical phase per lamella.

The first term describes the *adiabatic rotation (AR)* of the polarization vector along the twisted optic axis of CLC. The second term describes the polarization rotation caused by the *circular Bragg resonance (CBR)* and becomes the dominant contribution to polarization rotation in the spectral vicinity of the photonic bandgap $\lambda_0 \sim n_{\text{avg}}P$.

As shown in Fig. 16(a), near the band-edges on both sides of the photonic bandgap the optical rotation power is large, but the transmission in this regime for the σ_+ wave [which 'sees' the photonic bandgap] is low compared to the near unity transmission of the σ_- wave. Such uneven transmission produces ellipticity in the transmitted light, giving rise to poor PF (Polarization figure-of-merit) defined in Ref. 51. To preserve the linearity of the polarization, both σ_+ and σ_- waves must have equal (near-unity) transmission and PF of unity, and therefore it is necessary to operate in the off-CBR regime, i.e., away from the band-edges where the CBR and AR together yield the maximum ORP, and transmission for both σ_+ and σ_- waves is unity. Operation in the wavelength regime $<700\,\text{nm}$ would simultaneously yield *near-unity* transmission and PF value and large ORP ($>400°/\text{mm}$), cf. Fig. 16(b).

The polarization rotation angle, which is the ultimate factor characterizing the practical polarization rotation ability of a chiral photonic crystal, depends on the thickness of CLC. To get a polarization rotation of 90° at 600 nm where the ORP is $400°/\text{mm}$, the CLC has to be at least 250-μm thick ($N > 600$). CLCs of such thickness and beyond require special field-assisted techniques.[50–52] It is important to note here that many chiral

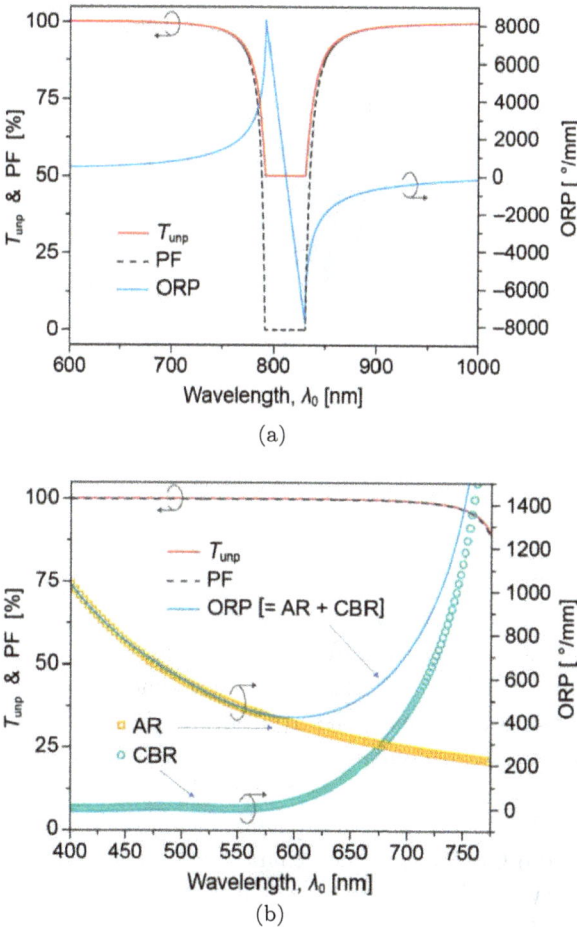

Figure 16. (a) Plot of total transmission, polarization figure of merit, and optical rotation power for (a) spectral regimes covering the short- and long-wavelength band-edges and (b) spectral regime below the short-wavelength band-edge of a $N > 500$ cholesteric liquid crystal.[50,51]

photonic crystals currently under development claim exceptionally large ORP, but they are unable to produce the needed π or 2π rotation for practical use due to their limited thickness and/or large optical losses.

Cholesteric liquid crystals of extraordinary thickness with $N>1000$ exhibit many properties impossible with other[110–113] existing or developing chiral photonic crystals. As shown in,[50–52] a $\sim$1 mm thick CLC is capable of >360 degrees polarization rotation of CW and ps-fs pulsed

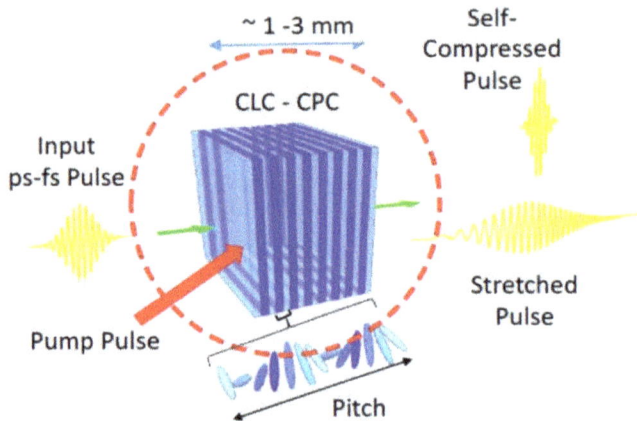

Figure 17.   Schematic representation of self-compression or pump-pulse induced polarization switching of ultrafast (fs and ps) laser pulse using a CLC-CPC.

lasers of conventional or complex polarization states in the entire visible to near- and mid-infrared (700–1700–5000 nm) regime. Dynamic switching of the polarization states and pulse modulations can also be performed by modulating the birefringence/index with a variety of mechanisms,[35–37,51,52,64,96,114–116] including electronic optical nonlinearities produced by femtosecond pump laser pulses, cf. Fig. 17.

In,[52] the theoretical simulation using measured CLC parameters shows that for orthogonal polarization switching, the required circular birefringence change (due to the dominant change in n_e) is ∼0.0068 for a 1.1-mm thick ($N \sim 5000$) CPC. Such index change requires a pump laser intensity of about 680 MW/cm^2. With the same material parameters, the simulations also show the feasibility of orthogonal polarization switching and pulse modulation operations throughout the near- to mid-infrared ($\lambda \sim$ 1-μm to 5-μm) spectral regime. Cholesteric liquid crystalline CPCs thus present themselves as promising alternatives to conventional bulky optical setups for such ultrafast photonic applications.

References

1. I.-C. Khoo, *Liquid Crystals* (3rd edn.), New Jersey: John Wiley & Sons (2022).
2. I.-C. Khoo and S.-T. Wu, *Optics and Nonlinear Optics of Liquid Crystals*, Hoboken, NJ: World Scientific (1993).
3. J. Li *et al.*, Infrared refractive indices of liquid crystals, *J. Appl. Phys.* **97**(7), 073501 (2005).

4. I. Khoo, The infrared optical nonlinearities of nematic liquid crystals and novel two-wave mixing processes, *J. Mod. Opt.* **37**(11), 1801–1813 (1990).

5. Y. Arakawa *et al.*, Design of an extremely high birefringence nematic liquid crystal based on a dinaphthyl-diacetylene mesogen, *J. Mater. Chem.* **22**(28), 13908–13910 (2012).

6. C.-S. Yang *et al.*, The complex refractive indices of the liquid crystal mixture E7 in the terahertz frequency range, *JOSA B* **27**(9), 1866–1873 (2010).

7. S. Mueller, A. Penirschke, C. Damm *et al.*, Broad-band microwave characterization of liquid crystals using a temperature-controlled coaxial transmission line, *IEEE Trans. Microwave The. Tech.* **53**(6), 1937–1945 (2005).

8. I. C. Khoo *et al.*, Synthesis and characterization of the multi-photon absorption and excited-state properties of a neat liquid 4-propyl 4'-butyl diphenyl acetylene, *J. Mater. Chem.* **19**(40), 7525–7531 (2009).

9. I. Khoo *et al.*, Nanosphere dispersed liquid crystals for tunable negative-zero-positive index of refraction in the optical and terahertz regimes, *Opt. Lett.* **31**(17), 2592–2594 (2006).

10. I. Khoo *et al.*, Liquid crystals for optical filters, switches and tunable negative index material development, *Mol. Cryst. Liq. Cryst.* **453**, 309–319 (2006).

11. S. Xiao, U. K. Chettiar, A. V. Kildishev *et al.*, Tunable magnetic response of metamaterials, *Appl. Phys. Lett.* **95**(3), Article Number: 033115 (2009).

12. D. H. Werner *et al.*, Liquid crystal clad near-infrared metamaterials with tunable negative-zero-positive refractive indices, *Opt. Express* **15**(6), 3342–3347 (2007).

13. X. Wang *et al.*, *Tunable optical negative-index metamaterials employing anisotropic liquid crystals*, *Appl. Phys. Lett.* **91**(14), 143122 (2007).

14. J. A. Bossard *et al.*, Tunable frequency selective surfaces and negative-zero-positive index metamaterials based on liquid crystals, *IEEE Trans. Antennas Propag.* **56**(5), 1308–1320 (2008).

15. J. Xu *et al.*, A review of tunable electromagnetic metamaterials with anisotropic liquid crystals, *Front. Phys.* **9**, 633104 (2021).

16. I. C. Khoo *et al.*, Liquid crystals tunable optical metamaterials, *IEEE J. Sel. Top. Quant. Electron.* **16**(2), 410–417 (2010).

17. G. Pawlik *et al.*, Liquid crystal hyperbolic metamaterial for wide-angle negative–positive refraction and reflection, *Opt. Lett.* **39**(7), 1744–1747 (2014).

18. A. Minovich *et al.*, Tunable fishnet metamaterials infiltrated by liquid crystals, *Appl. Phys. Lett.* **96**(19), 193103 (2010).

19. R. Pratibha *et al.*, Tunable optical metamaterial based on liquid crystal-gold nanosphere composite, *Opt. Express* **17**(22), 19459–19469 (2009).

20. F. Zhang *et al.*, Electrically controllable fishnet metamaterial based on nematic liquid crystal, *Opt. Express* **19**(2), 1563–1568 (2011).

21. O. D. Lavrentovich, Liquid crystals, photonic crystals, metamaterials, and transformation optics, *Proc. Natl. Acad. Sci.* **108**(13), 5143–5144 (2011).

22. I. C. Khoo, Nonlinear optics, active plasmonics and metamaterials with liquid crystals, *Prog. Quant. Electron.* **38**(2), 77–117 (2014).

23. I. C. Khoo, Nonlinear optics of liquid crystalline materials, *Phys. Rep.* **471**(5–6), 221–267 (2009).

24. I. C. Khoo, Cholesteric and blue-phase liquid photonic crystals for nonlinear optics and ultrafast laser pulse modulations, *Liq. Cryst. Rev.* **6**(1), 53–77 (2018).

25. M. Y. Shih, A. Shishido, P. H. Chen *et al.*, (2000) *All-optical image processing with a supranonlinear dye-doped liquid-crystal film, Opt. Lett.* **25**(13), 978–980.

26. A. Miniewicz, J. Girones, P. Karpinski *et al.*, Photochromic and nonlinear optical properties of azo-functionalized POSS nanoparticles dispersed in nematic liquid crystals, *J. Mater. Chem. C* **2**, 432–440 (2014).

27. H. Ono and Y. Harata, Higher-order optical nonlinearity observed in host–guest liquid crystals, *J. Appl. Phys.* **85**, 676 (1999). https://doi.org/10.1063/1.369202.

28. I. C. Khoo, Optical-thermal induced total internal reflection-to-transmission switching at a glass-liquid crystal interface, *Phys. Lett.* **40**, 645 (1982). https://doi.org/10.1063/1.93227.

29. S. V. Serak, H. J. Eichler, A. A. Kovalev, and T. A. Davidovich, Passive laser Q-switching using a dye-doped liquid-crystalline layer near total internal reflection, *Opt. Commun.* **196**, 1–6, 269–280 (2001).

30. P. Mormile, E. Casale, L. De Stefano, and M. Villiargio, Light switching at a prism-liquid crystal interface. A new sensor for magnetic fields, *Appl. Phys. B* **63**(4), 385–388 (1996).

31. I. C. Khoo, K. Chen, and Y. Z. Williams, Orientational photorefractive effect in undoped and CdSe nanorods-doped nematic liquid crystal — Bulk and interface contributions, *IEEE J. Selected Top. Quant. Electron.* **12**(3), 443–450 (2006).

32. I. C. Khoo, H. Li, and Y. Liang, Observation of orientational photorefractive effects in nematic liquid crystals, *Opt. Lett.* **19**(21), 1723–1725 (1994).

33. M. Peccianti, A. De Rossi, G. Assanto, A. De Luca, C. Umeton, I. C. Khoo, Electrically assisted self-confinement and waveguiding in planar nematic liquid crystal cells, *Appl. Phys. Lett.* **77**(1), 7–9 (2000).

34. I. C. Khoo, P. Y. Yan, G. M. Finn, T. H. Liu, and R. R. Michael, Low-power (10.6-μm) laser-beam amplification by thermal-grating-mediated degenerate four-wave mixing in a nematic liquid-crystal film, *JOSA B* **5**(2), 202–206 (1988).

35. J. Hwang *et al.*, Enhanced optical nonlinearity near the photonic bandgap edges of a cholesteric liquid crystal, *Opt. Lett.* **29**(22), 2644–2646 (2004).

36. J. Hwang and J. Wu, Determination of optical Kerr nonlinearity of a photonic bandgap structure by Z-scan measurement, *Opt. Lett.* **30**(8), 875–877 (2005).

37. L. Song *et al.*, Direct femtosecond pulse compression with miniature-sized Bragg cholesteric liquid crystal, *Opt. Lett.* **38**(23), 5040–5042 (2013).

38. I. C. Khoo and A. Diaz, Multiple-time-scale dynamic studies of nonlinear transmission of pulsed lasers in a multiphoton-absorbing organic material, *JOSA B* **28**(7), 1702–1710 (2011).

39. I. C. Khoo, Nonlinear organic liquid-cored fiber array for all-optical switching and sensor protection against short-pulsed lasers, *IEEE J. Selected Top. Quant. Electron.* **14**(3), 946–951 (2008).
40. I.-C. Khoo *et al.*, Passive optical limiting of picosecond-nanosecond laser pulses using highly nonlinear organic liquid cored fiber array. *IEEE J. Selected Top. Quant. Electron.* **7**(5), 760–768 (2001).
41. G. S. He *et al.*, Multiphoton absorbing materials: Molecular designs, characterizations, and applications, *Chem. Rev.* **108**(4), 1245–1330 (2008).
42. I.-C. Khoo, J.-H. Park, and J. D. Liou, Theory and experimental studies of all-optical transmission switching in a twist-alignment dye-doped nematic liquid crystal, *JOSA B* **25**(11), 1931–1937 (2008).
43. I. C. Khoo, A. Diaz, J. Ding, Nonlinear-absorbing fiber array for large-dynamic-range optical limiting application against intense short laser pulses. *JOSA B* **21**(6), 1234–1240 (2004).
44. I. C. Khoo and H. Li, Nonlinear optical propagation and self-limiting effect in liquid-crystalline fibers, *Appl. Phys. B* **59**, 573–580 (1994).
45. I. C. Khoo *et al.*, Nonlinear liquid-crystal fiber structures for passive optical limiting of short laser pulses, *Opt. Lett.* **21**(20), 1625–1627 (1996).
46. I. C. Khoo *et al.*, Nonlinear absorption and optical limiting of laser pulses in a liquid-cored fiber array, *JOSA B* **15**(5), 1533–1540 (1998).
47. I. C. Khoo *et al.*, Molecular photonics of a highly nonlinear organic fiber core liquid for picosecond–nanosecond optical limiting application, *Chem. Phys.* **245**(1-3), 517–531 (1999).
48. I. C. Khoo *et al.*, Nonlinear optical liquid cored fiber array and liquid crystal film for ps-cw frequency agile laser optical limiting application, *Opt. Express* **2**(12), 471–482 (1998).
49. D.-K. Yang and S.-T. Wu, *Fundamentals of Liquid Crystal Devices*, New Jersey: John Wiley & Sons (2014).
50. C.-W. Chen and I. C. Khoo, Extraordinary polarization rotation of vector beams with high-period-number chiral photonic crystals, *Opt. Lett.* **44**(21), 5306–5309 (2019).
51. C.-W. Chen and I. C. Khoo, Optical vector field rotation and switching with near-unity transmission by fully developed chiral photonic crystals, *Proc. Natl. Acad. Sci.* **118**(16), e2021304118 (2021).
52. C.-W. Chen *et al.*, Massive, soft, and tunable chiral photonic crystals for optical polarization manipulation and pulse modulation, *Appl. Phys. Rev.* **10**(1), 011413 (2023).
53. P. V. Dolganov, K. D. Baklanova, and V. K. Dolganov, Optical properties and photonic density of states in one-dimensional and three-dimensional liquid-crystalline photonic crystals, *Liq. Cryst.* **47**(2), 231–237 (2020).
54. T. J. White *et al.*, Phototunable azobenzene cholesteric liquid crystals with 2000 nm range, *Adv. Fun. Mater.* **19**(21), 3484–3488 (2009).
55. A. Ryabchun and A. Bobrovsky, Cholesteric liquid crystal materials for tunable diffractive optics, *Adv. Opt. Mater.* **6**(15), 1800335 (2018).
56. H. Coles and S. Morris, Liquid-crystal lasers, *Nat. Photonics* **4**(10), 676–685 (2010).

57. Y. Matsuhisa *et al.*, Cholesteric liquid crystal laser in a dielectric mirror cavity upon band-edge excitation, *Opt. Express* **15**(2), 616–622 (2007).
58. T.-H. Lin *et al.*, Cholesteric liquid crystal laser with wide tuning capability, *Appl. Phys. Lett.* **86**(16), 161120 (2005).
59. N. Wang *et al.*, Lasing properties of a cholesteric liquid crystal containing aggregation-induced-emission material, *Opt. Express* **23**(26), 33938–33946 (2015).
60. H. Khandelwal *et al.*, Application of broadband infrared reflector based on cholesteric liquid crystal polymer bilayer film to windows and its impact on reducing the energy consumption in buildings, *J. Mater. Chem. A* **2**(35), 14622–14627 (2014).
61. C.-W. Chen *et al.*, Normally transparent smart window based on electrically induced instability in dielectrically negative cholesteric liquid crystal, *Opt. Mater. Express* **8**, 691–697 (2018).
62. A. Jullien, U. Bortolozzo, S. Grabielle, J.-P. Huignard, N. Forget, and S. Residori, Continuously tunable femtosecond delay-line based on liquid crystal cells, *Opt. Express* **24**(13), 14483–14493 (2016).
63. C.-W. Chen *et al.*, Slowing sub-picosecond laser pulses with 0.55 mm-thick cholesteric liquid crystal, *Opt. Mater. Express* **7**(6), 2005–2011 (2017).
64. Y. Liu *et al.*, Ultrafast pulse compression, stretching-and-recompression using cholesteric liquid crystals, *Opt. Express* **24**(10), 10458–10465 (2016).
65. D. K. Yang, J. L. West, L. C. Chien, and J. W. Doane, Control of reflectivity and bistability in displays using cholesteric liquid crystals, *J. Appl. Phys.* **76**, 1331–1333 (1994).
66. C.-C. Li, H.-Y. Tseng, C.-W. Chen, C.-T. Wang, H.-C. Jau, Y.-C. Wu, W.-H. Hsu, and T.-H. Lin, Versatile energy-saving smart glass based on tristable cholesteric liquid crystals, *ACS Appl. Energy Mater.* **3**, 7601–7609 (2020).
67. J.-C. Lai, W.-F. Cheng, C.-K. Liu, and K.-T. Cheng, Optically switchable bistable guest–host displays in chiral-azobenzene-and dichroic-dye-doped cholesteric liquid crystals, *Dyes Pigm.* **163**, 641–646 (2019).
68. Z. Liu, C.-H. Lin, B.-R. Hyun, C.-W. Sher, Z. Lv, B. Luo, F. Jiang, T. Wu, C.-H. Ho, and H.-C. Kuo, Micro-light-emitting diodes with quantum dots in display technology, *Light Sci. Appl.* **9**, 83 (2020).
69. C. W. Hsu, B. Zhen, W. Qiu, O. Shapira, B. G. DeLacy, J. D. Joannopoulos, and M. Soljačić, Transparent displays enabled by resonant nanoparticle scattering, *Nat. Commun.* **5**, 3152 (2014).
70. D. Lanman and D. Luebke, Near-eye light field displays, *ACM Trans. Graphics (TOG)* **32**, 1–10 (2013).
71. C. Chang, K. Bang, G. Wetzstein, B. Lee, and L. Gao, Toward the next-generation VR/AR optics: A review of holographic near-eye displays from a human-centric perspective, *Optica* **7**, 1563–1578 (2020).
72. A. Maimone, A. Georgiou, and J. S. Kollin, Holographic near-eye displays for virtual and augmented reality, *ACM Trans. Graphics (Tog)* **36**, 1–16 (2017).

73. G. Jr. Paul Montgomery, J. L. West, and W. Tamura-Lis, Light scattering from polymer-dispersed liquid crystal films: Droplet size effects, *J. Appl. Phys.* **69**(3), 1605–1612 (1991).

74. P. Song *et al.*, Studies on the electro-optical and the light-scattering properties of PDLC films with the size gradient of the LC droplets, *Liq. Cryst.* **42**(3), 390–396 (2015).

75. I. Dierking, Polymer network–stabilized liquid crystals, *Adv. Mater.* **12**(3), 167–181 (2000).

76. A. Y. G. Fuh, S. Y. Chih, and S. T. Wu, Advanced electro-optical smart window based on PSLC using a photoconductive TiOPc electrode, *Liq. Cryst.* **45**(6), 864–871 (2018).

77. C.-C. Li *et al.*, Bistable cholesteric liquid crystal light shutter with multielectrode driving, *Appl. Opt.* **53**(22), E33–E37 (2014).

78. H. Ren and S.-T. Wu, Reflective reversed-mode polymer stabilized cholesteric texture light switches, *J. Appl. Phys.* **92**(2), 797–800 (2002).

79. R. Bao, C.-M. Liu, and D.-K. Yang, Smart bistable polymer stabilized cholesteric texture light shutter, *Appl. Phys. Express* **2**, 112401 (2009).

80. H. Xianyu, T.-H. Lin, and S.-T. Wu, Rollable multicolor display using electrically induced blueshift of a cholesteric reactive mesogen mixture, *Appl. Phys. Lett.* **89**(9), 091124 (2006).

81. M. Yu, H. Yang, and D.-K. Yang, Stabilized electrically induced Helfrich deformation and enhanced color tuning in cholesteric liquid crystals, *Soft Matter* **13**(46), 8728–8735 (2017).

82. K.-W. Lin *et al.*, Mechanism of scattering bistable light valves based on salt-doped cholesteric liquid crystals, *Opt. Express* **29**(25), 41213–41221 (2021).

83. C. F. Bohren, and D. R. Huffman, *Absorption and Scattering of Light by Small Particles*, John Wiley & Sons (2008).

84. W. T. Doyle, Optical properties of a suspension of metal spheres, *Phys. Rev. B* **39**(14), 9852–9858 (1989).

85. M. S. Wheeler, J. S. Aitchison, and M. Mojahedi, Coated nonmagnetic spheres with a negative index of refraction at infrared frequencies, *Phys. Rev. B* **73**(4), 045105 (2006).

86. G. Pawlik *et al.*, Field-induced inhomogeneous index distribution of a nano-dispersed nematic liquid crystal metamaterial near the Freedericksz transition: Monte Carlo studies, *J. Opt. Soc. Am. B* **27**(3), 567–576 (2010).

87. G. Pawlik *et al.*, Large gradients of refractive index in nanosphere dispersed liquid crystal metamaterial with inhomogeneous anchoring: Monte Carlo study, *Opt. Mater.* **33**, 1459–1463 (2011).

88. S. Wuestner *et al.*, Overcoming losses with gain in a negative refractive index metamaterial, *Phys. Rev. Lett.* **105**(12), 127401(2010).

89. A. De Luca *et al.*, Dispersed and encapsulated gain medium in plasmonic nanoparticles: A multipronged approach to mitigate optical losses, *ACS Nano* **5**(7), 5823–5829 (2011).

90. M. A. Naveed *et al.*, Novel spin-decoupling strategy in liquid crystal-integrated metasurfaces for interactive metadisplays, *Adv. Opt. Mater.* **10**(13), 2200196 (2022).

91. J. A. Dolan, H. Cai, L. Delalande *et al.*, Broadband liquid crystal tunable metasurfaces in the visible: Liquid crystal inhomogeneities across the metasurface parameter space, *ACS Photonics* **8**(2), 567–575 (2021).

92. A. Lininger *et al.*, Optical properties of metasurfaces infiltrated with liquid crystals, *PNAS* **117**(34), 20390–20396 (2020)

93. J. Wu *et al.*, Liquid crystal programmable metasurface for terahertz beam steering, *Appl. Phys. Lett.* **116**(13), 131104 (2020).

94. J. Sipe and R. W. Boyd, Nonlinear susceptibility of composite optical materials in the Maxwell Garnett model, *Phys. Rev. A* **46**(3), 1614–1629 (1992).

95. R. Del Coso and J. Solis, Relation between nonlinear refractive index and third-order susceptibility in absorbing media, *JOSA B* **21**(3), 640–644 (2004).

96. D. C. Kohlgraf-Owens and P. G. Kik, Numerical study of surface plasmon enhanced nonlinear absorption and refraction, *Opt. Express* **16**(14), 10823–10834 (2008).

97. I. C. Khoo *et al.*, Liquid-crystals-plasmonics for ultrafast broadband all-optical switching. MRS Online Proc. Lib. (OPL), **1293** (2011).

98. N. Podoliak *et al.*, High optical nonlinearity of nematic liquid crystals doped with gold nanoparticles, *J. Phys. Chem. C* **116**(23), 12934–12939 (2012).

99. E. Graugnard *et al.*, Electric-field tuning of the Bragg peak in large-pore TiO2 inverse shell opals, *Phys. Rev. B* **72**(23), 233105 (2005).

100. A. d'Alessandro *et al.*, All-optical intensity modulation of near infrared light in a liquid crystal channel waveguide, *Appl. Phys. Lett.* **97**(9), 192 (2010).

101. T. T. Larsen *et al.*, Optical devices based on liquid crystal photonic bandgap fibres, *Opt. Express* **11**(20), 2589–2596 (2003).

102. J. Ptasinski *et al.*, Optical tuning of silicon photonic structures with nematic liquid crystal claddings, *Opt. Lett.* **38**(12), 2008–2010 (2013).

103. J. Ptasinski, I.-C. Khoo, and Y. Fainman, Enhanced optical tuning of modified-geometry resonators clad in blue phase liquid crystals, *Opt. Lett.* **39**(18), 5435–5438 (2014).

104. Y. J. Liu, Y. B. Zheng, J. Liou *et al.*, All-optical modulation of localized surface plasmon coupling in a hybrid system composed of photoswitchable gratings and au nanodisk arrays, *J. Phys. Chem. C* **115**(15), 7717–7722 (2011).

105. H. Qingzhen, Z. Yanhui, J. B. Krishna *et al.*, Frequency-addressed tunable transmission in optically thin metallic nanohole arrays with dual-frequency liquid crystals, *J. Appl. Phys.* **109**, Article no. 084340 (2011).

106. Y. Zhao, Q. Z. Hao, Y. Ma *et al.*, Light-driven tunable dual band absorber with liquid-crystal-plasmonic asymmetric nanodisk array, *Appl. Phys. Lett.* **100**, Article no. 053119 (2012).

107. H. Hao *et al.*, Tunable enhanced spontaneous emission in plasmonic waveguide cladded with liquid crystal and low-index metamaterial, *Opt. Express* **25**(4), 3433–3444 (2017).

108. H. Hao *et al.*, High-contrast switching and high-efficiency extracting for spontaneous emission based on tunable gap surface plasmon, *Sci. Rep.* **8**(1), 11244 (2018).

109. S. Chandrasekhar and J. S. Prasad, Theory of rotatory dispersion of cholesteric liquid crystals, *Mol. Cryst. Liq. Cryst.* **14**(1–2), 115–128 (1971).

110. I. Hodgkinson, Q. H. Wu, B. Knight *et al.*, Vacuum deposition of chiral sculptured thin films with high optical activity, *Appl. Opt.* **39**(4), 642–649 (2000).

111. M. Kuwata-Gonokami *et al.*, Giant optical activity in quasi-two-dimensional planar nanostructures, *Phys. Rev. Lett.* **95**(22), 227401 (2005).

112. A. Y. Zhu *et al.*, Giant intrinsic chiro-optical activity in planar dielectric nanostructures, *Light Sci. Appl.* **7**, 17158 (2018).

113. S. Takahashi *et al.*, Giant optical rotation in a three-dimensional semiconductor chiral photonic crystal, *Opt. Express* **21**(24), 29905–29913 (2013).

114. Y. Liu, Y. Wu, C.-W. Chen, X. Xie, W. Hu, P. Chen, J. Wen, J. Zhou, T.-H. Lin, and I. C. Khoo, Ultrafast switching of optical singularity eigenstates with compact integrable liquid crystal structures, *Opt. Express* **26**(22), 28818–28826 (2018).

115. Y. Liu *et al.*, Ultrafast optical signal processing with Bragg structures, *Appl. Sci.* **7**(6), 556 (2017).

116. I. Muševič, Liquid-crystal micro-photonics, *Liq. Cryst. Rev.* **4**(1), 1–34 (2016) DOI: 10.1080/21680396.2016.1157768.

Chapter 6

Plasmonic-based Biosensors for the Rapid Detection of Harmful Pathogens

Francesca Petronella[*,‖], Daria Stoia[†,‡], Yasamin Ziai[§], Federica Zaccagnini[¶],
Viviana Scognamiglio[*], Dana Maniu[†,‡], Chiara Rinoldi[§], Monica Focsan[†],
Amina Antonacci[*], Filippo Pierini[§], and Luciano De Sio[¶,**]

[*]*National Research Council of Italy, Institute of Crystallography CNR-IC,*
Area della Ricerca Roma 1 Strada Provinciale 35d,
n. 9 - 00010 Montelibretti (RM), Italy
[†]*Nanobiophotonics and Laser Microspectroscopy Centre,*
Interdisciplinary Research Institute on Bio-Nano-Sciences,
Babes-Bolyai University, 42 Treboniu Laurian Street,
400271 Cluj-Napoca, Romania
[‡]*Biomolecular Physics Department, Faculty of Physics,*
Babes-Bolyai University, 1 M. Kogalniceanu Street,
400084 Cluj-Napoca, Romania
[§]*Department of Biosystems and Soft Matter,*
Institute of Fundamental Technological Research,
Polish Academy of Sciences, ul. Pawińskiego 5B,
Warsaw 02-106, Poland
[¶]*Department of Medico-Surgical Sciences and Biotechnologies*
Sapienza University of Rome, Latina, Italy
[‖]*francesca.petronella@ic.cnr.it*
[**]*luciano.desio@uniroma1.it*

The survival of humankind and the Earth is closely linked to a "one health" vision of scientific and technological progress. Our health and the health of our planet are bound to find equilibria and interplay also involving cohabitation with pathogenic microorganisms. The need to control human and environmental health is strictly related to the availability of tools able to achieve the accurate, precise, and fast detection of pathogens, such as the biosensors. Noble metal nanoparticles can offer unique opportunities to develop high-performing and user-friendly biosensors for pathogen monitoring. This chapter describes how noble metal nanoparticles, due to their plasmonic properties, can be used as a powerful and unique toolbox for developing biosensors for pathogen detection, in different biosensing systems. Specific sections are dedicated to refractometric optical biosensors, surface-enhanced Raman scattering, and electrochemical biosensors. Moreover, attractive opportunities for biosensors on flexible substrates are provided. A visionary perspective on the future challenges in the field of plasmonic biosensing is provided in the conclusive section of the chapter.

1. Introduction

Although in the last decades, much progress has been made in the medical world regarding the diagnosis and treatment of various illnesses, infectious diseases still remain one of the major causes of worldwide mortality.[1] More than one million people die every year because of nosocomial infections and bacterial antimicrobial resistance caused by the overuse and misuse of antibiotics.[2]

An unbalanced presence of some bacterial strains in the environment, often caused by extreme weather events, pollution, and improper water treatment, threatens ecological equilibria, with severe consequences on animal and human health. Furthermore, the release of pathogens in potable water can also be used as a biological weapon. The experience of the COVID-19 pandemic also demonstrated to what extent our globalized lifestyle promotes the uncontrollable spreading of pathogenic microorganisms. It follows that from the "one health" perspective, which is mandatory for the survival of our planet, pathogenic surveillance is gaining increasing importance.

The gold standard for the identification and quantification of pathogenic microorganisms is represented by the plate counting method. Further analytical techniques include the Matrix-assisted Laser Desorption/Ionization Time of Flight–Mass Spectrometry (MALDI-TOF-MS),[3] the real-time polymerase chain reaction (PCR), the enzyme-linked immunosorbent assay (ELISA), electrochemiluminescence, for immunoassays, and, lately, fluorescence detection has started to be primarily used for pathogenic microorganism detection.[4–6]

Unfortunately, all the mentioned methods lack sensitivity, are time-consuming, and require qualified and specialized personnel. Still, more importantly, they cannot be harnessed in miniaturized and portable devices, being limited — consequently — to their lab use.[7] This unavailability of direct on-site pathogens' detection, identification, and monitorization pushed material scientists to put nanotechnology-inspired solutions on the table for developing more and more performing biosensors for pathogen detection.[8]

In this framework, plasmonic nanoparticles (NPs) offer unique opportunities. The most commonly used plasmonic NPs are noble metal NPs (Au, Ag, Cu), characterized by free electrons on their surface. Such a feature determines the phenomenon of localized surface plasmon resonance (LSPR), which dictates how the opto-electronical properties of nanometals depend on their size, shape, and the chemical properties of their surrounding medium. These unique characteristics underlie the great interest in noble metal NPs for biosensing applications. The present chapter will overview the progress of the recent scientific literature in developing biosensors based on noble metal NPs for the detection of pathogenic microorganisms. It will deal with refractometric biosensors, surface-enhanced Raman scattering-based biosensors, and electrochemical biosensors. Moreover, a section will focus on plasmonic biosensing platforms fabricated on flexible substrates. A closing section will point out the future challenges in biosensing that plasmonic NPs have the credential to win successfully.

2. Refractometric Biosensors

LSPR-based biosensors are also identified as refractometric biosensors. Their working principle relies on the high sensitivity of plasmonic NPs to refractive index changes. As illustrated in Box 1, the frequency of the LSPR band is dependent on the refractive index of the medium, surrounding the plasmonic NPs. In other words, the alteration of chemical environment experienced by plasmonic NPs determines a shift of the maximum absorption wavelength of the LSPR band. This phenomenon makes plasmonic NPs extremely efficient optical transducers.

Moreover, the reduced dimensions of plasmonic NPs imply a high specific surface area, thus ensuring a huge number of surface sites available for interacting with the desired analyte.

Moreover, LSPR sensing requires simple instrumentation and guarantees high temporal resolution enabling real-time detection of analytes.

The quality of an LSPR-based biosensor can be essentially defined considering three parameters: the limit of detection (LOD), the bulk sensitivity (S), and the figure of merit (FOM).

The LOD is generally defined as the minimum single result which, with a stated probability, can be distinguished from a suitable blank value and indicates the point at which the analysis is reliable.

The bulk sensitivity (S) is defined as

$$S = \frac{\Delta\lambda_p}{dn}, \tag{1}$$

where $\Delta\lambda p$ is the shift of the plasmonic band (in nm) and dn is the refractive index variation. S is reported in nanometers of peak shift per refractive index unit (nm/RIU). It can be calculated as the slope obtained from the linear regression analysis of points resulting by plotting the $\Delta\lambda p$ vs the refractive index variation.

The FOM is defined as the ratio of the S over the full width at half maximum (FWHM) indicated as $\Delta\lambda$ in the following equation:

$$FOM = \frac{S}{\Delta\lambda}. \tag{2}$$

Essentially, the S indicates the ability of the biosensing system to detect change of the refractive index, while the FOM points out the ability of the system to discriminate very small alterations of the refractive index.[9]

To exploit the potential of plasmonic NPs as optical transducers, it is crucial to provide selectivity and specificity by functionalizing the NPs (or the NP-based system) with a suitable biorecognition element, namely, probe molecules that trigger the interaction between the pathogen, to be determined, and the NP surface. In biological assays, exploiting LSPR sensing, biotin-streptavidin and antibody-antigen are the most commonly used robust biomolecular interactions. However, other probe biomolecules include oligonucleotides, peptides, aptamers, and nucleic acids. Biotin-streptavidin forms a strong and specific bond with a huge variety of reagents.[10] Immunoassay sensors rely on the NPs' surface functionalization with capture antibodies, generating a self-assembled monolayer or standard bioconjugate linkers.[11]

The effectiveness of NP functionalization with the biorecognition element can be demonstrated by several techniques. The optical shift of the LSPR band in absorption spectra performed after promoting the interaction with pathogens accounts for a change of the local refractive index, which can be associated with a biorecognition event involving the pathogen and the functional NP. However, electron microscopy[12–14] and atomic force microscopy[15] are often employed to directly observe the accumulation of functionalized NPs on bacteria cell surface.

Antibodies are excellent biomolecules to be used as biorecognition elements for surface functionalization of plasmonic NPs. Antibodies, indeed, play pivotal physiological roles in the recognition of unknown entities such as antigens expressed by pathogenic agents. Their specificity, especially for monoclonal antibodies, enables their extended use in the detection of several targets ranging from proteins, lipid derivatives, drugs, aberrant cells to bacterial and fungal pathogens.

Spherical Gold NPs (Au NPs) synthesized by Turkevich's method where functionalized with polyclonal antibodies (IgY) derived from chicken. The bioconjugation procedure is essentially a passive absorption promoted by controlling the pH and the amount of NaCl in the colloidal dispersion of the bioconjugate. The nanoprobe was used to quantify the *Escherichia coli* 0157:H7, one of the more aggressive *E. coli* strains, by absorption spectroscopy. The assay performed by absorption spectroscopy was able to detect pathogens up to 10 CFU/ml. However, the specificity of the proposed system is determined by ELISA tests.[16]

In a recently published work, spherical Au NPs were successfully used for fabricating refractometric nanoplasmonic sensors for the early-stage diagnosis of periodontitis. In this case, the winning strategy consisted in functionalizing Au nanospheres, immobilized on a multiwell with casein. The Au NP immobilization was achieved by modifying the multiwell surface with a polyelectrolyte multilayer (PEM) that fostered the incorporation of nanospheres by immersion. The functionalization with casein, realized by physisorption, allowed to monitor the proteolytic activity of gingipains. The gingipains are biomarkers of the *Porphyromonas gingivalis* (key pathogen of the periodontitis) and their presence can also provide information on disease development. Therefore, in this biosensing system, the casein degradation, caused by the presence of gingipains, was translated in a blue-shift of the LSPR band of Au NPs. Therefore, a simple analytic technique, such as absorption spectroscopy, enabled the quantification of gingipains with an LOD <0.1 μg/ml, thus representing a very effective tool for the diagnosis and monitoring of a disease affecting about 40% of the adult population.[17]

Remarkably, the immobilization of plasmonic NPs on a substrate is a convenient and mandatory pre-requisite for realistic biosensing applications.

Indeed, Au NP arrays proved to be effective for the detection of *Salmonella typhimurium* in food samples. In this case, the LSPR biosensor was designed as a monolayer of Au NPs self-assembled on a glass substrate and then functionalized with an aptamer.[18] Fiber-optics are also used as a convenient substrate for realizing LSPR biosensors to be functionalized with plasmonic NPs. An Ω-shaped fiber-optic was modified with an array of Au NPs functionalized with a suitable aptamer for the detection of *S. Typhimurium*.[19] In a recent paper, a fiber-optics modified spherical Au NPs were functionalized by a bacteriophage used as a biorecognition element. By measuring the change in absorbance, the interactions between the bacteriophage and the target pathogen were detected and quantified.[20] The paper of Funari *et al.* reports the fabrication of a biochip consisting of Au mushroom-like nanostructures on a silicon substrate, used to monitor in real time the kinetic of *E. coli* biofilm formation, by recording the absorption spectrum. The blue-shift of the peak of Au nanomushrooms indicated the bacterial biofilm formation.[21]

Excellent sensing performances were achieved using anisotropic Au NPs as nanorods (Au NRs). Indeed, several researchers used Au NRs to prepare bioconjugates for spectroscopic detection of pathogens mainly in colloidal dispersions.[22,23] The anisotropic morphology, determines the occurrence of two LSPR bands in the absorption spectrum of Au NRs. The transverse plasmon is typically centered at 530 nm, while the longitudinal plasmon is located at higher wavelengths. The resonance frequency of the transverse plasmon can be tuned according to Au NR aspect ratio[11] and it is more sensitive to an alteration with respect to the transverse one.

Therefore, De Sio *et al.* recently developed a bioactive Au NRs array, for the detection of *E. coli* cells in water matrices (Fig. 1(a)). The Au NRs array was fabricated by the polyelectrolyte-assisted electrostatic layer by layer assembly, which shows several benefits such as (i) a user friendly, scalable, reproducible, and safe fabrication procedure, (ii) the realization of the Au NRs array on conformal surface, (iii) the possibility of conveying, on a rigid substrate, the optical properties of a colloidal dispersion. Indeed, the spectra of the Au NRs array clearly show the occurrence of both the LSPR band of the AuNRs and a FWMH similar to the one of AuNRs' dispersion. The innovation comes from the achievement of an LOD of 8.4 CFU/ml, well below the LOD values of previously reported LSPR-based biosensors, and, more important, from the possibility to reuse the bioactive Au NRs array. Indeed, the author exploited the photothermal properties of Au NRs to obtain the substrate disinfection and developed a suitable washing procedure to remove dead cells and reuse the substrate.[24]

It is well known that the intersection between nanotechnology and the domain of nucleic acids has opened up unpredictable scenarios. Advances in the field of LSPR biosensors arise not only from the high specificity associated with a genetic-level analysis, but also from the possibility of synthesizing nucleic acid-based probes with flexibility, higher reaction yields, and reduced costs, especially with respect to monoclonal antibodies.[25]

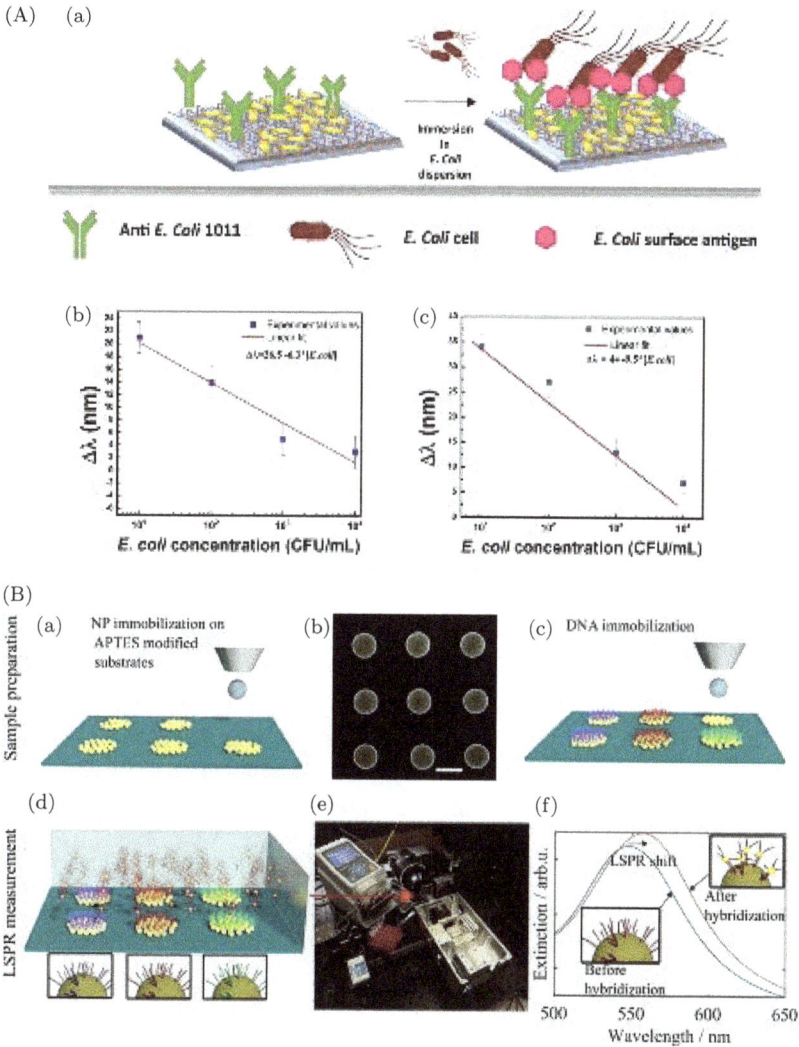

Figure 1. Schematic representation of two different Au NP-based platforms for the detection of bacteria. Panel A: A schematic representation of an AuNR-based plasmonic substrate fnctionalized with an antibody able to capture E. coli cells (a) providing an optical shift proportional to the E. coli cell concentration in absence (b) and in presence (c) of the fluorescent label SYTO 9™. Reproduced with permission from Ref. 24. Copyright 2023. Panel B: The schematic representation of the preparation and measurement principle for an array of spherical Au NPs, fabricated by a piezoelectric dispenser system (a and b) and functionalized with several capture DNA sequences (c and d). The system enables the multiplex detection of DNA from pathogens as the hybridization between the capture DNA and the target DNA determines an optical shift of the plasmon band (f). Reproduced with permission from Ref. 27. Copyright 2023.

Consequently, nucleic acid can be used as convenient and effective biorecognition element for developing LSPR-based biosensors.

Au NP-probes were prepared by functionalizing Au nanosphere with kinetoplast DNA (kDNA) sequences from *Leishmania major* for diagnosis and differentiation of *L. major* from *L. tropica* and *L. infantum* species. The interaction of the kDNA Au NPs-probe with the complementary DNA strand resulted not only in a shift of the LSPR band and in a color change, visible to the naked eye, but also in an increased colloidal stability. Interestingly, the introduction of HCl in kDNA Au NPs-probe dispersion caused aggregation in presence of the non-complementary strand, while the hybridization of the complementary strand with the kDNA probe prevented the kDNA Au NPs aggregation. The probe was demonstrated to be specific for quantifying *L. major* in clinical samples, with LOD of 0.34 ng/μL and an efficiency comparable to conventional Real Time PCR analysis.[26]

Zopf *et al.* fabricated an array of spherical Au NPs (80 nm in diameter), consisting of several spots composed of Au nanospheres. The array of Au NP spots was realized by using a piezoelectric dispenser, while the functionalization of spots with different thiol-modified DNA molecules (capture probes) was achieved by exploiting the affinity of thiol functionality toward Au surface (Fig. 1(b). The incubation of the DNA-activated array with the DNA sequence of the target pathogen promotes the binding between two DNA complementary strands, resulting in a change of the local refractive index and consequently a change of the LSPR peak. Such a strategy allows the multiplex detection and identification of five fungal pathogen DNA sequences.[27]

3. SERS-based Techniques for Pathogen Detection

In recent years, Raman spectroscopy, due to its ability to offer specific and readable structural information about the composition of pathogens of interest, has gained more and more attention in on-site monitoring of relevant pathogens, by identification of their specific molecular vibrations.[28] But, although this vibrational spectroscopy technique possesses many attractive properties, like specificity, high response speed, etc., its direct implementation in pathogen detection applications is limited because of its poor intrinsic sensitivity, the Raman effect being a very weak phenomenon. In this context, to overcome this major limitation, an enhanced Raman detection technique was developed,[29] which offers, over conventional Raman spectroscopy, ultrahigh sensitivity (i.e., down to single molecule level) and selectivity, together with rapid spectral fingerprinting recognition capabilities.[30] This particular case of Raman spectroscopy technique, called Surface-enhanced Raman spectroscopy (SERS), allows the enhancement of Raman-active vibrations of the molecules located in the proximity of a nanostructured plasmonic surface, integrating thus Raman spectroscopy with nanotechnology in a benefic manner.

In particular, SERS is a rapid, non-destructive, low-cost, sensitive, and direct detection strategy, based on the enhancement of the Raman signal of the pathologic molecule by its adsorption and immobilization on metal nanostructures (e.g., Au, silver (Ag)), that provides qualitative and quantitative data of the investigated sample.[31] By using metal NPs as active substrates, "hotspots" are created based on the strong electromagnetic field generated by the controlled aggregation or self-assembling of the NPs. These generated "hotspots" together with the localized surface plasmon resonance constitute the two enhancement phenomena that form the SERS detection method: (i) the electromagnetic field enhancement (EM) due to noble metal substrates, being independent of the adsorbate; and (ii) the surface chemical (CM) enhancement, which arises from the interaction of the target absorbate with the metallic surface/NPs, allowing detection of extremely low concentration of biomolecules in the range of pico to femto molar concentrations.[32–34]

Based on the two phenomena, the SERS signal produces an enhancement of several orders of magnitude, being reported even as a signal enhancement of 1014 over the normal Raman signal.[35,36]

Hence, SERS is able to identify various chemical species in a rapid way, from pigments, narcotic substances, biomaterials, and, most importantly, detects single molecules, viruses, various pathogen species as well as cancer cells.[37] Moreover, the signal-to-noise ratio is significantly improved by SERS-active molecules' ability to quench fluorescence, and a broad range of pathogens can be detected based on the unique molecular fingerprints provided by SERS analysis.[38] Therefore, keeping in mind all these abovementioned SERS' advantages, researchers have been continuously motivated to develop innovative microfluidic devices, point-of-care testing, or various SERS-based strategies capable of detecting pathogens in food, water, or different biofluids in a portable and easy-to-operate manner.

In this chapter, we highlight successful developments of the versatile SERS-based nano(bio)sensors implemented in the literature in the last five years for the ultrasensitive detection and identification of bacteria and virus.

3.1. *SERS-based plasmonic (nano)sensors for bacteria detection*

To date, SERS technique has been able to ease the direct detection of different pathogenic bacteria due to a good Raman scattering cross-section of bacteria. In this approach, the bacteria are captured on the metallic-nanostructured surfaces, and the enhanced Raman fingerprint of pathogens is analyzed. For example, Zhang *et al.* efficiently detected *Staphylococcus aureus* using a novel 3D SERS substrate fabricated by repeated immersion of polymer brush in colloidal Ag NPs. They reported a very good LOD of 8 CFU ml^{-1}.[39] Wang *et al.* used dendritic Ag nanostructures to detect *S. enterica* and *E. coli*.[40]

Wei *et al.* successfully detected *E. coli* and *S. aureus* using highly active SERS NPs prepared via the microwave heating method.[41] In most cases, for distinguishing different bacteria SERS signatures, principle component analysis (PCA) and hierarchical cluster analysis (HCA) were successfully employed.

In order to discriminate the SERS signal obtained from different strains of *Campylobacter (C. jejuni, C. coli, C. lari,* and *C. upsaliensis.)* from poultry samples, Witkowska *et al.* used polycrystalline silicon covered with silicon nitride and Ag nanostructures.[42] SERS measurements were repetitively performed, for dissimilar hotspots on the SERS Ag-Si substrate. The SERS signal discrimination was achieved based on PCA algorithms.

Recently, Liu *et al.* reported SERS label-free detection of 20 strains of common pathogenic bacteria using NR SERS substrate.[43] They used multivariate procedures for achieving high accuracy of SERS discrimination of pathogenic bacteria.

Bashir *et al.* collected seven SERS spectra of six different *E. coli* strains (three tigecycline-resistant and three tigecycline-sensitive) using plasmonic NPs, for bacterial signal enhancement.[44] The biochemical changes due to tigecycline resistance have been identified in SERS spectra, and a successful discrimination between resistant and sensitive strain has been completed using PCA, HCA, and Partial Least Square-Discriminant Analysis (PLS-DA).

Metallic NPs were also used for the SERS detection of bacteria in food matrices. *E. coli* was detected by SERS in apple juice[45] exploiting Ag NPs, while Au NPs were employed for the detection of *S. Aureus* and *Shigella flexneri* in pork mince samples.[46] To this end, Zhao *et al.* developed a flexible and portable SERS platform presented in Fig. 2.[46] The SERS

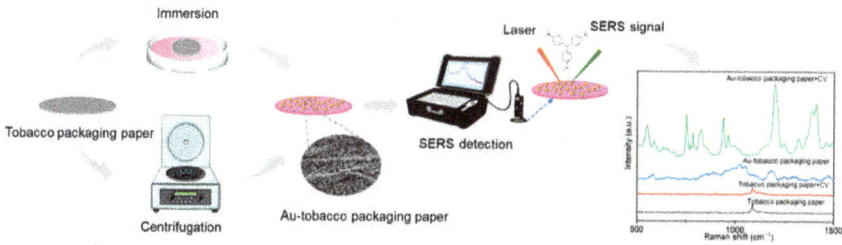

Figure 2. Schematic illustration of the preparation of the Au-TPP substrate and the SERS detection process. (Tobacco packaging paper (TPP)). Reproduced with permission from Ref. 46. Copyright 2023.

platform consisted of commercial tobacco packaging paper loaded with Au NPs by immersion. The approach for the preparation of the SERS platform and the analysis is displayed in Fig. 2.

Specimens from humans such as urine,[45] wound fluids[47] and also environmental matrices[48] were analyzed by using Ag NPs, Au NPs, and a microfluidic device with Ag NPs,[48] respectively. As shown in Fig. 3,[48] a microfluidic device composed of two cross-stack microfluidic channels (sample and extraction) with a hydrophilic porous membrane in between, was employed for *E. coli and Pseudomonas taiwanensis* detection in tap water. The fluid containing bacteria and NPs flows through the porous membrane from sample channel to the extraction channel due to the potential difference between the two microfluidic channels, as presented in Fig. 3. The membrane acts as a trap for the bacteria and the NPs, producing a sensitive SERS platform for bacteria detection.

Deb *et al.* demonstrate that capping agents of the Au NPs influenced the pattern and functionality of metallic nanostructures, and consequently, the efficiency of direct SERS bacteria identification.[49] Indeed, from the three investigated capping agents (thioglucose, polyvinylpyrrolidone, and citrate) for the SERS detection of *Cutibacterium acnes, E. coli,* and *S. aureus,* thioglucose proved good efficiency and highest accuracy in bacterial detection and discrimination.

In recent years, a variety of indirect SERS techniques have been developed in order to rapidly detect different types of bacteria, with high sensitivity and specificity. In this respect, various SERS-based plasmonic biosensors have been explored. A SERS plasmonic biosensor combines specific bacterial recognition elements with resourceful functionalized NPs, as SERS tags, hence it has the advantage of having the capacity to achieve a high specificity and performant LOD.[50–53]

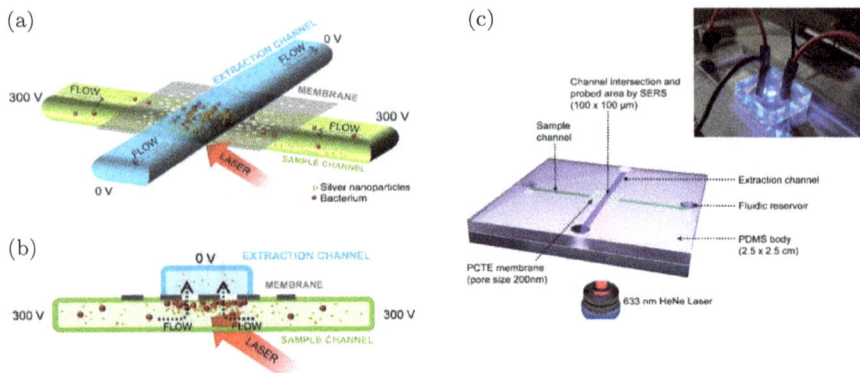

Figure 3. The working principle of the microfluidic device for bacteria enrichment and SERS analysis: (a) side view, (b) cross-section, (c) the overall appearance of the lab-on-a-chip device for the concentration of bacteria followed by SERS detection (inset shows a photo of the microfluidic chip with electrical connections during fluorescence measurement). Reproduced with permission from Ref. 48. Copyright 2023.

Moreover, SERS technique was combined with different methods (i.e., magnetic separation,[54–57] lateral flow assay (LFA) and immunochromatography,[58–60] microfluidics,[61] etc.) to gain sensitive and selective SERS biosensors with high capability of pathogenic bacteria detection, in short time and at low cost.

The combination of magnetic separation and SERS leads to the development of sensitive biosensor detection of methicillin-resistant *S. aureus*, *E. coli*, and *S. typhimurium*.[62] By functionalizing NPs with magnetic nanomoieties, specific *antibodies*, and exclusive SERS reporters, authors successfully detected very low concentrations (herein 10 CFU/ml) of the aforementioned bacteria.

In a similar manner, Zhang *et al.* used Au-coated magnetic NPs functionalized with vancomycin for capturing and quantifying gram-positive and gram-negative bacteria via SERS, resulting in a capturing efficiency of 88.89% for *S. aureus* and 74.96% for *E. coli*).[63]

Several dual SERS-LFA biosensors were also developed. He *et al.*, proposed a dual SERS-LFA sensor for the detection of *Campylobacter jejuni*.[64] The conjugation pad of a lateral flow assay strip was immersed in a solution of platinum-coated Au NRs (AuNR@Pt) functionalized with a specific antibody (mouse anti-*Campylobacter jejuni* monoclonal antibody) and Raman reporter (4-mercaptobenzoic acid, MBA). A wide range of *Campylobacter jejuni* concentrations (from 10^2 CFU/ml to 5×10^6 CFU/ml) was detected, with a linear relationship between the SERS signal intensities and the bacteria concentrations and an LOD of 50 CFU/mL for SERS and 75 CFU/ml for the LFA test. Moreover, authors tested the proposed system for detection of other six foodborne pathogens (*E. coli, S. aureus, C. freundii, V. parahaemolyticus, P. aeruginosa,* and *S. typhimurium*), achieving a good accuracy even for real sample test (milk and chicken).

Using the same approach, Gao *et al.* detected the *Cronobacter sakazakii* using a dual SERS immunochromatographic test strip.[59]

The LOD using visual inspection was 10^5 CFU/ml. Much better LOD (201 CFU/ml) was establish using the SERS signal obtained from the test line (due to the presence of PATP reporter in Au NPs decoration). Good reproducibility, specificity, and selectivity of the SERS strip was also reported.

You *et al.* reached a single cell detection limit, using starch magnetic beads coated with Au NPs and functionalized with specific antibody for capturing and concentrating the target *E. coli.*[57]

Duan *et al.* developed a SERS aptamer platform using Au NPs, fixed on flexible and transparent PDMS film, functionalized with two specific aptamers as bacteria traps and some Au NPs, decked with the same two aptamers and with other two small molecules (MBA, and Nile blue A), as SERS tags.[65]

The system was successfully employed for determining the *Vibrio parahaemolyticus* and *S. typhimurium* even at very low pathogen concentrations of 18 CFU/ml and 27 CFU/ml, respectively.

Very recently, Shen *et al.*, proved the multiplex analysis of pathogenic bacteria by means of a SERS-immunoassay.[60]

As SERS amplifying substrate, they used graphene oxide nanosheets coated with Au shells and functionalized with 4-MBA (as Raman label) and specific antibodies, as illustrated in Fig. 4. The proposed SERS labels adhere efficiently to bacterial cells and supply plentiful hotspots for SERS signal enrichment. The biosensor is completed with a sample pad (for sample solution charging), absorption pad (for capillarity), and NC membrane with three test lines (each containing different capture antibodies) and one control line (and Goat anti-mouse IgG). The projected SERS-immunoassay was fruitfully applied for rapid (20 min) and simultaneous recognition of *S. aureus, E. coli,* and *S. typhimurium*, with low detection limits (8, 10, and 10 cells/ml, respectively).

Rippa *et al.* developed a portable SERS device based on Au aperiodic nanostructure functionalized with 4-aminothiophenol and Tbilisi bacteriophages for the on-site detection of Brucella abortus, a pathogenic bacterium

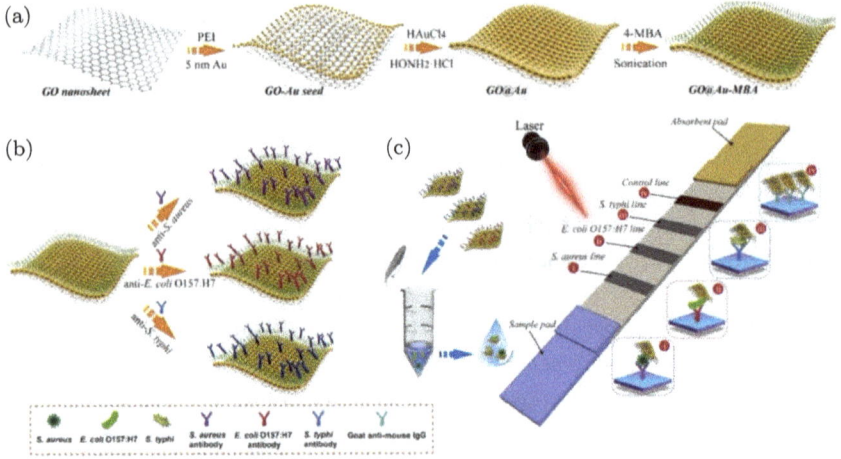

Figure 4. Schematic representation of (a) the fabrication procedure of flexible GO@Au nanosheets, (b) the preparation of immuno-GO@Au SERS tags, and (c) GO@Au tags-based ICA for multiplex and ultrasensitive analysis of three target foodborne pathogens. Reproduced with permission from Ref. 60. Copyright 2023.

affecting healthcare and sometimes the economy, due to the contamination of food and drinks.[66]

This new SERS-sensing platform combines high bacterial capture capacity of bacteriophages with a valuable SERS marker — the $1322\,cm^{-1}$ peak, related to the vibrational stretching of the diazo-bond formed in the process of bacteriophage immobilization on Au surface via 4-aminothiophenol. The as-made SERS-sensing device was tested with good results on real food samples (micro-filtrated milk).

Wang *et al.* engineered another complex SERS-immunoassay for simultaneous detection of four pathogens (*L. mono, E. coli, S. aureus,* and *S. typhi*) on two test lines.[67]

The SERS-immunoassay relies on graphene oxide nanosheets coated with Au shells and grafted with NPs (in order to increase the efficiency of bacteria binding and create additional SERS hotspots). MBA and L. mono DTNB were employed as Raman labels. The discrimination between the four types of bacteria and the concentration quantification are assured by SERS intensity of the typical signature of DTNB (at $1328\,cm^{-1}$) and 4-MBA ($1585\,cm^{-1}$). In addition to the valuable LOD of the proposed GO@Au/Ag nanosticker-based SERS-LFA biosensor of only 9 cell/ml, the multiple tests performed in real biofluids indicate high accuracy and stability.

Ma *et al.* implemented a different SERS method for the detection of *S. aureus*.[68] Firstly, they fabricated SERS substrates by self-assembly properties if Au@Ag NPs. Then, they modify the surface of Au@Ag NPs substrate with cysteamine hydrochloride in order to increase the immobilization of ROX-aptamer. The ROX-aptamer possesses dual functions: it performs the selective detection of *S. aureus*, and provides a valuable Raman fingerprint for SERS detection. In the presence of *S. aureus*, with the ROX-aptamers connected to its target bacteria, the corresponding SERS signal decreases noticeably, as we can observe in Fig. 5. Evidently, a detection limit of 6 CFU/ml was obtained. The novel SERS method could be adapted to another pathogen detected by simply changing the aptamer.

In summary, although remarkable progress has been made recently in the SERS detection of the pathogens, there is still plenty of room for future work in order to obtain easy operation of SERS-based plasmonic nanosensors with high specificity, sensitivity, and reproducibility for as many types of bacteria as possible.

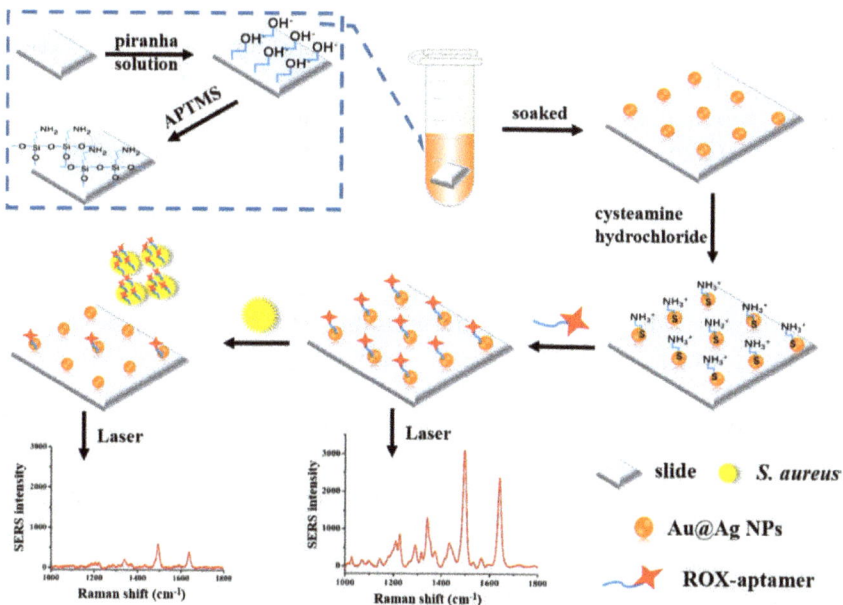

Figure 5.   Schematic illustration of the developed SERS detection method. Reproduced with permission from Ref. 68. Copyright 2023.

3.2. SERS-based plasmonic (nano)sensors for virus detection

Viruses are the source of the greatest genetic diversity in the world, with a rapid transmission occurring via aerosols, droplets, even fecal matter.[69]

They represent a real threat to human health. In this view, early diagnosis of viral infections, such as influenza virus, Ebola virus, human immunodeficiency virus (HIV), and more recently the severe acute respiratory syndrome coronavirus 2 (SARS-CoV-2) has always played a crucial role. Plasmonic-based-SERS virus detection techniques are a valuable alternative for viruses' detection, being rapid, portable, simple and highly sensitive, as Lu *et al.* have recently summarized with respect to food-borne viruses.[70]

In the context of the recent global outbreak of coronavirus disease in 2019,[71] portable SERS detection has started to be an alternative emerging technique that can be successfully employed not only for the quantification of the SARS-CoV-2, but also for influenza virus.[72]

In this pandemic context, Gao *et al.* proposed an interesting and accurate triple-mode biosensor (i.e., colorimetric/SERS/fluorescence) for SARS-CoV-2 detection, based on Au nanospheres for rapid and selective RNA detection in only 40 minutes. As SERS enhancers, the authors used citrate-stabilized spherical Au NPs. Using a micro Raman spectrometer, the fabricated nanosensor achieved an LOD of 395 fM in simulated sample conditions.[73]

The Wawrousek group developed a very simple and ultrasensitive SERS assay to detect SARS-CoV-2 spike protein directly in human saliva.[74] Concretely, the rapid diagnostic in this specific case implies a sandwich assay, illustrated schematically in Fig. 6, between the saliva sample incubated with a single-chain variable fragment (scFv) selected herein due to its capability to bind to the receptor binding domain of the targeted SARS-CoV-2 spike protein, labeled with Malachite green (MG) — employed as a Raman-active dye with scFv conjugated to the Au-coated magnetic NPs' surface. Then a magnetic field was employed to concentrate the MG dye. The proposed point-of-care (POC) test system achieved an LOD of 1.94×10^3 genomes ml^{-1} and it demonstrated being more sensitive than commercially available COVID-19 diagnostic tests.

The urgent need for the fast and ultrasensitive source detection of norovirus (NoV) infection, (causing gastritis and colitis) motivated Achadu *et al.* to construct a dual-mode detection platform based on core-satellite

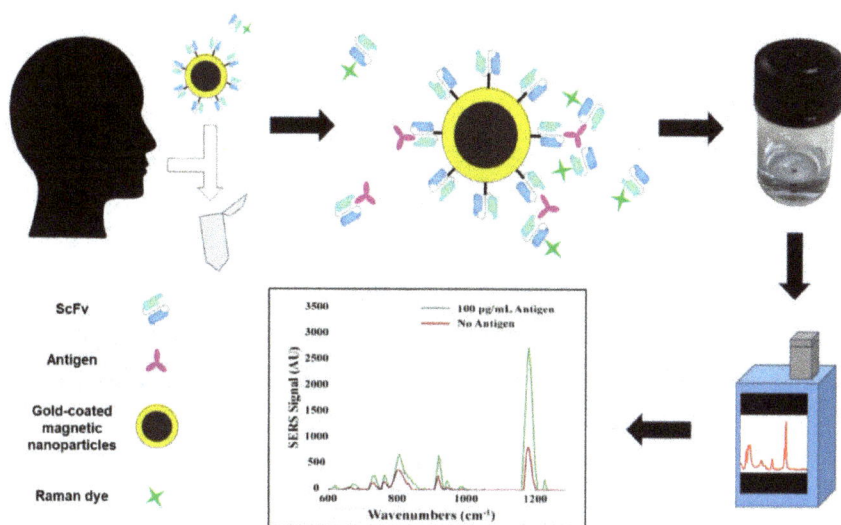

Figure 6. Schematic illustration of the designed single SERS immunoassay, realized by mixing the virus-binding scFv antibody fragments labeled with malachite green with Raman dye and scFv-conjugated magnetic Au nanoshells. Reproduced with permission from Ref. 74. Copyright 2023.

immunocomplex between anti-NoV antibody conjugated sulfur-doped agarderivate carbon dots (as a single active probe) and novel polydopamine-functionalized magnetic nanocubes. Based on the specific antigen–antibody immunoreaction, an excellent LOD of 0.1 fg ml^{-1} was determined via the ultrasensitive SERS detection.[75]

A quite similar LOD of 5.2 fg ml^{-1} in human fecal samples was obtained by the same group, employing in this different case a novel plasmonic @ magnetic derivative of MoO_3 nanocubes (as SERS nanotags) and a single-layer graphene oxide (as a solid capture platform). This new formed sandwich-type immunocomplex[76] was then clinically validated, using human fecal samples, obtaining an LOD of ~60 copies/ml, proving the potential transfer of the nanoplatform in point-of-care diagnostic testing (Fig. 7).

In order to detect hepatitis B virus, the main reason for liver cirrhosis and liver cancer, an interesting SERS-based strategy was successfully developed.[77] This sandwich immunoassay employed for the sensitive and selective SERS detection of antigen HBsAg, in spiked and real samples from nine patients, allows an LOD of 0.05 pg/m, which is a lower value compared to the ELISA method.

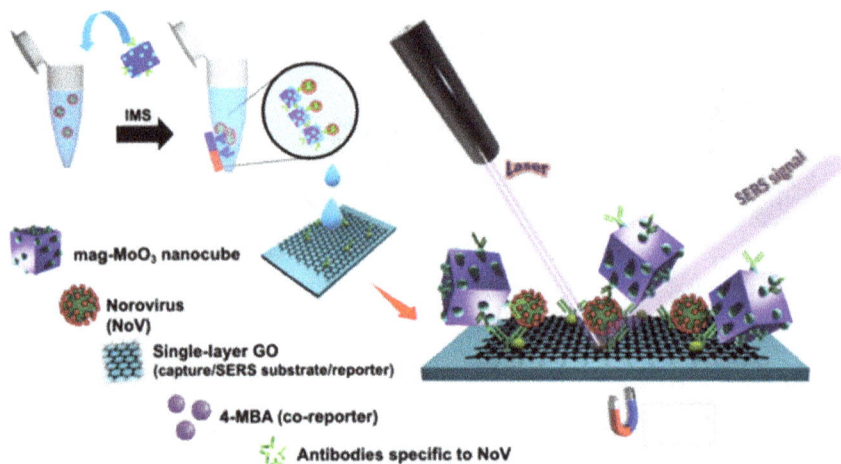

Figure 7. Schematic representation of the Norovirus detection method in human feces samples. The sandwich-type immunocomplex, consisting of plasmonic magnetic MoO_3 nanocubes and a single-layer graphene oxide platform, owns the ability to capture and separate the specific Norovirus, making use of the SERS effect. Reproduced with permission from Ref. 76. Copyright 2023.

Although great amount of progress was achieved in the design of highly efficient SERS-based nanosensors for virus detection, there are still concerns that remain to be addressed in order to fabricate innovative detection nanoplatforms with great stability, reproducibility, and efficiency.

4. Electrochemical Biosensors

Electrochemical detection possesses several advantages in the design of biosensors, including the simple design of the final device and its low manufacturing cost, robustness, easy miniaturization, and excellent detection limits in addition to the possibility of handling small analyte volumes. Many electrochemical biosensors have been described in the literature for the detection of pathogens, showing very diverse sensing configurations based on both natural and artificial biological components, as well as combining nanomaterials,[78] nucleic acid amplification methods,[79,80] and microfluidics.[81]

An example of an electrochemical immunosensor was described by Guner and colleagues,[82] based on a pencil graphite electrode (PGE) modified with a PPy/Au NP/MWCNT/Chi bionanocomposite, for the detection

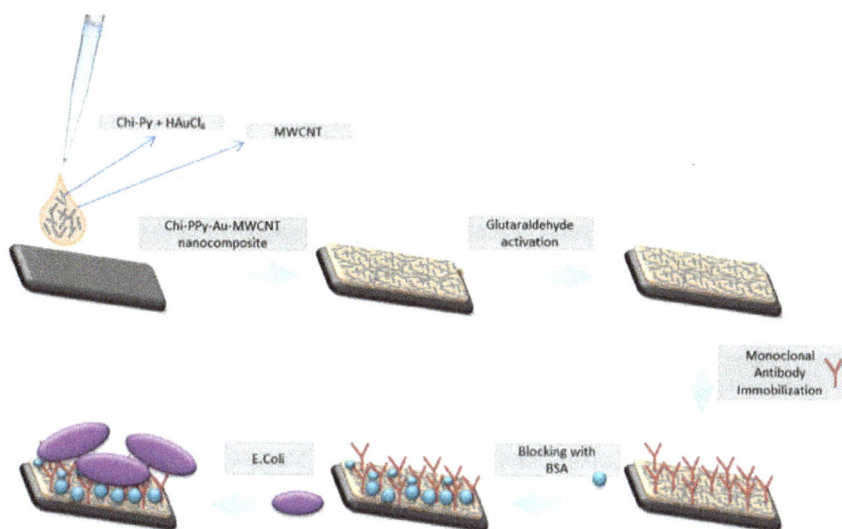

Figure 8.   Schematic representation of the experimental setup of the *E. coli* immunosensor fabrication. Reproduced with permission from Ref. 82.

of *E. coli* O157:H7 exploiting specific monoclonal antibodies. Figure 8 reports the different steps for the preparation and the functionalization of the electrode, including the immobilization of the monoclonal antibodies. The setup of the immunosensing bionanocomposite electrode was followed by cyclic voltammetry (CV). Under the optimum conditions, this biosensor was able to detect *E. coli* O157:H7 by amperometry in a concentration range from 3×10^1 to 3×10^7 CFU/ml with an LOD of $\sim$30 CFU/ml. Gram negative reference strains, including *E. coli* O124:NM (ATCC 43893), *P. aeruginosa* (ATCC 27853), *S. sp.* (ATCC 11126), *Burkholderia cepacia* (ATCC 25416), and *S. enteritidis* (ATCC 4931), were exploited to evaluate the selectivity of the proposed biosensor, showing that *E. coli* O157:H7 was highly detected as compared to the other analyzed bacteria. Moreover, the biosensor showed good reproducibility (with an RSD value of 8.9%) and stability (89.3% activity within the first 2 weeks), demonstrating to be useful for the specific detection of *E. coli* O157:H7 contamination in food quality and safety control.

Besides antibodies, the combination of electrochemical techniques with aptamers also led to the development of simple, fast, sensitive, and selective biosensors that act as powerful tools for pathogen detection in many application fields. An aptasensor based on an Au NPs/CNPs/CNFs functionalized glassy carbon electrode was designed by Ranjbar *et al.*[83] for the rapid detection of *S. aureus*. In this case, the cellulose nanofibers (CNF)-based hybrid nanocomposite was used as a durable scaffold for the immobilization of aptamers designed ad hoc considering the specific target. This nanocomposite was able to provide high surface area, excellent conductivity, and good biocompatibility. Indeed, CNFs, which consist of β-1,4 linked anhydro-d-Glucose units, are a natural polymer capable of furnishing an excellent platform for the immobilization of many biological components. On the other hand, Au NPs are used thanks to their excellent conductivity and high surface area for capturing aptamers as well as for their capability of easily and effectively conjugating the biological components with thiolated motifs via the formation of the self-assembled monolayers. Thanks to a similar configuration, the designed aptasensor showed a wide linear dynamic range $(1.2 \times 10^1$ to $1.2 \times 10^8)$ CFU ml^{-1} and an LOD of 1 CFU ml^{-1} for the detection of *S. aureus* by electrochemical impedance spectroscopy (EIS) in human blood serum. The selectivity of the biosensor was evaluated using different bacteria strains (i.e., *P. aeruginosa*, *E. coli* O157, *S. typhimurium*, and *S. flexneri*), highlighting values of ΔRct for other strains much lower than the value obtained for *S. aureus*.

Electrochemical biosensors for pathogen detection have been associated in the last years also with nucleic acid amplification systems, to provide higher sensitivity. This is the case of the genosensor described by Yu and co-workers,[84] who developed a portable multiplexed electrochemical platform targeting multiple DNA regions of interest. In detail, a dry-reagent mLATE-PCR mix that is compatible with a downstream detection method using an electrochemical genosensor was setup. The dry-reagent mLATE-PCR mix simplifies the whole workflow of the preparation of the amplification reaction mix into a two-step assay to which ultrapure water and the DNA template are added. In this system, the presence of target DNAs enabled generating corresponding single-stranded DNA (ssDNA) amplicons tagged with a fluorescein label. The fluorescein-labeled ssDNA amplicons were thus analyzed using capture probe-modified screen-printed Au electrodes. Enzymatic amplification of the hybridization event is achieved through the catalytic production of electroactive α-naphthol by anti-fluorescein-conjugated alkaline phosphatase. The whole process was depicted in Fig. 9. *Vibrio cholerae* serogroups O1 and O139 were analyzed using the proposed platform, which showed excellent sensitivity (10 CFU/ml) and specificity (100%) when challenged with 168 spiked stool samples. The genosensor

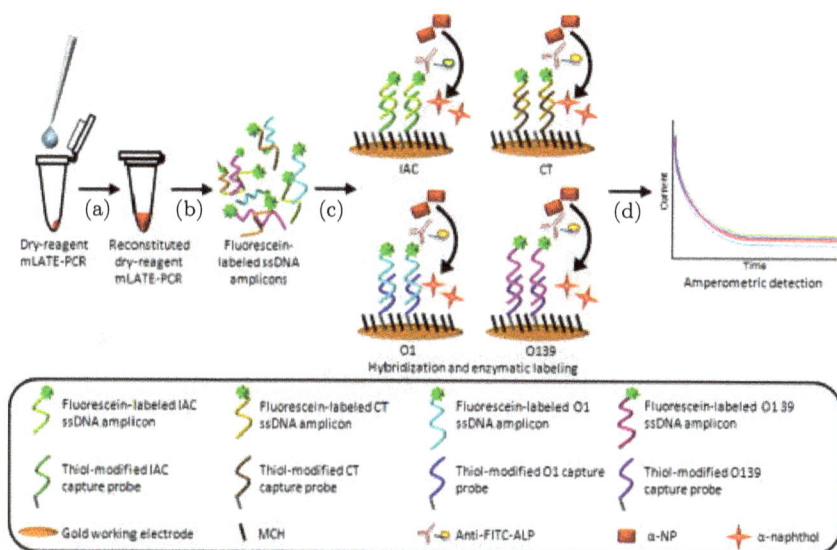

Figure 9. Schematic illustration of the procedures involved in nucleic acid detection with a dry-reagent mLATE-PCR-coupled multiplex electrochemical genosensor; (a) two-step protocol for setting up the mLATE-PCR mix by adding ultrapure water and the DNA template; (b) generation of fluorescein-labeled ssDNA amplicons during amplification; (c) hybridization between ssDNA amplicons and complementary capture probes that have been self-assembled on the Au working electrodes followed by binding of anti-FITC-ALP to fluorescein-labeled ssDNA amplicons and conversion of electroinactive α-NP into electroactive α-naphthol by ALP; and (d) amperometric detection of α-naphthol at a working potential of 0.25 V. Reproduced with permission from Ref. 84. Copyright 2023.

was tested with a panel of bacterial strains, showing no cross-reactivity with other non-target pathogens.

The convergence of crosscutting technologies including microfluidic and electrochemical sensing supported in the last years the development of forefront POC devices to realize fast, automatic, and affordable genetic analysis. A crucial example was proposed by Park *et al.*[85] which described the development of a single-chip biosensor integrating a film-based PCR module and an electrode module for gene amplification and electrochemical analysis as well as a polydimethylsiloxane (PDMS)-based finger-actuated microfluidic module for sample migration and mixing. Such integrated pumpless microfluidic chips can replace external pumping systems and provide analysis of multiple samples by square wave voltammetry.

Figure 10 reports the fabrication of the integrated microfluidic chip. Adhesion films were used to assemble the injection, PCR, mixing, and

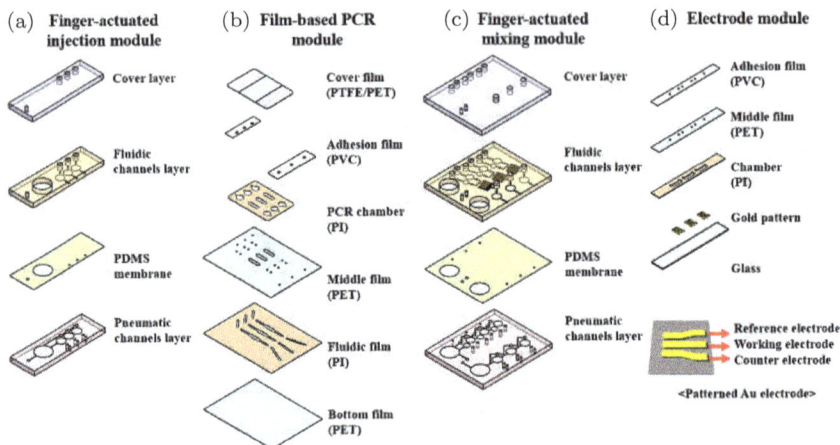

Figure 10. Configuration of each module of the integrated pumpless microfluidic chip. (a) Finger-actuated microfluidic module for injection of reagents. (b) Film-based PCR module for PCR. (c) Finger-actuated microfluidic module for mixing electrolyte and amplified genes. (d) Au electrode module for electrochemical detection, including film-based chambers. Reproduced with permission from Ref. 85. Copyright 2023.

electrode modules onto a polycarbonate (PC) sheet. Using this platform, E. coli O157:H7 was sensitively analyzed with an LOD of 10^2 CFU/ml.

To evaluate the usability of the integrated chip, E. coli O157:H7, S. enteritidis, and Bacillus cereus pathogens spiked into real samples of milk were tested and the result indicated that the applied pathogenic nucleic acids were successfully amplified and analyzed by the integrated chip with good selectivity.

5. Biosensors on Flexible Substrates

Soft and flexible substrates play a vital role in detecting, POC testing, and biosensing. These powerful tools made a remarkable advance in monitoring and detection in various areas such as environment, food safety, and healthcare diagnosis. The unique flexibility feature of these materials allows them to have efficient contact with the system even with complicated surfaces, resulting in enhanced signal transduction. Due to these properties, the measurements are in real time; thus, more reliable and accurate data can be achieved.[86]

In recent years, a lot of research has been done on the flexible substrates for biosensing applications, and excellent works have been published which have discussed this field from different aspects, such as materials, detection methods, and applications. Substrate selection is of crucial significance when designing a biosensor, as the platform's characteristics are dependent on the support substrate and can be requirements that can be applied in this stage. Recently, biosensors based on electrospun nanofibers alongside hydrogels, films, microgels, and other 3D structures have been explored. These all provide key essentials of a flexible substrate for biosensing, with enhancements in sensitivity, integration, and efficient performance.[87,88]

5.1. *Electrospun nanofibers*

Nanomaterials have been widely used with analytical systems as they can result in a faster mass transfer rate, thus achieving lower LOD compared to larger-size materials.[89] Electrospinning is known as a valuable and flexible technique to fabricate fibers in the micro- and nanodimensions, used massively for various applications. Nanofibers fabricated via this method can be adjusted to desirable sizes and mechanical properties, while the wide selection of materials and electrospinning parameters can alter the chemical functionalities. A typical electrospinning setup consists of a pump, a collector, and a high voltage source, which is the system's driving force, guiding the polymeric solution from the tip of the spinneret to the collector.[90,91]

One of the most vital aspects of electrospun nanofibers in biosensing applications is the extremely high surface-area-to-volume ratio. Sensing elements (e.g., plasmonic NPs) can be introduced into the nanofibers via three approaches: Plasmonic NPs incorporated into the electrospinning solution (pre-mix), metal precursors dispersed into the polymer solution followed by a post-treatment on fibers to transfer metal ions to metal NPs (in situ), and introduction of plasmonic particles by surface modifications after electrospinning (ex situ).[92]

As a simple and easy process, the pre-mix approach is commonly used for biosensor fabrication. Surface coverage of the particles can be easily tuned by modulating the concentration in the polymer solution, and the desired shape and size of the particles can be easily determined as they are pre-prepared. Wang *et al.* fabricated electrospun Poly(*N*-isopropylacrylamide) (PNIPAAm) mat embedded with NPs (Ag NPs) for the detection of adenosine in the urea solution. The presence and alignment of the Ag NPs generated hotspots upon laser-induced heating, resulting in a very low LOD.[93] Ag NPs were also used by Shi *et al.* with poly(ε-caprolactone) (PCL) nanofibrous mats. The homogeneous dispersion of the particles in PCL fibers was checked with scanning electron microscopy (SEM), as shown in Fig. 11(a). This platform was used for sensing melamine, and with this technique, a detection limit was achieved, which was much lower than FDA standards.[94] Polyvinyl alcohol (PVA) embedded nanofibers with Au NPs were designed and fabricated to detect methylene blue and methyl salicylate.[95] Au NPs were also used with PVA nanofibers with high concentrations. This composition has provided an excellent matrix for the fibers and stabilized the NPs. Detection of the 4-MBA (4-MBA) molecules was significantly enhanced by this method.[96]

The in-situ approach consists of two main steps of fabrication. In the first step, metal ions dispersed in the polymer solution undergo electrospinning, and in the next step, a reduction of the ions into particles results in the final platform. Metal NPs can be obtained through processes such as thermal, chemical, plasma etching, and photo-irradiation. Chen *et al.* designed an electrospun mat of electrospun polyvinylidene fluoride (PVDF) nanofibers while the Au seeds were deposited on top of the fibers. The seeds were reduced into NPs using a chemical method by introducing I^- ion to the platform. This system was used to detect thiram using a SERS method.[97] Au NPs were also grown in situ in the work of Severyukhina *et al.* using chitosan nanofibers as substrate and trisodium citrate in heated water as a reducing agent. Rhodamine 6G and D-glucose, as famous biomarkers for diabetes, were detected by this platform successfully.[98] There is also a possibility of adjusting the shape and size of the particles throughout the process. This is a system proposed by Jia *et al.* to grow Ag nanoplates and polyhedra on polyacrylonitrile (PAN) nanofibers. By controlling the experimental parameters of the system, it is possible to manipulate the size and shape of the particles. The sensitivity of the biosensor to detect 4-MBA was shown to be significantly increased due to getting more hotspots and surface area.[99] The simple scheme of the procedure can be seen in Fig. 11(b).

NPs can also be independently synthesized and immobilized on the surface of the nanofibers with covalent or non-covalent bonds. This process depends on the topology of the nanofibers, NPs, and their surface functional groups. Fathi *et al.* fabricated a platform to detect *S. enterica* using an Au-based aptasensor. Chitosan electrospun with carbon nanofibers was decorated with Au NPs and placed on a graphite electrode to make an electrochemical sensor.[100] Figure 11(c) shows the schematic representation of the fabrication process and the sensing procedure with the help of other molecules, such as methylene blue. A SERS-based biosensor was

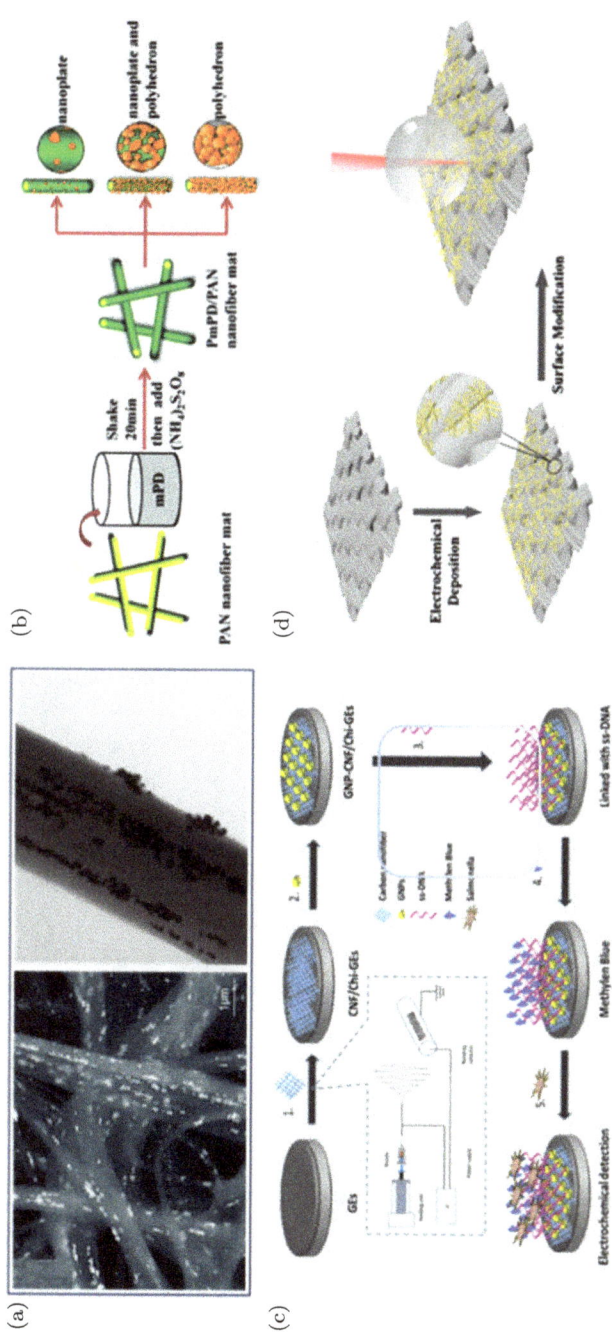

Figure 11. (a) Backscattering SEM and TEM images of Ag NPs/PCL nanofiber membrane. (b) Schematic diagram showing fabrication of nanofiber decorated with Ag NCs in situ, resulting in different morphologies. (c) Schematic illustration of the aptasensor fabrication steps. The surface of the GE was modified with aptamer-Au NP/carbon nanofiber-chitosan. (d) Schematic diagram of Ag NDs deposited on flexible carbon fiber cloth for direct droplet detection. Reproduced with permission from Ref. 101. Copyright 2023.

designed by the use of poly (acrylic acid) (PAA)/poly (vinyl alcohol) (PVA)
electrospun nanofibers, where Au NPs were assembled on the structure by
simple electrostatic interaction. This platform has proved to be a precise
biosensor to detect small molecules such as 4-ATP and R6G.[101] To secure
the immobilization of the NPs, chemical procedures have been suggested.
A biosensor with high sensitivity for 4-ATP was synthesized using this
method. PCL nanofibrous mats were grafted with acrylic acid under UV
light, generating a negative charge at their surface. The functionalized NPs
(cysteamine-capped Au NPs) bearing a positive charge can form a hydrogen
bonding with the substrate.[102]

5.2. *Other substrates*

Various flexible substrates based on different structures and techniques than
electrospinning have also been reported in the literature. In this frame,
researchers have explored and designed several other different kinds of
flexible substrates for biosensing applications, including films, membranes,
hydrogel networks, and 3D structures.[88,103] These systems can be covered,
coated, or loaded with plasmonic particles such as Au or Ag NPs to obtain
high-performance SERS substrates for the detection and monitoring of
molecules.[104,105]

Wang *et al.* designed a ternary film composed of a poly(methyl
methacrylate) (PMMA) template film, a fluorescent quantitative polymerase chain reaction adhesive film, and a polyethylene terephthalate
(PET) covering film. Additionally, Ag-coated Au NPs were obtained in a
monolayer array and self-assembled on the PMMA layer of the ternary-film
package. The highly active and stable SERS substrate revealed the ability
to sense residues of thiabendazole in fruit juices.[106]

Another PMMA-based flexible SERS substrate covered by Au NPs was
proposed by Zhong *et al.*, who fabricated a self-assembled system for the efficient detection of a model molecule (i.e., malachite green isothiocyanate) at
low concentrations.[107] Au NPs were also used for producing pseudo-paper
films to detect residues of pesticides on surfaces (e.g., apple peel). The NPs
were synthesized in situ by iteratively seeding on polyethyleneimine grafted
onto microcrystalline cellulose. The substrate showed several "hotspots"
and high SERS activity. Furthermore, Sun *et al.* investigated the potential
of polyvinylidene fluoride membrane where Au NPs were immobilized on

the surface for efficient monitoring of the pH of liquids and SERS detection of gas molecules in the atmosphere.[108]

On the other hand, Ag particles were used to fabricate a bimetallic Tantal/Ag film by applying the magnetron sputtering technique. The substrate showed high chemical and structural stability and proved its potential as an active SERS sensor for microbial cells like *E. coli* bacteria.[109] In another study, the thermal deposition of Ag NPs was performed on Kapton tape to guarantee the SERS stability in the long term (up to 4 months).[110]

Hydrogel structures have also been exploited for biosensing applications by loading plasmonic particles into their networks. The resulting systems can efficiently and stably incorporate Ag particles for efficient sensing of biomolecules, as well as pH and temperature variations.[105,111]

Finally, in a study published by Lu *et al.*, the possibility of developing 3D flexible SERS substrates was assessed. The system was formed by electrochemical deposition of Ag nanodendrites on a 3D carbon fiber cloth, augmenting the "hotspot" effect and showing a superhydrophobic surface (Fig. 11(d)). The authors highlighted the great advantages of the proposed substrate, such as the short detection time and the capability of simultaneously detecting at least three different molecules.[112]

6. Future Perspectives

The reduced dimensions of metallic NPs are responsible for their peculiar interaction with electromagnetic radiation. These properties generate opportunities for several technological applications, including interactions with biomolecules and microorganisms such as biosensors. The sections of the present chapter have demonstrated the versatility of metallic NPs for developing biosensors for pathogen detection. As illustrated in the previous sections, several configurations, transducing strategies, and detection systems are possible. However, the next technological challenges will push a paradigm shift that must embrace the "one health" vision of the future. Accordingly, nanomaterials and nanotechnology must adopt green solutions to address the next challenges. The next-generation biosensors should be designed to minimize their ecological footprint in each step of their lifecycle. The preparation approaches for the next biosensing nanoplatforms should be cost-effective, safe, and versatile. Original strategies must be adopted to reuse and recycle nanobiosensing platforms to amortize their production costs. Another challenge for plasmonic biosensor development involves new functionalization strategies of the metallic nanostructure to perform multiplex sensing, and multivariate analysis, especially at the molecular level. This progress will be possible by precisely controlling the functionalization procedure by integrating, for instance, microfluidic circuits or applying inkjet printing or microcontact printing to immobilize biomolecules. The NPs and biomolecules' immobilization procedure should be improved to realize effective flexible and wearable pathogen biosensors to be applied on personal protective equipment to be used by medical personnel, by professionals that perform sampling in a particular site, or by biological warfare inspectors. Deep learning and artificial intelligence should establish a more robust dialog with nanotechnological biosensors to enrich the information that can be obtained from a single measurement, perform multiplex analysis, and control the biosensors' efficiency. Moreover, the reduced dimension of plasmonic NPs should be exploited to introduce NPs' insight into the cross-talk mechanism of pathogenic cells to detect the antimicrobial resistance mechanism at the molecular level. Finally, plasmonic NPs should be suitably assembled to fabricate bio-inspired metasurfaces or metamaterials interacting with pathogens, thus generating an ultrasensitive predictive biosensing system for pathogen detection.

Appendix: Optical Properties of Plasmonic Nanoparticles

The interaction of an electromagnetic (EM) radiation with metallic nanomaterials can produce an optical phenomenon called localized surface plasmon resonance (LSPR). It occurs when the frequency of the impinging EM wave is in resonance with the oscillations of the conductive electrons of the nanomaterial. The electronic oscillations are simply called surface plasmon resonances (SPR) in case of flat metallic surfaces and they propagate as an evanescent wave, while in nanostructured materials, such as NPs, they are localized on their surface, and therefore they are extremely localized, so that this phenomenon is called LSPR.

Gans theory depicts the solutions of Maxwell equations in case of ellipsoid NPs interacting with EM waves. The analytical expressions for the extinction cross-section (Eq. (B1)), within the dipole approximation, can be synthetized in the absorption contribution (Eq.(B2)), neglecting the scattering one, in case of NRs:

$$\sigma_{ext} = \sigma_{abs} + \sigma_{sca} \approx \sigma_{abs}, \tag{B1}$$

$$\sigma_{abs} = \frac{2\pi V}{3\lambda} \varepsilon_m^{\frac{3}{2}} \sum_{j=1}^{3} \frac{\left(\frac{1}{P_j^2}\right)\varepsilon_2}{\left(\varepsilon_1 + \frac{(1-P_j)\varepsilon_m}{P_j}\right)^2 + \varepsilon_2^2}. \tag{B2}$$

NRs have spheroidal shape with a major (A) and two minor axes (B, C) and P_j, which are called depolarization factors along those axes, as reported in the Eqs. (B3–B5), respectively. They are dependent on the size of the NRs through their aspect ratio (length-to-width ratio) that is contained in the definition of eccentricity, e(see Eq. (B3)).

$$P_A = \left(\frac{1-e^2}{e^2}\right)\left(\frac{1}{2e}\ln\left(\frac{1+e}{1-e}\right) - 1\right), \tag{B3}$$

$$P_B = P_C = \frac{1-P_A}{2}, \tag{B4}$$

$$e = \sqrt{\left(1 - \left(\frac{1}{AR}\right)\right)}. \tag{B5}$$

Electronic oscillations on transverse and longitudinal axes of NRs generate two corresponding peaks in the absorption spectrum. The longitudinal mode is stronger and has lower energy compared to the transverse one; moreover, it is strictly dependent on geometrical and environmental factors. The two modes for the electrons' oscillation are depicted in Figs. B1 and B2: the smaller oscillation of the longitudinal mode makes this peak more sensitive to system changes.

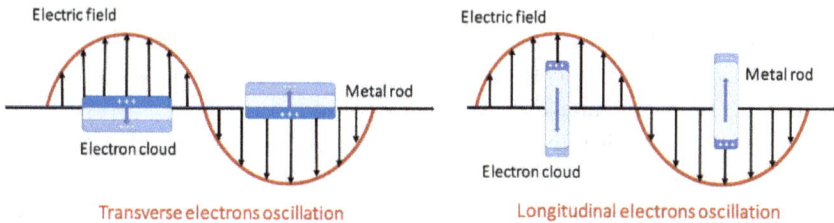

Figure B1.   Transverse electrons' oscillation. Longitudinal electrons' oscillation.

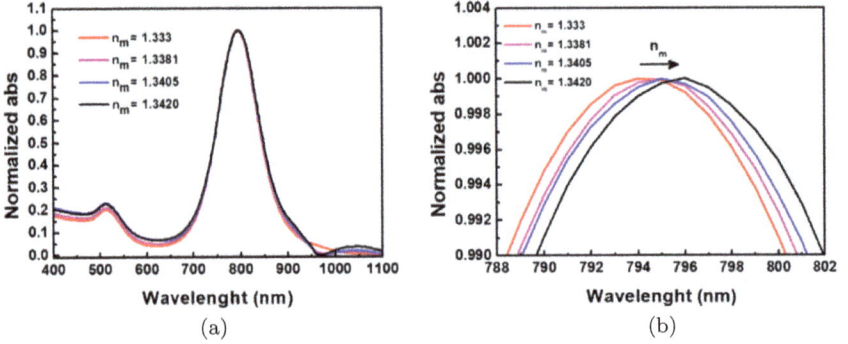

Figure B2. (a) Longitudinal plasmon peak shift according to the refractive index variation. (b) Magnification of the panel (a) in the wavelength range from 788 nm to 802 nm.

The LSPR condition is given by the Frolich condition (Eq. (B6)) that defines the real part of the dielectric function for NRs as follows:

$$\varepsilon_1 = -\left((1 - P_j) \frac{\varepsilon_m}{P_j} \right), \tag{B6}$$

where the absorption is strongly enhanced and confined close to the surface.

Geometrical factors such as shape and dimension appear in the absorption cross-section through the depolarization factors. As the aspect ratio increases, the wavelength position of the longitudinal peak is red-shifted to higher values: the corresponding colorimetric variation of the colloidal solutions of NRs having different aspect ratios is sensitive enough to be visible to the naked eye.

Since LSPR phenomenon occurs on the surface of NRs, the surrounding medium has a significant impact. Gans theory includes the refractive index of the surrounding medium through its permittivity $\varepsilon_m = n_m^2$. As n_m gets higher, the longitudinal peak position shifts to longer wavelengths.

The latter property constitutes the reason for the popular application of metal NRs in sensing. Figure B1 shows the red-shift of the wavelength position of the longitudinal peak in the absorption spectra of a colloidal dispersion of Au NRs, in mixture of water and methanol at a different methanol percentage as reported in Table B1. The different mixtures of methanol and water show different values of the refractive index, shown in Table B1, at a temperature of 19°C, as M. Martens et al. have demonstrated.[1]

Table B1.   Au NRs solutions with different methanol–water percentages in the solvent.

Methanol (%)	0	20	30	40
Refractive index	1.333	1.3381	1.3405	1.3420
λLSPRl (nm)	794.00	794.87	795.18	795.82

The dispersion of Au NRs in the four mixtures produced absorption spectra characterized by longitudinal LSPR bands centered at different wavelengths, according to the refractive index of the surrounding medium. As summarized in Fig. B2(a), and in Table B1, a very small increase of refractive index results in a small (but detectable) red shift.

Acknowledgments

Francesca Petronella, Viviana Scognamiglio, and Amina Antonacci acknowledge project INF-ACT, National Recovery, and Resilience Plan (NRRP), Mission 4 Component 2 Investment 1.3 — Call for tender No. 0000341 of 15/03/2023 of Italian Ministry of University and Research funded by the European Union — NextGenerationEU (D.D. MUR Prot n. 00015553 of 11/10/2022).

Monica Focsan and Daria Stoia acknowledge the financial support of the Romanian National Authority for Scientific Research, CNCS-UEFISCDI, project number PN-III-P2-2.1-PED-2021-3342. Filippo Pierini acknowledges the National Science Centre (NCN) SONATA BIS Project No. 2020/38/E/ST5/00456 and the National Agency for Academic Exchange (NAWA) grant no. PPI/APM/2018/1/00045/U/001.

F. Zaccagnini and L. De Sio acknowledge the NATO Science for Peace and Security Program (G5759) and the European Project EOARD 2022/2025 supported by the Air Force Office of Scientific Research, Air Force Material Command, U.S. Air Force. "Digital optical network encryption with liquid-crystal grating metasurface perfect absorbers" FA8655-22-1-7007.

References

1. World Health Organization (WHO), The top 10 causes of death, https://www.who.int/news-room/fact-sheets/detail/the-top-10-causes-of-death#:~:text=The%20top%20global%20causes (accessed 5 December 2022).

2. World Health Organization (WHO), World Antimicrobial Awarness Week 2022, https://www.who.int/campaigns/world-antimicrobial-awareness-week/2022 (accessed 5 December 2022).

3. X.-F.Chen, X. Hou, M. Xiao, L. Zhang, J.-W. Cheng, M.-L. Zhou, J.-J. Huang, J.-J. Zhang, Y.-C. Xu, and P.-R.Hsueh, Matrix-assisted laser desorption/ionization time of flight mass spectrometry (MALDI-TOF MS) analysis for the identification of pathogenic microorganisms: A review, *Microorganisms* **9**(7), 1536 (2021).

4. T. Lee, M. Mohammadniaei, H. Zhang, J. Yoon, H. K. Choi, S. Guo, P. Guo, and J.-W. Choi, Single functionalized pRNA/gold nanoparticle for ultrasensitive microRNA detection using electrochemical surface-enhanced Raman spectroscopy, *Adv. Sci.* **7**, 1902477 (2020).

5. L. J. Carter, L. V. Garner, J. W. Smoot, Y. Li, Q. Zhou, C. J. Saveson, J. M. Sasso, A. C. Gregg, D. J. Soares, T. R. Beskid, S. R. Jervey, and C. Liu, Assay techniques and test development for COVID-19 diagnosis, *ACS Central Sci.* **6**(5), 591–605 (2020).

6. E. J. Kim, H. Kim, E. Park, T. Kim, D. R. Chung, Y.-M. Choi, and M.Kang, Paper-based multiplex surface-enhanced Raman scattering detection using polymerase chain reaction probe codification, *Anal. Chem.* **93**(8), 3677–3685 (2021).

7. N. E. Dina, M. A. Tahir, S. Z. Bajwa, I. Amin, V. K. Valev, and L. Zhang, SERS-based antibiotic susceptibility testing: Towards point-of-care clinical diagnosis, *Biosens. Bioelectron.* **219**, 114843 (2023).

8. M. Potara, A. Campu, D. Maniu, M. Focsan, I. Botiz, and S. Astilean, Advanced nanostructures for microbial contaminants detection by means of spectroscopic methods. In L. Baia, Z. Pap, K. Hernadi and M. Baia (eds.), *Advanced Nanostructures for Environmental Health* (pp. 347–384). Elsevier (2020).

9. S. Zhang, C. L. Wong, S. Zeng, R. Bi, K. Tai, K. Dholakia, and M. Olivo, Metasurfaces for biomedical applications: Imaging and sensing from a nanophotonics perspective, *Nanophotonics* **10**(1), 259–293 (2021).

10. X. Li, L. Jiang, Q. Zhan, J. Qian, and S.He, Localized surface plasmon resonance (LSPR) of polyelectrolyte-functionalized gold-nanoparticles for bio-sensing, *Colloids Surf. A Physicochem. Eng. Aspects* **332**, 172–179 (2009).

11. K. M.Mayer, and J. H.Hafner, Localized surface plasmon resonance sensors, *Chem. Rev.* **111**(6), 3828–3857 (2011).

12. W.Liao, Q. Lin, Y.Xu, E.Yang, and Y. Duan, Preparation of Au@Ag core–shell nanoparticle decorated silicon nanowires for bacterial capture and sensing combined with laser induced breakdown spectroscopy and surface-enhanced Raman spectroscopy, *Nanoscale* **11**(12), 5346–5354 (2019).

13. Y. Fan, A. C. Pauer, A. A. Gonzales, and H. Fenniri, Enhanced antibiotic activity of ampicillin conjugated to gold nanoparticles on PEGylated rosette nanotubes, *Int. J. Nanomed.* **14**, 7281–7289 (2019).

14. K. Whang, J.-H. Lee, Y. Shin, W. Lee, Y. Kim, D. Kim, L. Lee, and T. Kang, Plasmonic bacteria on a nanoporous mirror via hydrodynamic trapping for

rapid identification of waterborne pathogens, *Light Sci. Appl.* **7**, Article number: 68 (2018).

15. A. P. V. Ferreyra Maillard, S. Gonçalves, N. C. Santos, B. A. López de Mishima, P. R. Dalmasso, and A. Hollmann, Studies on interaction of green silver nanoparticles with whole bacteria by surface characterization techniques, *Biochimica et Biophysica Acta (BBA) — Biomembranes* **1861**(6), 1086–1092 (2019).

16. F. Yaghubi, M. Zeinoddini, A. R. Saeedinia, A. Azizi, and A. Samimi Nemati, Design of localized surface plasmon resonance (LSPR) biosensor for immunodiagnostic of E. coli O157:H7 using gold nanoparticles conjugated to the chicken antibody, *Plasmonics* **15**(5), 1481–1487 (2020).

17. A. Svärd, J. Neilands, E. Palm, G. Svensäter, T. Bengtsson, and D. Aili, Protein-functionalized gold nanoparticles as refractometric nanoplasmonic sensors for the detection of proteolytic activity of porphyromonas gingivalis, *ACS Appl. Nano Mater.* **3**(10), 9822–9830 (2020).

18. S. Y. Oh, N. S. Heo, S. Shukla, H.-J. Cho, A. T. E. Vilian, J. Kim, S. Y. Lee, Y.-K. Han, S. M. Yoo, and Y. S. Huh, Development of gold nanoparticle-aptamer-based LSPR sensing chips for the rapid detection of Salmonella typhimurium in pork meat, *Sci. Rep.* **7**(1), 10130 (2017).

19. Y. Xu, Z. Luo, J. Chen, Z. Huang, X. Wang, H. An, and Y. Duan, Ω-Shaped fiber-optic probe-based localized surface plasmon resonance biosensor for real-time detection of Salmonella typhimurium, *Anal. Chem.* **90**(22), 13640–13646 (2018).

20. P. Halkare, N. Punjabi, J. Wangchuk, S. Madugula, K. Kondabagil, S.Mukherji, and Label-free detection of escherichia coli from mixed bacterial cultures using bacteriophage T4 on plasmonic fiber-optic sensor, *ACS Sens.* **6**(7), 2720–2727 (2021).

21. R. Funari, N. Bhalla, K.-Y. Chu, B. Söderström, and A. Q. Shen, Nanoplasmonics for real-time and label-free monitoring of microbial biofilm formation, *ACS Sens.* **3**(8), 1499–1509 (2018).

22. Q. Chen, L. Zhang, Y. Feng, F. Shi, Y. Wang, P. Wang, and L. Liu, Dual-functional peptide conjugated gold nanorods for the detection and photothermal ablation of pathogenic bacteria, *J. Mater. Chem. B* **6**(46), 7643–7651 (2018).

23. J. Zhou, F. Tian, R. Fu, Y. Yang, B. Jiao, and Y. He, Enzyme–nanozyme cascade reaction-mediated etching of gold nanorods for the detection of Escherichia coli, *ACS Appl. Nano Mater.* **3**(9), 9016–9025 (2020).

24. F. Petronella, D. De Biase, F. Zaccagnini, V. Verrina, S.-I. Lim, K.-U.Jeong, S. Miglietta, V. Petrozza, V. Scognamiglio, N. P. Godman, D. R. Evans, M. McConney, and L. De Sio, Label-free and reusable antibody-functionalized gold nanorod arrays for the rapid detection of Escherichia coli cells in a water dispersion, *Environ. Sci. Nano* **9**(9), 3343–3360 (2022).

25. J. A. Park, C. Amri, Y. Kwon, J.-H. Lee, and T. Lee, Recent advances in DNA nanotechnology for plasmonic biosensor construction, *Biosensors* **12**(6), 418 (2022).

26. N. Sattarahmady, A. Movahedpour, H. Heli, and G. R. Hatam, Gold nanoparticles-based biosensing of Leishmania major kDNA genome: Visual and spectrophotometric detections, *Sens. Actuators B: Chem.* **235**, 723–731 (2016).

27. D. Zopf, A. Pittner, A. Dathe, N. Grosse, A. Csáki, K. Arstila, J. J. Toppari, W. Schott, D. Dontsov, G. Uhlrich, W. Fritzsche, and O. Stranik, Plasmonic nanosensor array for multiplexed DNA-based pathogen detection, *ACS Sen.* **4**(2), 335–343 (2019).

28. L. Wang, W. Liu, J.-W. Tang, J.-J. Wang, Q.-H. Liu, P.-B. Wen, M.-M. Wang, Y.-C. Pan, B. Gu, and X. Zhang, Applications of Raman spectroscopy in bacterial infections: Principles, advantages, and shortcomings, *Front. Microbiol.* **12**(2021).

29. Z. Wu, H. Pu, and D.-W. Sun, Fingerprinting and tagging detection of mycotoxins in agri-food products by surface-enhanced Raman spectroscopy: Principles and recent applications, *Trends Food Sci Technol.* **110**, 393–404 (2021).

30. A. I. Pérez-Jiménez, D. Lyu, Z. Lu, G. Liu, and B. Ren, Surface-enhanced Raman spectroscopy: benefits, trade-offs and future developments, *Chem. Sci.* **11**(18), 4563–4577 (2020).

31. S. Efrima, and L. Zeiri, Understanding SERS of bacteria, *J. Raman Spectrosc.* **40**(3), 277–288 (2009).

32. X. Zhou, Z. Hu, D. Yang, S. Xie, Z. Jiang, R. Niessner, C. Haisch, H. Zhou, and P. Sun, Bacteria detection: From powerful SERS to its advanced compatible techniques, *Adv. Sci.* **7**(23), 2001739 (2020).

33. R. Prucek, A. Panáček, Ž. Gajdová, R. Večeřová, L. Kvítek, J. Gallo, and M. Kolář, Specific detection of Staphylococcus aureus infection and marker for Alzheimer disease by surface enhanced Raman spectroscopy using silver and gold nanoparticle-coated magnetic polystyrene beads, *Sci. Rep.* **11**(1), 6240 (2021).

34. T. Lemma, J. Wang, K. Arstila, V. P. Hytönen, and J. J. Toppari, Identifying yeasts using surface enhanced Raman spectroscopy, *Spectrochimica Acta Part A: Mol. Biomol. Spectrosc.* **218**, 299–307 (2019).

35. C. J. L. Constantino, T. Lemma, P. A. Antunes, and R. Aroca, Single-molecule detection using surface-enhanced resonance Raman scattering and Langmuir–Blodgett monolayers, *Anal. Chem.* **73**(15), 3674–3678 (2001).

36. K. Kneipp, H. Kneipp, G. Deinum, I. Itzkan, R. R. Dasari, and M. S. Feld, Single-molecule detection of a cyanine dye in silver colloidal solution using near-infrared surface-enhanced Raman scattering, *Appl. Spectrosc.* **52**(2), 175–178 (1998).

37. E. Witkowska, K. Niciński, D. Korsak, T. Szymborski, and A. Kamińska, Sources of variability in SERS spectra of bacteria: Comprehensive analysis of interactions between selected bacteria and plasmonic nanostructures, *Anal. Bioanal. Chem.* **411**(10), 2001–2017 (2019).

38. M. Kahraman, M. M. Yazıcı, F. Şahin, and M. Çulha, Convective assembly of bacteria for surface-enhanced Raman scattering, *Langmuir* **24**(3), 894–901 (2008).

39. Q. Zhang, X.-D. Wang, T. Tian, and L.-Q. Chu, Incorporation of multilayered silver nanoparticles into polymer brushes as 3-dimensional SERS substrates and their application for bacteria detection, *Appl. Surf. Sci.* **407**, 185–191 (2017).

40. P. Wang, S. Pang, J. Chen, L. McLandsborough, S. R. Nugen, M. Fan, L. He, Label-free mapping of single bacterial cells using surface-enhanced Raman spectroscopy, *Analyst* **141**(4), 1356–1362 (2016).

41. C. Wei, M. Li, and X. Zhao, Surface-enhanced Raman scattering (SERS) with silver nano substrates synthesized by microwave for rapid detection of foodborne pathogens, *Front. Microbiol.* **9** (2018).

42. E. Witkowska, K. Niciński, D. Korsak, B. Dominiak, J. Waluk, and A. Kamińska, Nanoplasmonic sensor for foodborne pathogens detection. Towards development of ISO-SERS methodology for taxonomic affiliation of Campylobacter spp, *J. Biophotonics* **13**(5), e201960227 (2020).

43. S. Liu, Q. Hu, C. Li, F. Zhang, H. Gu, X. Wang, S. Li, L. Xue, T. Madl, Y. Zhang, and L. Zhou, Wide-range, rapid, and specific identification of pathogenic bacteria by surface-enhanced Raman spectroscopy, *ACS Sens.* **6**(8), 2911–2919 (2021).

44. S. Bashir, H. Nawaz, M. Irfan Majeed, M. Mohsin, A. Nawaz, N. Rashid, F. Batool, S. Akbar, M. Abubakar, S. Ahmad, S. Ali, and M. Kashif, Surface-enhanced Raman spectroscopy for the identification of tigecycline-resistant E. coli strains, *Spectrochimica Acta Part A: Mol. Biomol. Spectrosc.* **258**, 119831 (2021).

45. A. B. Nowicka, M. Czaplicka, T. Szymborski, and A. Kamińska, Combined negative dielectrophoresis with a flexible SERS platform as a novel strategy for rapid detection and identification of bacteria, *Anal. Bioanal. Chem.* **413**(7), 2007–2020 (2021).

46. H. Zhao, D. Zheng, H. Wang, T. Lin, W. Liu, X. Wang, W. Lu, M. Liu, W. Liu, Y. Zhang, M. Liu, and P. Zhang, In Situ collection and rapid detection of pathogenic bacteria using a flexible SERS platform combined with a portable Raman spectrometer *Int. J. Mol. Sci.* **23**(13), 7340 (2022).

47. Y. Tanaka, E. H. Khoo, N. A. b. M. Salleh, S. L. Teo, S. Y. Ow, L. Sutarlie, and X. Su, A portable SERS sensor for pyocyanin detection in simulated wound fluid and through swab sampling, *Analyst* **146**(22), 6924–6934 (2021).

48. B. Krafft, A. Tycova, R. D. Urban, C. Dusny, and D. Belder, Microfluidic device for concentration and SERS-based detection of bacteria in drinking water, *Electrophoresis* **42**(1–2), 86–94 (2021).

49. M. Deb, R. Hunter, M. Taha, H. Abdelbary, and H. Anis, Rapid detection of bacteria using gold nanoparticles in SERS with three different capping agents: Thioglucose, polyvinylpyrrolidone, and citrate, *Spectrochimica Acta Part A: Mol. Biomol. Spectrosc.* **280**, 121533 (2022).

50. X.-Y. Wang, J.-Y. Yang, Y.-T. Wang, H.-C. Zhang, M.-L. Chen, T. Yang, and J.-H. Wang, M13 phage-based nanoprobe for SERS detection and inactivation of Staphylococcus aureus, *Talanta* **221**, 121668 (2021).

51. X. Gao, Y. Yin, H. Wu, Z. Hao, J. Li, S. Wang, and Y. Liu, Integrated SERS platform for reliable detection and photothermal elimination of bacteria in whole blood samples, *Anal. Chem.* **93**(3), 1569–1577 (2021).

52. B. Xie, Z.-P. Wang, R. Zhang, Z. Zhang, and Y. He, A SERS aptasensor based on porous Au-NC nanoballoons for Staphylococcus aureus detection, *Anal. Chim. Acta* **1190**, 339175 (2022).

53. S. Zhou, X. Guo, H. Huang, X. Huang, X. Zhou, Z. Zhang, G. Sun, H. Cai, H. Zhou, and P. Sun, Triple-function Au–Ag-stuffed nanopancakes for SERS detection, discrimination, and inactivation of multiple bacteria, *Anal. Chem.* **94**(15), 5785–5796 (2022).

54. M.-C. Yang, A. Hardiansyah, Y.-W. Cheng, H.-L. Liao, K.-S. Wang, A. Randy, C. Harito, J.-S. Chen, R.-J. Jeng, and T.-Y. Liu, Reduced graphene oxide nanosheets decorated with core-shell of Fe3O4-Au nanoparticles for rapid SERS detection and hyperthermia treatment of bacteria, *Spectrochimica Acta Part A: Mol. Biomol. Spectrosc.* **281**, 121578 (2022).

55. Z. Zhou, R. Xiao, S. Cheng, S. Wang, L. Shi, C. Wang, K. Qi, and S. Wang, A universal SERS-label immunoassay for pathogen bacteria detection based on Fe(3)O(4)@Au-aptamer separation and antibody-protein A orientation recognition, *Anal. Chim. Acta* **1160**, 338421 (2021).

56. E. Yang, D. Li, P. Yin, Q. Xie, Y. Li, Q. Lin, and Y. Duan, A novel surface-enhanced Raman scattering (SERS) strategy for ultrasensitive detection of bacteria based on three-dimensional (3D) DNA walker, *Biosens. Bioelectron.* **172**, 112758 (2021).

57. S.-M. You, K. Luo, J.-Y. Jung, K.-B. Jeong, E.-S. Lee, M.-H. Oh, Y.-R. and Kim, Gold Nanoparticle-Coated Starch Magnetic Beads for the Separation, Concentration, and SERS-Based Detection of E. coli O157:H7, *ACS Appl. Mater. & Interf.* **12**(16), 18292–18300 (2020).

58. D. He, Z. Wu, B. Cui, E. Xu, and Z. Jin, Establishment of a dual mode immunochromatographic assay for Campylobacter jejuni detection, *Food Chem.* **289**, 708–713 (2019).

59. S. Gao, J. Wu, H. Wang, S. Hu, and L. Meng, Highly sensitive detection of Cronobacter sakazakii based on immunochromatography coupled with surface-enhanced Raman scattering, *J. Dairy Sci.* **104**(3), 2748–2757 (2021).

60. W. Shen, C. Wang, S. Zheng, B. Jiang, J. Li, Y. Pang, C. Wang, R. Hao, and R. Xiao, Ultrasensitive multichannel immunochromatographic assay for rapid detection of foodborne bacteria based on two-dimensional film-like SERS labels. *J. Hazard Mater.* **437**, 129347 (2022).

61. S. Asgari, R. Dhital, A. Mustapha, and M. Lin, Duplex detection of foodborne pathogens using a SERS optofluidic sensor coupled with immunoassay, *Int. J. Food Microbiol.* **383**, 109947 (2022).

62. H. Kearns, R. Goodacre, L. E. Jamieson, D. Graham, and K. Faulds, SERS detection of multiple antimicrobial-resistant pathogens using nanosensors, *Anal. Chem.* **89**(23), 12666–12673 (2017).

63. C. Zhang, C. Wang, R. Xiao, L. Tang, J. Huang, D. Wu, S. Liu, Y. Wang, D. Zhang, S. Wang, and X. Chen, Sensitive and specific detection of clinical

bacteria via vancomycin-modified Fe3O4@Au nanoparticles and aptamer-functionalized SERS tags, *J. Mater. Chem B* **6**(22), 3751–3761 (2018).

64. D. He, Z. Wu, B. Cui, E. Xu, and Z. Jin, Establishment of a dual mode immunochromatographic assay for Campylobacter jejuni detection, *Food Chem.* **289**, 708–713 (2019).

65. N. Duan, M. Shen, S. Qi, W. Wang, S. Wu, and Z. Wang, A SERS aptasensor for simultaneous multiple pathogens detection using gold decorated PDMS substrate, *Spectrochimica Acta Part A: Mol. Biomol. Spectrosc.* **230**, 118103 (2020).

66. M. Rippa, R. Castagna, D. Sagnelli, A. Vestri, G. Borriello, G. Fusco, J. Zhou, and L. Petti, SERS biosensor based on engineered 2D-aperiodic nanostructure for in-situ detection of viable brucella bacterium in complex matrix, *Nanomaterials* **11**(4), 886 (2021).

67. C. Wang, C. Wang, J. Li, Z. Tu, B. Gu, and S. Wang, Ultrasensitive and multiplex detection of four pathogenic bacteria on a bi-channel lateral flow immunoassay strip with three-dimensional membrane-like SERS nanostickers, *Biosens. Bioelectron.* **214**, 114525 (2022).

68. X. Ma, S. Xu, L. Li, and Z. Wang, A novel SERS method for the detection of Staphylococcus aureus without immobilization based on Au@Ag NPs/slide substrate, *Spectrochimica Acta Part A: Mol. Biomol. Spectrosc.* **284**, 121757 (2023).

69. E. Mauriz, Recent progress in plasmonic biosensing schemes for virus detection, *Sensors* [Online] (2020).

70. L.-L. Qu, Y.-L. Ying, R.-J. Yu, and Y.-T. Long, In situ food-borne pathogen sensors in a nanoconfined space by surface enhanced Raman scattering, *Microchimica Acta* **188**(6), 201 (2021).

71. World Health Organization (WHO) Coronavirus disease (COVID-19) pandemic https://www.who.int/emergencies/diseases/novel-coronavirus-2019 (accessed 12 December 2022).

72. F. Saviñon-Flores, E. Méndez, M. López-Castaños, A. Carabarin-Lima, K. A. López-Castaños, M. A. González-Fuentes, and A. Méndez-Albores, A review on SERS-based detection of human virus infections: Influenza and coronavirus, *Biosensors* **11**(3), 66 (2021).

73. Y. Gao, Y. Han, C. Wang, L. Qiang, J. Gao, Y. Wang, H. Liu, L. Han, and Y. Zhang, Rapid and sensitive triple-mode detection of causative SARS-CoV-2 virus specific genes through interaction between genes and nanoparticles, *Anal. Chim. Acta* **1154**, 338330 (2021).

74. M. Mohammadi, D. Antoine, M. Vitt, J. M. Dickie, S. Sultana Jyoti, J. G. Wall, P. A. Johnson, and K. E. Wawrousek, A fast, ultrasensitive SERS immunoassay to detect SARS-CoV-2 in saliva, *Anal. Chim. Acta* **1229**, 340290 (2022).

75. O. J. Achadu, F. Abe, F. Hossain, F. Nasrin, M. Yamazaki, T. Suzuki, and E. Y. Park, Sulfur-doped carbon dots@polydopamine-functionalized magnetic silver nanocubes for dual-modality detection of norovirus, *Biosens. Bioelectron.* **193**, 113540 (2021).

76. O. J. Achadu, F. Abe, T. Suzuki, and E. Y. Park, Molybdenum triox-
    ide nanocubes aligned on a graphene oxide substrate for the detection
    of norovirus by surface-enhanced Raman scattering, *ACS Appl. Mater.
    Interfaces* **12**(39), 43522–43534 (2020).

77. M. Liu, C. Zheng, M. Cui, X. Zhang, D. P. Yang, X. Wang, and
    D. Cui, Graphene oxide wrapped with gold nanorods as a tag in a SERS
    based immunoassay for the hepatitis B surface antigen, *Mikrochimica Acta*
    185(10), 458 (2018).

78. F. Petronella, A. Antonacci, and V. Scognamiglio, Nanoparticles in biosen-
    sor design for the agrifood sector. In L. Fernandes Fraceto, H. W. Pereira de
    Carvalho, R. de Lima, S. Ghoshal, C. Santaella (eds.), *Inorganic Nanopes-
    ticides and Nanofertilizers: A View from the Mechanisms of Action to
    Field Applications* (pp. 213–251). Cham: Springer International Publishing
    (2022).

79. M. De Felice, M. De Falco, D. Zappi, A. Antonacci, and V. Scognamiglio,
    Isothermal amplification-assisted diagnostics for COVID-19, *Biosens. Bio-
    electron.* **205**, 114101–114101 (2022).

80. M. De Falco, M. De Felice, F. Rota, D. Zappi, A. Antonacci, and
    V. Scognamiglio, Next-generation diagnostics: Augmented sensitivity in
    amplification-powered biosensing. *TrAC Trends Anal. Chem.* **148**, 116538
    (2022).

81. C. Carrell, A. Kava, M. Nguyen, R. Menger, Z. Munshi, Z. Call, M. Nuss-
    baum, and C. Henry, Beyond the lateral flow assay: A review of paper-based
    microfluidics, *Microelectron. Eng.* **206**, 45–54 (2019).

82. A. Güner, E. Çevik, M. Şenel, and L. Alpsoy, An electrochemical
    immunosensor for sensitive detection of Escherichia coli O157:H7 by using
    chitosan, MWCNT, polypyrrole with gold nanoparticles hybrid sensing
    platform, *Food Chem.* **229**, 358–365 (2017,).

83. S. Ranjbar, and S. Shahrokhian, Design and fabrication of an electro-
    chemical aptasensor using Au nanoparticles/carbon nanoparticles/cellulose
    nanofibers nanocomposite for rapid and sensitive detection of Staphylococ-
    cus aureus, *Bioelectrochemistry* **123**, 70–76 (2018).

84. C. Y. Yu, G. Y. Ang, K. G. Chan, K. K. Banga Singh, and Y. Y. Chan,
    Enzymatic electrochemical detection of epidemic-causing Vibrio cholerae
    with a disposable oligonucleotide-modified screen-printed bisensor coupled
    to a dry-reagent-based nucleic acid amplification assay, *Biosens. Bioelectron.*
    70, 282–288 (2015).

85. Y. M. Park, J. Park, S. Y. Lim, Y. Kwon, N. H. Bae, J.-K. Park,
    and S. J. Lee, Integrated pumpless microfluidic chip for the detection of
    foodborne pathogens by polymerase chain reaction and electrochemical
    analysis, *Sens. Actuators B: Chem.* **329**, 129130 (2021).

86. H. Liu, Y. He, and K. Cao, Flexible surface-enhanced Raman scattering
    substrates: A review on constructions, applications, and challenges, *Adv.
    Mater. Interf.* **8**(21), 2100982 (2021).

87. M. Xu, D. Obodo, and V. K. Yadavalli, The design, fabrication, and
    applications of flexible biosensing devices, *Biosensors and Bioelectronics*
    124–125, 96–114 (2019).

88. Y. Ziai, C. Rinoldi, P. Nakielski, L. De Sio, and F. Pierini, Smart plasmonic hydrogels based on gold and silver nanoparticles for biosensing application, *Curr. Opin. Biomed. Eng.* **24**, 100413 (2022).

89. L. Matlock-Colangelo, and A. J. Baeumner, Recent progress in the design of nanofiber-based biosensing devices, *Lab Chip* **12**(15), 2612–2620 (2012).

90. A. Zakrzewska, M. A. Haghighat Bayan, P. Nakielski, F. Petronella, L. De Sio, and F. Pierini, Nanotechnology transition roadmap toward multifunctional stimuli-responsive face masks *ACS Appl. Mater.* **14**(41), 46123–46144 (2022).

91. M. A. Haghighat Bayan, F. Afshar Taromi, M. Lanzi *et al.*, Enhanced efficiency in hollow core electrospun nanofiber-based organic solar cells. *Sci. Rep.* **11**, 21144 (2021).

92. M. A. Haghighat Bayan, Y. J. Dias, C. Rinoldi, P. Nakielski, D. Rybak, Y. B. Truong, A. L. Yarin, and F. Pierini, *J. Polym. Sci.* **61**(7), 1–13 (2023).

93. L. Wang, Y. Zhang, W. Zhang, T. Ren, F. Wang, and H. Yang, Laser-induced plasmonic heating on silver nanoparticles/poly(N-isopropylacrylamide) mats for optimizing SERS detection, *J. Raman Spectrosc.* **48**(2), 243–250 (2017).

94. J. Shi, T. You, Y. Gao, X. Liang, C. Li, and P. Yin, Large-scale preparation of flexible and reusable surface-enhanced Raman scattering platform based on electrospinning AgNPs/PCL nanofiber membrane, *RSC Adv.* **7**(75), 47373–47379 (2017).

95. M. S. S. Bharathi, C. Byram, D. Banerjee, D. Sarma, B. Barkakaty, and V. R. Soma, Gold nanoparticle nanofibres as SERS substrate for detection of methylene blue and a chemical warfare simulant (methyl salicylate), *Bull. Mater. Sci.* **44**(2), 103 (2021).

96. M. Cao, S. Cheng, X. Zhou, Z. Tao, J. Yao, and L.-J. Fan, Preparation and surface-enhanced Raman performance of electrospun poly(vinyl alcohol)/high-concentration-gold nanofibers, *J. Polym. Res.* **19**(1), 9810 (2011).

97. D. Chen, L. Zhang, P. Ning, H. Yuan, Y. Zhang, M. Zhang, T. Fu, and X. He, In-situ growth of gold nanoparticles on electrospun flexible multilayered PVDF nanofibers for SERS sensing of molecules and bacteria, *Nano Res.* **14**(12), 4885–4893 (2021).

98. A. N. Severyukhina, B. V. Parakhonskiy, E. S. Prikhozhdenko, D. A. Gorin, G. B. Sukhorukov, H. Möhwald, and A. M. Yashchenok, Nanoplasmonic Chitosan nanofibers as effective SERS substrate for detection of small molecules, *ACS Appl. Mater. Interf.* **7**(28), 15466–15473 (2015).

99. P. Jia, J. Qu, B. Cao, Y. Liu, C. Luo, J. An, and K. Pan, Controlled growth of polyhedral and plate-like Ag nanocrystals on a nanofiber mat as a SERS substrate, *Analyst* **140**(15), 5190–5197 (2015).

100. S. Fathi, R. Saber, M. Adabi, R. Rasouli, M. Douraghi, M. Morshedi, and R. Farid-Majidi, Novel competitive voltammetric aptasensor based on electrospun carbon nanofibers-gold nanoparticles modified graphite electrode for Salmonella enterica serovar detection, *Biointerface Res. Appl. Chem.* **11**, 8702–8715 (2020).

101. Z. Liu, Z. Yan, L. Jia, P. Song, L. Mei, L. Bai, and Y. Liu, Gold nanoparticle decorated electrospun nanofibers: A 3D reproducible and sensitive SERS substrate, *Appl. Surf. Sci.* **403**, 29–34 (2017).
102. l. Wang, Y. Sun, J. Wang, and Z. Li, Assembly of gold nanoparticles on electrospun polymer nanofiber film for SERS applications, *Bull. Korean Chem. Soc.* **35**, 30–34 (2014).
103. L. Zhang, X. Li, W. Liu, R. Hao, H. Jia, Y. Dai, M. Usman Amin, H. You, T. Li, and J. Fang, Highly active Au NP microarray films for direct SERS detection, *J. Mater. Chem. C* **7**(48), 15259–15268 (2019).
104. M. Shen, N. Duan, S. Wu, Y. Zou, and Z. Wang, Polydimethylsiloxane gold nanoparticle composite film as structure for aptamer-based detection of vibrio parahaemolyticus by surface-enhanced Raman spectroscopy, *Food Anal. Methods* **12**(2), 595–603 (2019).
105. X. Liu, X. Wang, L. Zha, D. Lin, J. Yang, J. Zhou, and L. Zhang, Temperature- and pH-tunable plasmonic properties and SERS efficiency of the silver nanoparticles within the dual stimuli-responsive microgels, *J. Mater. Chem. C* **2**(35), 7326–7335 (2014).
106. K. Wang, D.-W. Sun, H. Pu, Q. Wei, and L. Huang, Stable, flexible, and high-performance SERS chip enabled by a ternary film-packaged plasmonic nanoparticle array, *ACS Appl. Mater. Interf.* **11**(32), 29177–29186 (2019).
107. L.-B. Zhong, J. Yin, Y.-M. Zheng, Q. Liu, X.-X. Cheng, and F.-H. Luo, Self-assembly of au nanoparticles on PMMA template as flexible, transparent, and highly active SERS substrates, *Anal. Chem.* **86**(13), 6262–6267 (2014).
108. J. Sun, Z. Zhang, C. Liu, X. Dai, W. Zhou, K. Jiang, T. Zhang, J. Yin, J. Gao, H. Yin, and H. Li, Continuous in situ portable SERS analysis of pollutants in water and air by a highly sensitive gold nanoparticle-decorated PVDF substrate, *Anal. Bioanal. Chem.* **413**(21), 5469–5482 (2021).
109. D. Chen, P. Ning, Y. Zhang, J. Jing, M. Zhang, L. Zhang, J. Huang, X. He, T. Fu, Z. Song, G. He, D. Qian, and X. Zhu, Ta@Ag porous array with high stability and biocompatibility for SERS sensing of bacteria, *ACS Appl. Mater. Interf.* **12**(17), 20138–20144 (2020).
110. X. Zhou, H. Li, G. Yu, Y. Chen, Y. Wang, Z. Zeng, and L. Chi, A highly-efficient, stable, and flexible Kapton tape-based SERS chip, *Mater. Chem. Front.* **5**(17), 6471–6475 (2021).
111. Y. Ziai, F. Petronella, C. Rinoldi, P. Nakielski, A. Zakrzewska, T. A. Kowalewski, W. Augustyniak, X. Li, A. Calogero, I. Sabała, B. Ding, L. De Sio, and F. Pierini, Chameleon-inspired multifunctional plasmonic nanoplatforms for biosensing applications, *NPG Asia Mater.* **14**(1), 18 (2022).
112. S. Lu, T. You, N. Yang, Y. Gao, and P. Yin, Flexible SERS substrate based on Ag nanodendrite–coated carbon fiber cloth: simultaneous detection for multiple pesticides in liquid droplet. *Anal. Bioanal. Chem.* **412**(5), 1159–1167 (2020).

Chapter 7

A Double Plasmonic/Photonic Approach for Multilevel Anticounterfeit and Food Safety Applications

Antonio De Luca*,†,**, Vincenzo Caligiuri*,†,††, Aniket Patra*,‡,
Maria P. De Santo*,†, Agostino Forestiero§, Giuseppe Papuzzo§,
Dante M. Aceti¶, Giuseppe E. Lio‖, and Riccardo Barberi*,†

*Department of Physics, University of Calabria,
via P. Bucci, 31C, 87036, Rende (CS), Italy
†CNR Nanotec UOS Rende, via P. Bucci, 31D,
87036, Rende (CS), Italy
‡Istituto Italiano di Tecnologia, via Morego 30,
16163 Genova (GE), Italy
§CNR-ICAR, Institute for High Performance and Networking,
via P. Bucci 8-9c, 87036 Rende (CS), Italy
¶Institute of Electronics, Bulgarian Academy of Sciences,
72, Tsarigradsko Chaussee blvd., 1784 Sofia, Bulgaria
‖CNR-INO and European Laboratory for Non Linear Spectroscopy (LENS),
Via Nello Carrara, 1 - 50019,
Sesto Fiorentino, Firenze (FI), Italy
**antonio.deluca@unical.it
††vincenzo.caligiuri@unical.it

Cutting-edge strategies used for authenticating goods are of fundamental importance considering the increasing counterfeiting levels. Physical Unclonable Functions (PUFs) constitute the frontier approach to face such a problem. PUFs are typically engineered in the framework of electronics, but alternative and more sophisticated strategies are taking hold, especially involving optics and nanotechnologies. In this chapter, we propose a hybrid plasmonic/photonic PUF working at the nanoscale. This approach combines the unique and non-reproducible morphology of a functional nanostructured surface with a resonant response to achieve a three-level challenge–response paradigm. The structure consists of a resonant cavity where the top mirror is replaced with a layer of plasmonic silver nano-islands. The naturally random spatial distribution of clusters and nanoparticles formed by this deposition technique constitutes the manufacturer-resistant nanoscale morphological fingerprint of the proposed PUF. The presence of Ag nano-islands allows to tailor the interplay between photonic and plasmonic modes to achieve two additional security levels. The first one is constituted by the chromatic response and broad iridescence of our structures, while the second, by their rich spectral response, accessible even through a common smartphone light-emitting diode (LED). The proposed architectures could also be used as an irreversible and quantitative temperature exposure label in the framework of food safety applications. They are inexpensive, scalable, and can be deposited over different substrates, envisioning the way toward a plethora of possible applications.

1. Introduction

In a study called "Trends in Trade in Counterfeit and Pirated Goods", the Organization for Economic Co-operation and Development (OECD) certified that the trade in counterfeit goods settles around 3.3% of global world trade, in 2019. Such a trend is rapidly increasing. The new emerging technological scenarios, like 5G, the internet of things, and distributed ledger technologies, require a new approach to secure intellectual properties as well as the authenticity of the traded goods. In this respect, a great promise is held by the Physical Unclonable Functions (PUFs).[1-4] PUFs are individual physical signatures whose intrinsic unpredictability produces a unique and unclonable response to a specific challenge. In general, a PUF is a physical system whose internal structure manifests random disorder and uncontrollable manufacturing variations which make it unique and "unclonable" for users, counterfeiters, and even for the original PUF-manufacturers. PUFs can be interrogated following a challenge–response scheme. Depending on the number of independent challenge–response pairs it offers, a PUF can be categorized as "weak" or "strong", according to a recently proposed taxonomy.[5]

The fulcrum over which a PUF leverages is constituted by the unavoidable randomness introduced in manufacturing processes, over which the operator or the designer has no control. In this respect, micro- and nanofabrication processes are the ideal reservoirs in which to look for user-independent randomness since, despite their reliability in terms of device operation functionalities, common lithographic techniques introduce a certain level of unpredictability.[6,7] A PUF respecting such prerequisites is often regarded as a manufacturer-resistant PUF and holds the highest level of reliability.[4] As a result, Silicon-based Metal-Oxide-Semiconductor (MOS) technology offered historically the first examples of PUFs.[8,9]

Field-Effect Transistors (FET), even if equipped with profound structural innovations, still remain the platform of election for the embodiment of PUFs.[10-16] It is however true that the always increasing necessity to encrypt and authenticate goods in which transistors, ring-oscillators, or electronic systems embedding CMOS-related PUFs do not find place makes it paramount to look for innovative materials and strategies.[17] In this respect, the interaction between light and matter constitutes a fertile ground over which new approaches to PUFs are being developed, sometimes called "optical PUFs".[18] Optical weak PUFs have been proposed employing different materials and relevant examples include nanostructures like nanoparticles,[19] random silver nanowires,[20] or organic nanolaser arrays.[21] Countless chemical processes have been harnessed as well to generate tags of different natures as optical PUFs.[22,23] Fluorescence has been widely used as an optical anti-counterfeit function.[24-27] Photoluminescence occurring in perovskite.[28] and inkjet-printed quantum dots[29] has been recently used as a PUF. Biological systems consisting of colonized populations of T-cells have also been demonstrated as high-security level biological unclonable functions.[30]

Colors represent a powerful tool for labeling, sometimes univocally characterizing the identity of the goods they are associated with. Red-Ferrari, for example, is perhaps the most recognizable feature of the brand, representing a strong authentication feature de facto. Colors, iridescence, and holography have been recognized to be so effectively strong in protecting and authenticating that practically the totality of banknotes adopts a color-related authentication label as one of the security gate-keepers.[31] The world of nanotechnologies offers, in this respect, unique possibilities. In particular, fields of nanotechnology in which peculiar chromatic effects stem from unusual properties of nanometric resonators are of interest for this purpose.[32-45] They, indeed, intrinsically offer the possibility of a double signature: (i) the chromatic one, represented by the specific colorful response of the system, and (ii) its spectral response.[46,47] While the former is very easy to challenge by a quick visual investigation, the latter requires slightly more sophisticated approaches, which often rely on spectroscopic analysis of the label. To this family, plasmonic materials belong. The capability of plasmonic nanoparticles (PNPS) to be used as anti-counterfeit labels has, indeed, been recently investigated. Localized coherent oscillations of electrons occurring in PNPS endow them with a resonant and colorful optical response that has been widely used all across human history, more or less consciously. The chromatic response of PNPS, together with their size and random positioning over a substrate, constitutes a powerful triple approach to producing a robust PUF.[47-49] However, to optimize the performance as PUFs and introduce a significant aleatory effect in their disposition over a substrate, very dispersed PNPS solutions have to be used, so that their inter-distance and clustering remain random. Under these conditions, the macroscopic chromatic response results in faint colors that necessitate a dark-field-microscopy investigation to be fully appreciated, ruling out the possibility of carrying out a simple and quick naked-eye challenge.[47]

In this chapter, we build on the concept of exploiting the plasmonic properties of nanostructured metallic surfaces in a hybrid plasmonic system engineered to be a three-level strong PUF. The proposed architecture is sketched in Fig. 1(a). The plasmonic PUF is structured to provide three different authentication levels: (i) chromatic (Fig. 1(b)), (ii) spectral (Fig. 1(c)), and (iii) morphological (Fig. 1(d)). This represents also one of the main advancements offered by our approach with respect to the state-of-the-art since small-scale devices have usually been thought to be

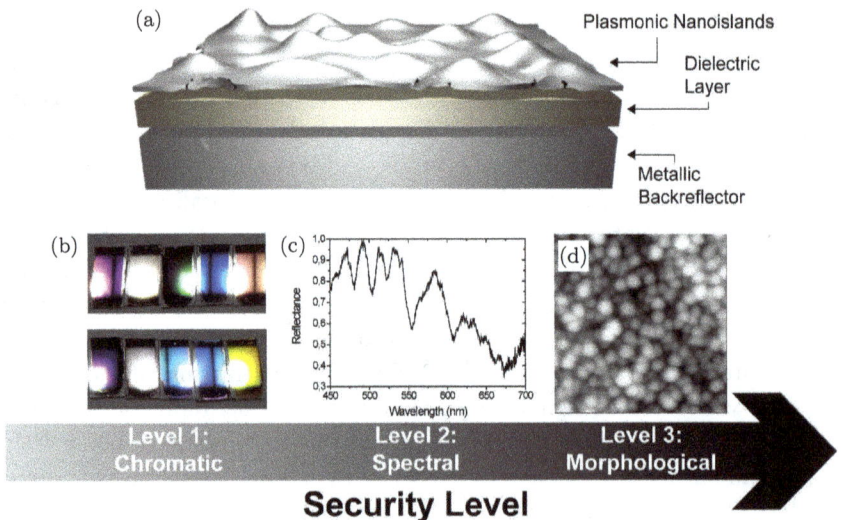

Figure 1. (a) Sketch of the Plasmonic/Photonic PUF architecture. (b) Chromatic and lightness effect, used as the "Level 1" chromatic signature. (c) Typical spectral response of a sample resembling the architecture shown in (a). This constitutes the "Level 2" spectral signature. (d) Ag nano-island morphology revealed via Atomic Force Microscopy analysis, constituting the "Level 3" nanoscale morphological signature. Image has been reproduced from the original work by V. Caligiuri et al., ACS Appl. Mater. Interfaces 2021, 13, 49172–49183, under the Creative Commons License.

unsuitable as strong PUF due to the intrinsic difficulty in introducing multilevel challenge–response pairs in a nano-metric device.

Here, the paradigm shift resides in scaling the challenge–response pairs from macroscopic to nanometric effects rather than the device itself. The technological core of the idea proposed hereafter lies, indeed, in Ag nano-islands obtained by depositing Ag layers below their percolation threshold, via a DC magnetron sputtering technique.[50–55] Such a procedure allows to obtain Ag nanoclusters whose density and size depend on the sputtering deposition parameters, but whose spatial disposition and clustering effect are random. There is, indeed, no possibility for the operator to gain control over this feature, which, therefore, constitutes the ideal morphological fingerprint. Moreover, Ag nano-islands unify a marked chromatic response with a peculiar plasmonic behavior. Here we demonstrate that, when used as the top layer in a multilayered configuration involving also one or more metal/insulator/metal (MIM) resonators, the interplay between photonic cavity resonances and plasmonic modes gives rise to a glaring and distinct chromatic response. We characterized it in both the CIE 1931 and the CIELAB color space, demonstrating that color tints with a broad variety of lightness-hue-saturation values can be obtained. Such a property, combined with a marked iridescence, gives rise to a chromatic response that is unique and cannot be reproduced by commercial paints, representing level 1 of authentication of the proposed PUF. This first level is also the one thought to be readily accessible to the customer. The exceptional chromatic properties manifested by our samples stem from their spectrally rich optical response (Fig. 1(c)), which we characterized both by sophisticated ellipsometric measurements and by using the LED flash lamp of a smartphone in a simple spectroscopic setup. This latter investigation demonstrates the capability of engineering portable spectroscopic systems with which to equip a smartphone and provide the customer easy access also to the spectral fingerprint. In the end, the strongest and most innovative authentication level is given by what we called the morphological fingerprint, which is constituted by the spatial disposition of Ag nano-islands (Fig. 1(d)). The morphology of a predetermined precise area of the surface of a sample on top of which Ag nano-islands have been deposited via sub-percolation threshold sputtering deposition has been investigated through Atomic Force Microscopy (AFM) measurements. The outcome image has been registered as a fingerprint to be compared via a neuromorphic imaging recognition algorithm with those of different areas. We found out that the number of tags recognized by the algorithm by comparing two images of the same area, acquired with two different AFM measurements (positive recognition), is more than two of magnitude larger than the number of tags recognized by comparing AFM measurements belonging to two different areas of the patterned surface (negative recognition). The distance between the two expectation values of the related distributions is one of the largest reported so far, confirming exceptional robustness of the proposed technique to false positive and false negative recognitions. We also simulate the critical experimental conditions in which the AFM operator could, by mistake, non-concentrically rotate the sample before performing the AFM measurement. Even in this case, our technique ensures an expectation value of recognized tags in the positive recognition process more than one order of magnitude larger than the negative recognition case. In the end, we demonstrate that our plasmonic PUF could work as a thermal exposure label to be used, for example, in the framework of food safety. We show, indeed, that when the sample is exposed to temperatures higher than a certain threshold, its color and optical response change dramatically and irreversibly in a deterministic and quantitatively measurable way.

The new concept reported in this chapter could potentially impact the PUF blockchain, introducing a new way of securing a broad variety of goods, since this architecture could potentially be evaporated over every kind of metallic substrate, acting as a back-reflector, or with the possibility of placing all the layers on a flexible substrate. The deposition technique is readily scalable from nano-to-wafer size, opening to the labeling of every kind of substrate, from chip-embedded to commercial electronics device size, up to plastics and bio-polymers used for food storage. The possibility of producing plasmonic nano-islands is not a prerogative of Ag. All kinds of metals could be used, such as Al and Cu, for example, to achieve inexpensive plasmonic labeling. In the end, the exceptional randomness of the proposed surfaces holds great promise in the field of cryptography and could be potentially used as a morphological encrypting matrix to carry out morpho-cryptography.

2. The Double Plasmonic/Photonic Approach

Leveraging on the unique features of nanotechnology is one of the most promising approaches to providing substantial progress to PUFs. Under this point of view, plasmonics offer several advantages. First of all, plasmonics is synonym for colors. Although their nature is completely different from that of natural colors and tints, plasmonic colors (sometimes also called "structural" colors) offer exceptional customizability in terms of hue, saturation, and brightness with the incomparable advantage of non-degradability over time. In particular, plasmonic colors emerge from resonance mechanisms occurring in a single plasmonic element or as a consequence of the interactions between neighboring elements as well as between plasmonic particles and other resonant systems like photonic cavities. The latter case is the one we consider. Such a hybrid approach takes the best of the two worlds: on the one hand, it benefits from the collective response of many, randomly placed, plasmonic resonators (Ag nano-islands), and, on the other hand, takes advantage of the high-quality-factor characteristics typical of photonic cavities.

Such a sophisticated electromagnetic scenario allows to consider three main features as candidates for PUF encryption levels: (i) the obtained structured color together with its iridescence, (ii) a spectral response with rich well-recognizable features, and (iii) a random morphology, perfect as a morphological bar-code. In the following subsections, a description of each of these three features as a PUF level will be provided in detail.

2.1. *The first physical unclonable function level: The chromatic signature*

Silver nano-islands manifest a marked size-dependent chromatic response. The size of Ag nano-islands can be easily controlled by changing the sputtering deposition time, as shown in the next section. The chromatic response of Ag nano-islands stems from their plasmonic properties. As such, the larger their size, the more the resonance shifts toward the red band of the optical spectrum. Despite the broad tunability of the plasmonic resonance of Ag nano-islands, the hue and saturation levels of the obtained colors are quite limited. Such a feature, together with a very low lightness, contributes to providing a "matte" color nuance. A far more variegated color gamma, together with a broad angle-dependence (iridescence), can be obtained by composite resonant structures made of one or many metal/insulator/metal (MIM) multilayered cavities in which the top metallic layer is replaced with Ag nano-islands (Fig. 2(a)). Metal/Dielectric/Nano-islands (MIN) are multilayered structures that allow to tailor the interplay between proper Fabry–Pérot modes, hosted by MIM cavities, and plasmonic modes sustained by Ag nano-islands. Such an interaction can be controlled by acting on both the Fabry–Pérot mode and the plasmonic sides. The former

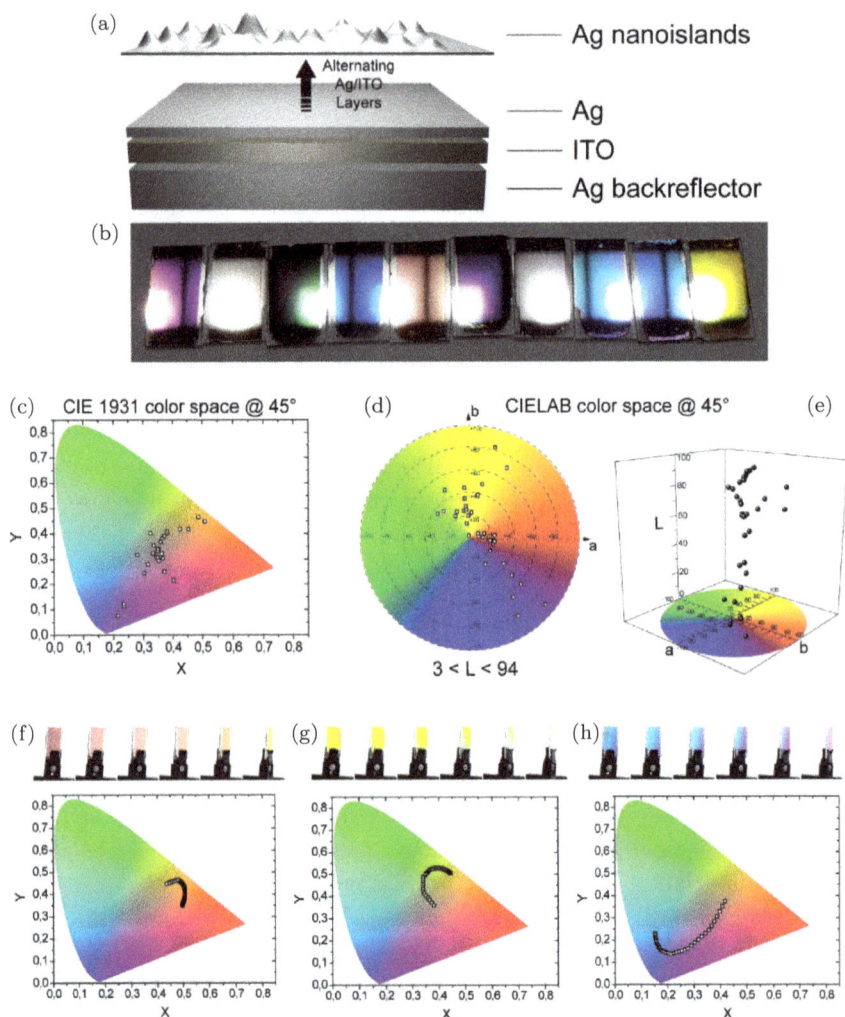

Figure 2.   (a) Sketch of the Metal/Insulator/Nano-Islands (MIN) multilayered configuration. (b) Picture showing several MIN structures selected to highlight the broad
chromatic palette achievable with this architecture. The picture highlights also the
metalized character of the obtained colors. The features appreciable in (b) can be
quantified by means of (c) CIE 1931 and (d) 2D and (e) 3D CIELAB colorspace analysis.
(d) and (e) allow to realize the vast hue-saturation-lightness combinations obtainable
by means of MIN structures. (f–h) Angle-dependent chromaticity (iridescence) of three
significant MIN structures whose values are centered in the red (f), the yellow (g), and
the blue (h) color gamma, together with the CIE 1931 colorspace analysis of their angle-
dependent chromaticity (iridescence). Image has been reproduced from the original work
by V. Caligiuri *et al.*, *ACS Appl. Mater. Interfaces* 2021, 13, 49172–49183, under the
Creative Common License.

task can be accomplished by playing with: (i) the number of resonators (MIM cavities by which the entire structure is composed),[56] (ii) thickness, and (iii) the refractive index of the employed dielectric layers.[57-59] The latter aim could be pursued by changing the sputtering deposition time of the nano-islands. A variation of the thickness and/or refractive index of the dielectric layer induces a blue-shift of the resonance of the single MIM resonator.[57,59-64]

A variation of the number of multilayers by which the MIN multilayered is composed, enriches the chromatic response, fostering the insurgence of novel resonant modes through cavity modes hybridization.[56,58,65,66]

In the end, longer sputtering deposition times induce a red-shift of the plasmonic resonance of the Ag nano-islands (see the following section, Fig. 3(h)).

The interplay between Fabry–Pérot and plasmonic resonances endows the multilayered MIN structures with a variegated chromatic response, which is perfectly captured in both the CIE 1931 (Fig. 2(c)) and CIELAB color space (Figs. 2(d) and 2(e)), united to unique iridescence (Figs. 2(f)–2(h)). The CIELAB diagram (Figs. 2(d) and 2(e)) is particularly effective in capturing the chromatic versatility of the proposed nanostructures, demonstrating the possibility to cover practically the entire saturation (radius of the circle) and luminescence (Z-Axis of Fig. 2(e)) range. It is, on the contrary, more complicated to obtain a green tint. This is mainly due to the physics governing the formation of resonances in such samples. To obtain a green tint, indeed, full absorption in both the blue and red range has to be provided together with sharp reflectance in the green range. However, even though obtaining a sharp reflectance peak in the green range is quite easy, it is very difficult to suppress the red and the blue component simultaneously. As a result, it is very easy to obtain hues in the blue, violet, and yellow range while obtaining pure green tints is rather difficult. The sophisticated chromatic scenario just described constitutes a distinct yet very hard to counterfeit signature of a potentially labeled object. The proposed structures, however, intrinsically possess a deeper security level which is constituted by their iridescence. Iridescence can be understood as the dependence of the chromaticity of an object on the observation angle. In our structures, this feature immediately translates into a modification of the reflectivity as a function of the angle. Figures 2(f)–2(h) show how the perceived color of three representative structures changes with the observation angle. To showcase such a feature, we selected three structures. The one shown in Fig. 2(f) manifests color hue in the red range, while the ones in Figs. 2(g) and 2(h) in the yellow and blue range, respectively. All of them undergo a marked color hue change while being rotated (see the stop motion sequence, made of six images for each sample, on the top of the corresponding color space map). To quantify such an effect, we drafted the CIE 1931 color space characterization for each sample. In such an analysis, each point corresponds to the CIE value of the color of the sample obtained at a particular angle, from 0° to 90° angle. The broadly customizable chromatic response, together with their iridescence, constitutes the first level of the complete plasmonic PUF. Reproducing with a common paint the particular lightness-hue-saturation characteristics of the colors obtained via our plasmonic architectures is, indeed, extremely difficult and it would be impossible to obtain the same iridescence.

2.2. Morphological and plasmonic characterization of Ag nano-islands

The Ag nano-islands can be obtained by depositing, via DC Magnetron sputtering, layers whose thickness is below the percolation threshold of Ag. This prevents the formation of a smooth and uniform film, while fostering the formation of randomly organized clusters of Ag nanoparticles we called nano-islands. Increasing the sputtering deposition time brings to denser packing of nano-islands, eventually forming a uniform Ag layer. Figures 3(a)–3(f) show Scanning Electron Micrograph (SEM) analysis of Ag nano-islands layers obtained via DC magnetron sputtering, with deposition time of (a) 5 s, (b) 10 s, (c) 15 s, (d) 30 s, (e) 40 s and (f) 60 s. The interaction between neighboring particles, together with the specific plasmonic resonance manifested by different sized clusters, contribute to provide a cooperative macroscopic plasmonic response whose signature is constituted by the reflectance dips shown in Fig. 3(h), occurring at the wavelengths highlighted in Fig. 3(i). From such a macroscopic plasmonic effect, a peculiar chromatic response is generated. The plasmonic response is, however, quite weak, leading to a recognizable but limited color gamut, as evidenced in the CIE 1931 diagram of Fig. 3(j), with very low saturation and lightness levels, as revealed by the CIELAB analysis shown in Fig. 3(k).

2.3. The second physical unclonable function level: The spectral signature

The unique chromatic response the multilayered MIN systems are endowed with stems from their scattering spectral response. This means that their

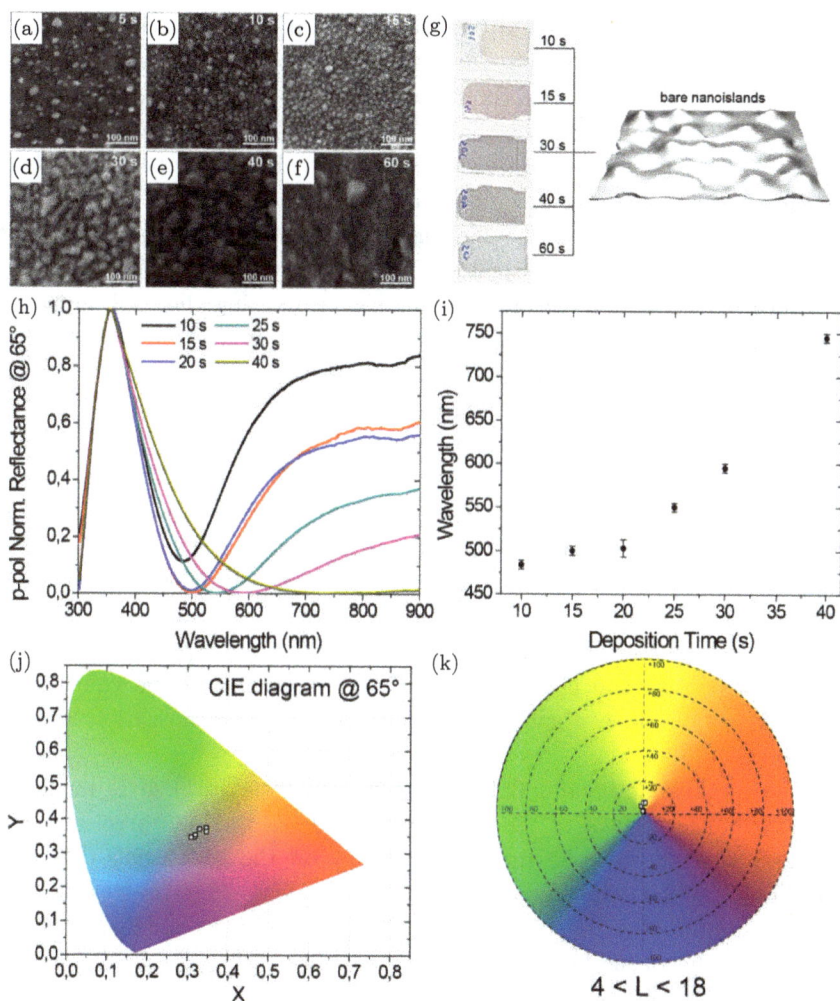

Figure 3. (a–f) SEM analysis of six different Ag nano-island layers at deposition times equal to 5 s, 10 s, 15 s, 30 s, 40 s, and 60 s, respectively. (g) Chromatic response of some of the samples analyzed before together with a 3D sketch of the Ag nano-islands layer. (h) Normalized p-polarization reflectance for the Ag nano-islands at 10 s, 15 s, 20 s, 25 s, 30 s, and 40 s measured at 65° angle of incidence together with (i) the spectral position of the reflectance dips associated to them. (j) CIE 1931 and (k) CIELAB hue–saturation diagram, together with an indication of the lightness range of all the Ag nano-island layers analyzed before. Image has been reproduced from the original work by V. Caligiuri *et al.*, *ACS Appl. Mater. Interfaces* 2021, 13, 49172–49183, under the Creative Commons License.

far-field scattering parameters (transmittance and reflectance) assume a rich and unique spectral shape. Each color corresponds to a specific spectral signature, constituting the second level of the complete plasmonic PUF. The level 2 signature could be challenged by simply illuminating it with a known light source which, for this purpose, could also simply be constituted by the flash lamp of a commercial smartphone, as shown in the sketch of Fig. 3(a). The response to the challenge is given by its spectral signature. In our experiments, we investigated the spectral response of the produced multilayered MIN architecture by illuminating it using the flash lamp of a smartphone and compared the obtained results with those acquired by a rigorous analysis carried out via spectroscopic ellipsometry. Examples of the most representative architectures are shown in Figs. 4(a)–4(f).

In particular, the reflectance spectra collected by an ellipsometric characterization of six representative samples are reported in the upper box of each panel of Figs. 4 (a1)–4(f1), compared with the spectra acquired by using a smartphone LED flash as a light source, as illustrated in the bottom figure of the same panels (black curves in Figs. 4(a2)–4(f2). The resonant behavior of the MIN multilayers holds a remarkable dependence on the angle of the impinging light, as described in the previous section. In addition, both the plasmonic and photonic nature of the compound multilayers introduce a marked dependence on the polarization of the impinging light.

To simplify the handling of this concept, in Figs. 4(a1)–4(f1) we show the reflectance spectra of the considered structures at one precise angle (being confident that the reader would rely on the iridescence of these structures provided in the previous section to figure out the resonances of each sample blue-shift while increasing the impinging angle). A quick comparison between the ellipsometric analyses and those carried out by illuminating with the LED flash lamp reveals that, as expected, in the case of the LED illumination, the spectral response is the result of the convolution of the p- and s-polarization reflectances (which have been rigorously discerned by the ellipsometric analysis). This endows the samples with a unique and rich spectral response, in which reflectance dips and peaks acquired at a precise angle are positioned at well-determined wavelengths. Replicating such a spectral signature with an opaque, commercially available paint would be impossible since their opacity and very scattering surface would compromise the measure itself. In the case of metallic paints, the reflectance of which could, on the contrary, be experimentally measured, it would however be impossible to reproduce the typical resonant response

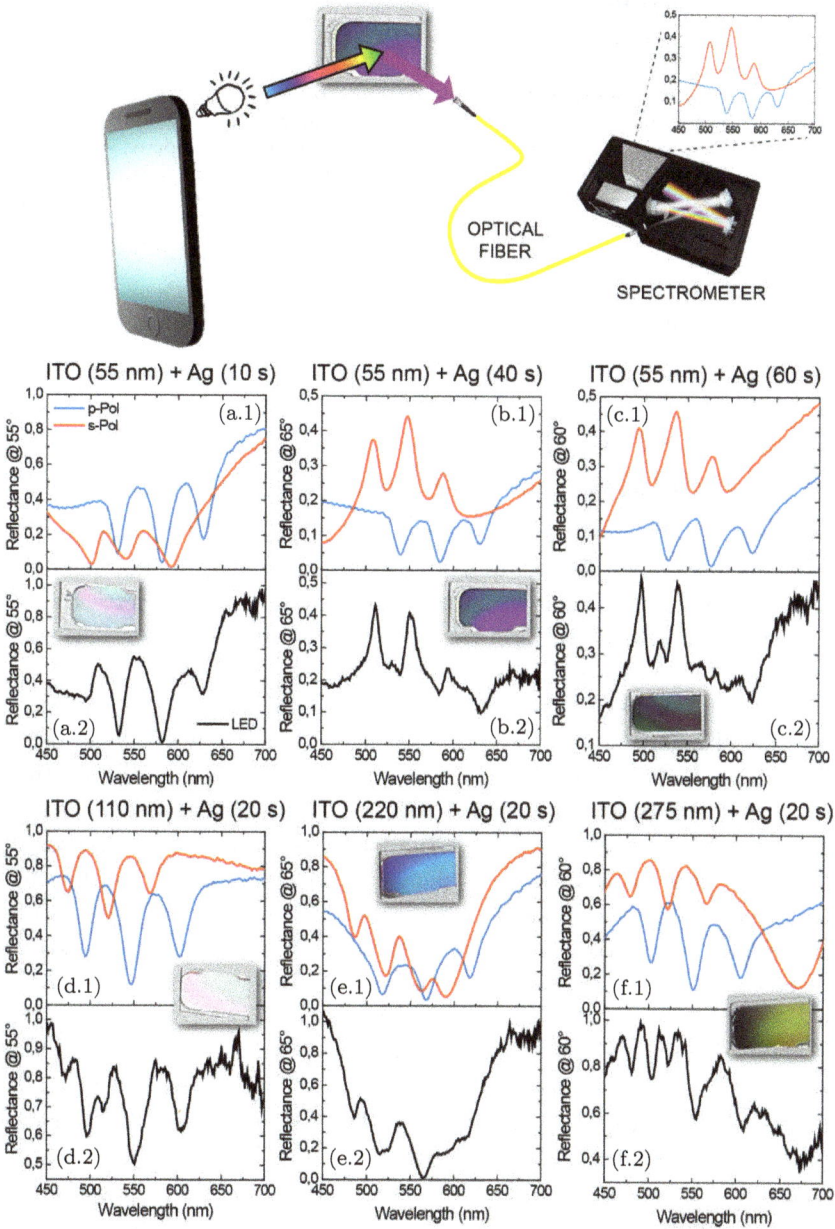

Figure 4.   (*Continued*)

of the MIN multilayers since metallic paints do not leverage on a resonant mechanism to produce their tints.

2.4. *The third physical unclonable function level: The morphological signature*

The third and strongest level of the complete plasmonic PUF is represented by the nanoscale morphology of the Ag nano-islands layer deposited on top of each MIN multilayer. As mentioned earlier, when an Ag layer with sub-percolation threshold size is deposited via DC magnetron sputtering, it does not assume the features of a smooth film but, rather, its morphology turns into a rough assembly of nano-islands, manifesting peculiar collective resonant plasmonic properties. Both the density and the average size of the nano-islands depend on the deposition time and the same is valid for their chromatic and plasmonic response, which comes out as the macroscopic effects of the cooperation between neighboring nano-islands. In other terms, even though the number of nano-islands per μm^2 and, therefore, the chromatic and plasmonic responses are determined by the sputtering parameters and remain repeatable, the morphological disposition of Ag nano-islands largely varies across the surface of the sample at a nanoscale level. What makes this signature unclonable is the fact that the operator has no control over the nanoscale morphology of a precise area of the sample, whose characteristics are not deterministically reproducible. A specific area of the sample, characterized by a peculiar nanoscale morphology, can, therefore, be taken as a manufacturer-resistant morphological barcode. The nanoscale morphology of a sample surface can be analyzed with great accuracy by means of Atomic Force Microscopy (AFM) investigations. Such a technique is a standard investigation tool and, differently from electron- or ion-beam-based imaging techniques does not require any particularly stringent experimental constraint, such as high vacuum chambers, metallization, or high-voltage exposure. Here, AFM analysis is used to challenge the PUF while its morphology constitutes the response. To prove the validity of Ag nano-islands as a morphological PUF, we sorted the top surface of an MIN multilayer as a grid, by lithographing a matrix of $3 \times 3\,\mu$m squares with i rows and j columns, so that each investigation area is well defined and recognizable (inset of Fig. 5(e)). The specific arrangement of the Ag nano-islands in each square determines its morphological barcode. Each square of the matrix has then been characterized via AFM measures both in trace and retrace mode. In the end, all the AFM measures have been digitalized to allow the conversion of salient morphological features into tags to be recognized while comparing two different morphological images. In the end, all the morphologies of the squares have been compared with each other by means of an open-access image recognition software (see Sec. 3). In particular, the AFM images are digitized and analyzed to evaluate relevant features for the comparison procedure (highlighted with colored circles in Figs. 5(a.1) and 5(a.2)).

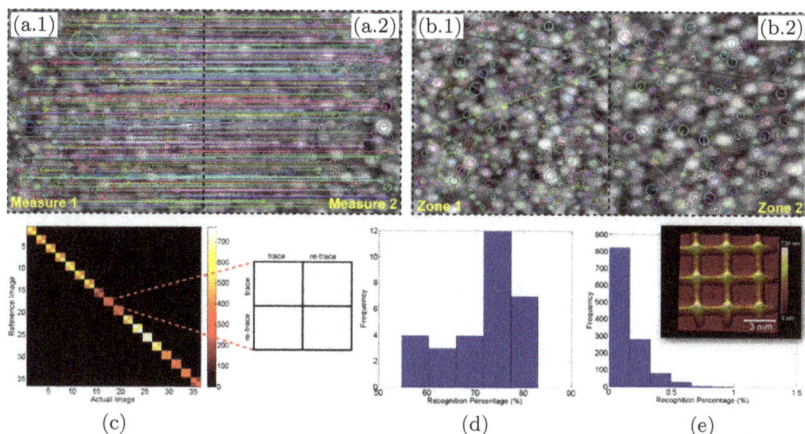

Figure 5. Comparison between (a.1) trace and (a.2) re-trace AFM measure of the same area. (b.1, b.2) Comparison between AFM measurements of two different areas. (c) Score-Match matrix representing all the possible comparisons between all the analyzed areas. In the zoom, a sketch of the on-diagonal element of the Score-Match matrix constituted by a 2×2 matrix accounting for the trace/trace, trace/re-trace, re-trace/trace, and re-trace/re-trace comparison of the same area. (d) Histogram of the distribution of the comparison between AFM images belonging to the same area together with (e) that belonging to comparisons of AFM measurements of different areas. In the inset, AFM measure of a portion of the $3 \times 3\,\mu$m grid into which the surface of an Ag nano-islands layer has been divided. Image has been reproduced from the original work by V. Caligiuri *et al.*, *ACS Appl. Mater. Interfaces* 2021, 13, 49172–49183, under the Creative Commons License.

The software then tries to recognize the same features between the two images of interest. Every time a match is confirmed, the software connects the two features with a line whose steepness carries information over the rotation of the image. A correlation matrix (sometimes called "Match-Score matrix"[17]) listing all the matching results in terms of recognized morphological features (tags) is then produced. The Match-Score matrix is one of the easiest and more widely used visual techniques to evaluate the validity of a PUF.[17] An example of software comparison between two AFM measurements over the same area is provided in Fig. 5(a) (a.1 and a.2), while an example of the comparison between two AFM measurements carried out over two different areas is shown in Fig. 5(b) (b.1 and b.2).

The number of features recognized in the case of the comparison between two measures carried out over the same area (Fig. 5(a.1), AFM trace, Fig. 5(a.2) AFM re-trace) is very high as demonstrated by the large number of connecting lines in Fig. 4(a). On the contrary, a comparison between two AFM images belonging to two different areas carries very few recognitions, as stated by the paltry number of connecting lines occurring between the two images of Figs. 5(b.1) and 5(b.2). The Match-Score matrix associated with the recognition procedure carried out over all the measured areas is reported in Fig. 5(c). The matrix has to be read considering that equal ith and jth elements correspond to a trace AFM measure of the same area if i and j are odd numbers and to a re-trace AFM measurement if they are even. Therefore, the $i + 1$ to j comparison corresponds to a re-trace to trace comparison of the same area and, finally, i to $j + 1$ comparison represents a trace to re-trace comparison of the same area. The elements on the diagonal of the Match-Score matrix are, therefore, 2×2 sub-matrices structured as shown in the zoom of Fig. 5(c). Off-diagonal elements correspond to comparisons between two different areas. The color of each i–j pair corresponds to the number of recognized features. Noticeably, the features recognized on the diagonal of the Match-Score matrix are more than two orders of magnitude larger than the number of features recognized in off-diagonal comparison.

Starting from the information acquired with the Match-Score matrix, we can define as a figure of merit the "tag recognition expectation value" (TREV), corresponding to the expectation value of the distribution of the number of tags recognized while comparing two different morphological areas. This parameter characterizes the capability of the nanoscale morphological fingerprint of working as a strong PUF. The distribution of the recognized tags for the cases in which a trace (retrace) square is compared with itself and/or with its retrace (trace) counterpart is shown in Fig. 5(d). The TREV of the distribution is about 76%, while the minimum of recognized tags in these cases is 55%. The number of recognized tags while comparing two different areas, no matter for the trace/retrace acquisition, does not exceed 1%, with a TREV of 0.11% (Fig. 5(e)). The two distributions do not show any overlap, highlighting the solidity of such a technique. The robustness of the Ag nano-islands to false-positive recognitions becomes glaring when comparing the two TREVs. The one inherent to the comparison of two different AFM scans of the same area is 690 times larger than that inherent to a comparison between two different areas, leaving practically no margins for false positive recognitions. Envisioning a practical application as a PUF, the most critical event corresponds to the case in which the correct area has been accidentally non-coaxially rotated by the operator while performing the AFM measure. This could potentially lead to a missed recognition due to an error of the operator (a false negative). To simulate this case, we carried out AFM investigations while rotating the same area from 0° to 90°. As for the previous case, four possibilities are salient: (i) a trace-to-trace comparison, in which the reference AFM analysis has been carried out as a trace measure (AFM tip moves left-to-right) and the compared image corresponds to another trace AFM measure, (ii) a trace-to-retrace comparison, where the compared image corresponds to a retrace AFM (AFM tip moves right-to-left), (iii) a retrace-to-retrace comparison, where both the reference and compared images correspond to retrace AFM measurements and, finally, (iv) a retrace-to-trace comparison in which a retrace AFM image is taken as a reference, compared to a trace AFM measure. Cases (i) and (iii) correspond to the homogeneous cases, in which higher recognition values are expected. Cases (ii) and (iv) represent the mixed cases and, for them, lower recognition values should be expected. The imaging recognition algorithm is however able to discriminate if the same image has been rotated with respect to the original one, being able to correlate an acceptable number of features also in this case. Fig. 6(a) shows the case of a 60° rotation of the same area (Fig. 6(a.1) for $\theta = 0°$, Fig. 6(a.2) for $\theta = 60°$), together with the connection lines between tags that the algorithm recognizes as equal.

We found that the number of recognized tags, $R(\vartheta)$, decreases with the exponential law $R(\vartheta) = A_0 e^{-\vartheta/\tau} + p_0$ with respect to the rotation angle. Here, A_0 corresponds to the tags recognition percentage at $\vartheta = 0°$, ϑ is the rotation angle, τ is the decay constant, measured in (degrees)-1, and p_0 is the offset of the exponential law, corresponding to the tags recognition

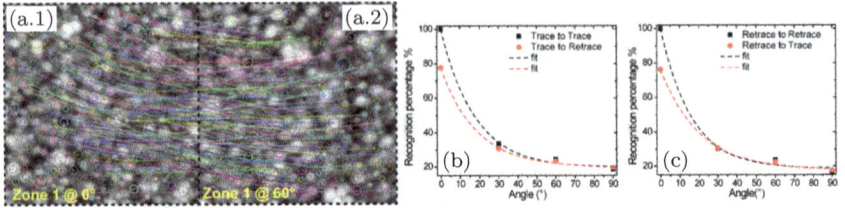

Figure 6.   Example of a comparison carried out by the imaging recognition software between the AFM image of a precise area of an Ag nano-island surface (a.1) and that of the same area tilted by 60° (a.2). Percentage of features recognized while rotating the sample in all the four possibilities: (b) Trace-to-trace and trace-to-retrace, (c) retrace-to-trace and retrace-to-retrace. Image has been reproduced from the original work by V. Caligiuri *et al.*, *ACS Appl. Mater. Interfaces* 2021, 13, 49172–49183, under the Creative Common License.

value toward which the exponential law tends to settle. The aforementioned cases (i) and (ii), in which the comparison is carried out using a trace AFM measure, are shown in Fig. 5(b), while cases (iii) and (iv), in which a retrace AFM measure constitutes the benchmark, are analyzed in Fig. 6(c). The starting points of the homogeneous cases correspond to comparing the same image and, of course, the tags recognition percentage is equal to 100%. Surprisingly, the exponential law describing the tags recognition percentage as a function of the rotation angle shows an offset around 20% ($p_0 \approx 20$). In the worst case of a 90° rotation, the tags recognition percentage does not go below 18%. In the following table (Table 1), the values of the tags recognition exponential decrease law are reported:

Table 1.   Fitting parameters (confidence interval $\pm$ 10%) inherent to the exponential law describing the decrease of recognized tags while rotating the sample under the AFM apparatus.

Comparison Type	A_0	τ	p_0
Trace to trace	80.39	17.73	19.55
Trace to retrace	57.87	18.45	19.95
Retrace to trace	81.33	16.17	18.61
Retrace to retrace	58.53	20.40	17.81

2.5. *Irreversible thermal switching in the field of food safety applications*

A feature that is often demanded in security labels is the capability to track temperature exposure of the labeled products. The multilayered MIN

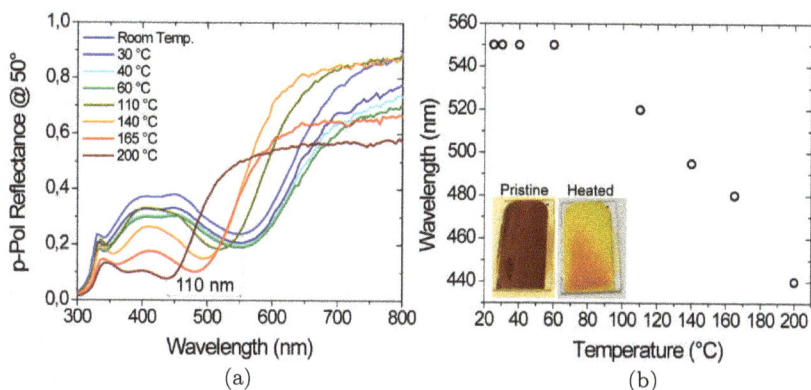

Figure 7. (a) Blue-shifting of the p-polarization reflectance response of an MIN sample made, from bottom to top, of Ag (100 nm)/ITO (110 nm)/Nano-islands (30 s), while increasing its temperature from room-temperature to 200°C. (b) Blue-shift of the reflectance dip shown in (a) as a function of the temperature of the sample. In the inset, a photo of the pristine and heated MIN showing a marked color change. Image has been reproduced from the original work by V. Caligiuri *et al.*, *ACS Appl. Mater. Interfaces* 2021, 13, 49172–49183, under the Creative Commons License.

architectures proposed in this work are the perfect candidates for the accomplishment of such a task since Ag nano-islands undergo an irreversible oxidation process when exposed to high temperatures. As a result, the resonance of the MIN multilayer experiences a significant blue-shift, as a function of the exposed temperature. In Fig. 6(a), we show the temperature-dependent *p*-polarization reflectance spectrum, acquired via temperature-varying ellipsometry (TVE) measurements, inherent to a MIN structure made of Ag (100 nm)/ITO (110 nm)/Nano-islands (30 s), from bottom to top.

As summarized in Fig. 7(b), where the spectral position of the mode of the MIN structure is plotted against the temperature while heating the sample from room temperature to 200°C, the main mode of the structure red-shifts by about 110 nm, with a knee at about 70°C. The oxidation process of Ag nano-islands induces a remarkable refractive index change of the plasmonic elements, resulting in a dramatic and irreversible color change. Such a mechanism can be easily used as a visual irreversible marker of product exposure to high temperatures.

In fact, if heated to ambient room temperature, Ag nano-islands undergo an oxidation process. Such a phenomenon is responsible for the marked color change and blue-shift of the plasmonic resonance

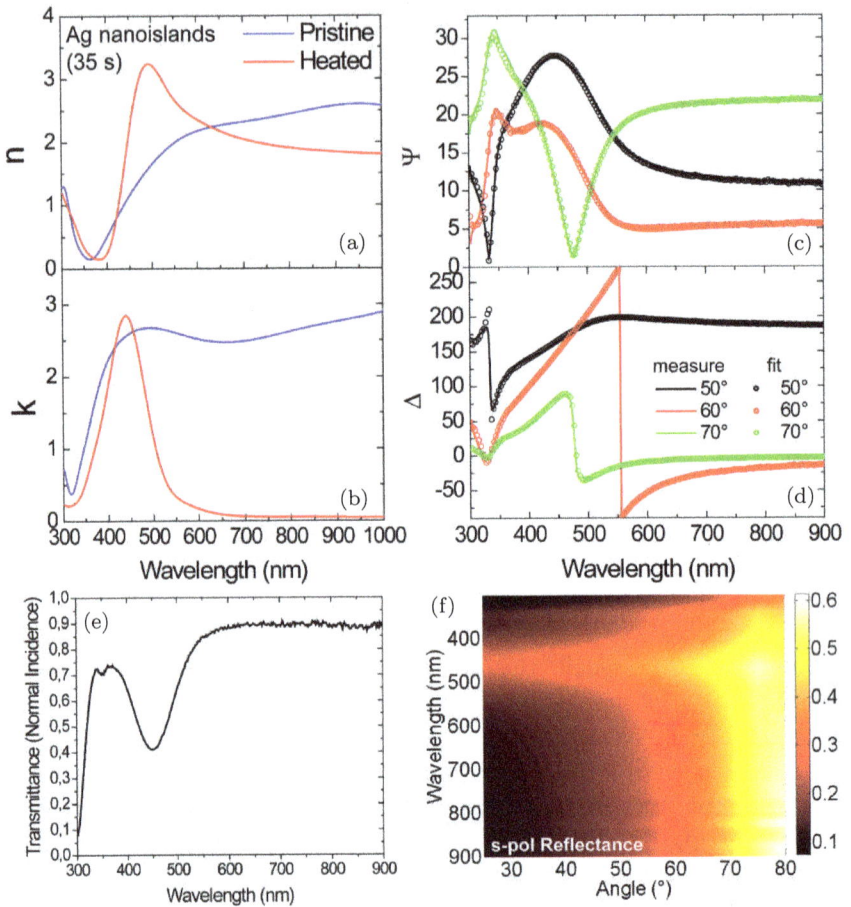

Figure 8.   (a, b) Real and imaginary refractive index of pristine (blue curve) and 200°C heated (red curve) of a nano-islands layer (deposition time 35 s), together with the measured and fitted ellipsometrical angles (c) Ψ and (d) Δ. (e) Transmittance of the Ag nano-islands layer (deposition time 35 s) together with (f) s-polarization angular reflectance map. Image has been reproduced from the original work by V. Caligiuri *et al.*, *ACS Appl. Mater. Interfaces* 2021, 13, 49172–49183, under the Creative Commons License.

shown earlier, which can be used as a temperature exposure labeling. The oxidation of Ag nano-islands can be confirmed ellipsometrically. In Figs. 8(a) and 8(b), a comparison between the measured refractive index of an Ag nano-islands layer, obtained with deposition time equal to 35 s before (blue curve) and after (red curve) the heating process at 200°C, is shown.

The occurring of a Lorentzian peak in the imaginary part of the heated sample around 450 nm can be immediately noted. This corresponds to the bandgap of Ag_2O, formed as a consequence of the oxidation process of Ag nano-islands. In Figs. 8(c) and 8(d), we report also the ellipsometric analysis and fit of the ellipsometrical angles Ψ and Δ, to highlight the good quality of the fitting procedure. To provide additional insights on the presence of a semiconductor bandgap, in Fig. 8(e) we present the transmittance spectrum of the heated Ag nano-islands layer. A marked dip at 450 nm is visible. Moreover, from Fig. 8(f), which shows the s-polarization angular transmittance 2D map of the same sample, it is possible to see that the wavelength of the transmittance dip does not shift as a function of the angles. Being dispersion-less, we can confirm that it corresponds to the bandgap of a semiconductor, Ag_2O in the case in point.

The properties described above are of special interest for food safety if we consider that these kinds of hybrid photonic/plasmonic structures can be easily fabricated over cellulose or flexible substrates, envisioning their printing ability over food paper packs. From this perspective, the proposed structures can constitute simple tags that irreversibly change color and simultaneously lose iridescence. Such a dramatic change in the optical properties can be easily detected *via* naked-eye investigations and can be thoroughly confirmed in a laboratory environment.

3. Experimental Methods

In the following subsections, we report a synthesis of the experimental methods used to produce the results that appeared in this chapter.

3.1. *Ag and Ito sputtering deposition parameters*

Multilayer Ag and ITO thin film were deposited via DC magnetron sputtering on glass substrates using the following sputtering parameters (Table 2).

Table 2.   Sputtering parameters.

Material	Power (W)	Rate (nm/s)	Pre-sputtering chamber pressure (mBar)	Sputtering chamber pressure (mBar)
Ag	20	0.25	3×10^{-5}	4.6×10^{-2}
ITO	40	0.16	3×10^{-5}	4.6×10^{-2}

3.2. Ag nano-islands sputtering deposition

Ag nano-islands have been obtained via DC magnetron sputtering deposition of Ag layers with sub-percolation threshold size. The obtained layers manifest a marked plasmonic response. The sputtering deposition parameters are as follows (Table 3):

Table 3.   Ag nano-islands sputtering parameters.

Deposition time (s)	Power (W)	Pre-deposition chamber pressure (mBar)	Deposition chamber pressure (mBar)
$5 \div 60$	12	3×10^{-5}	4.6×10^{-2}

3.3. CIE 1931 and CIELAB colorspace analysis

CIE 1931 and CIELAB color space analysis have been carried out via a customized MatLAB code. We begin by measuring both p- and s-polarized measurements and considering the convolution of both spectra. The electromagnetic signal reaching the human eye after being scattered by the sample is, indeed, the result of the convolution of both the polarizations. In particular, after having introduced the tri-stimulus values functions, we determine both the XYZ and xyz coordinate of the CIE 1931 color space. We then plot the related diagram. Then, once the CIE 1931 coordinates have been determined, we translate them into CIELAB parameters to determine, for each sample and at the desired angles, the values of hue, saturation, and lightness.

3.4. Ellispometrical and smartphone LED flashlamp characterization of the spectral response of MIN structures

P- and S-polarization reflectance measurements have been carried out over an ellipsometric setup (M2000 from Woollam), using an Xe lamp as a source. We then carried out a reflectance spectroscopic investigation using the LED flash lamp of a smartphone as a light source and collected the unpolarized reflectance spectrum at many angles (we have shown only the most significant of them) using an Ocean Optics Flame spectrometer.

3.5. *Temperature varying ellipsometry*

Temperature Varying Ellipsometry measurements have been carried out by equipping a VVase ellipsometer from Woollam with a custom hot stage produced by CaLCTec (Calabria Liquid Crystal Technology). We performed scattering and spectroscopic ellipsometry at each temperature specified in Fig. 7 of (i) bare Ag nano-islands, (ii) bare ITO layer, and (iii) the entire MIN structure. This allowed us to quantify the refractive index variation of both ITO and Ag nano-islands and evaluate their contributions in the spectral shift shown in Fig. 7(a). The pristine and heated refractive indices of both ITO and Ag nano-islands have been reported in Sec. 2.5.

3.6. *Image recognition algorithm*

Image recognition and feature matching have been carried out by means of a software application based on the Scale Invariant Feature Transform (SIFT) algorithm. SIFT is a computer vision algorithm for pattern recognition in 2D-image invariants to rotation, scale zooming, brightness changing.[67] The main steps are: (i) Constructing a Scale-Space, to make sure that features are scale-independent; (ii) Keypoint Localization, to identify the suitable features or keypoints; (iii) Orientation Assignment, to guarantee keypoints are rotation-invariant; and (iv) Keypoint Descriptor, to assign a unique fingerprint to each keypoint.[68]

The keypoint detection consists of identifying locations and scales that can be assigned many times to the same object from different points of view. Locations, invariant to scale change of the image, can be detected by searching for features with stable values for all possible scales. A continuous function of scale, the scale space, defined as a function, $L(x, y, \sigma)$ and formulated as the convolution of a variable-scale Gaussian, $G(x, y, \sigma)$, with an image, $I(x, y)$, is exploited to detect stable keypoint locations in scale space. Each keypoint is featured with a consistent orientation based on local image properties, enabling the generation of a keypoint descriptor related to this orientation and therefore invariant to image rotation. The pixel difference allows computing the gradient magnitude, $m(x, y)$, and orientation, $\theta(x, y)$. The correspondence among feature points in the original image and feature points in the input image can be evaluated, identifying, for each feature point, the nearest neighbor in feature vectors of the input image. The nearest neighbor is the feature point with a lower

Euclidean distance for the invariant descriptor vector. A software library implementing the SIFT algorithm written in C++, available on https://opencv.org/, was employed to design a Java tool for analyzing and comparing the acquired images, allowing to obtain the results reported in this chapter.

3.7. Atomic force microscopy measurements

Atomic Force Microscopy measurements have been carried out using a Multimode 8 equipped with a Nanoscope V controller (Bruker). Data were acquired in tapping mode, using silicon cantilevers (model TAP150, Bruker). Surfaces were imaged in air in a scan size of $1 \times 1\,\mu m$ and at different scan angles.

4. Conclusions

In this chapter, we engineered a strong Physically Unclonable Function based on a multilevel validity check process, all of which was included in the same platform. The architecture around which the PUF is designed consists of a composite multilayer in which a plasmonic element represented by Ag nano-islands interacts with a resonant one made of one or more metal/insulator/metal cavities. Such a hybrid plasmonic/photonic architecture manifests exceptional chromatic versatility, characterized by means of both CIE 1931 and CIELAB diagrams, that, united to a peculiar iridescence, constitutes the first recognition checkpoint which we call "chromatic signature". The specific chromatic response each structure is endowed with stems from a rich and detailed spectral response, resulting from the convolution of both the P- and S-polarized reflectance of the sample. The spectral response of the structure constitutes also the second recognition checkpoint, which we identify as the "spectral signature". We demonstrate that, despite the sophisticated nature of such an investigation, it could be readily and easily carried out by illuminating the sample employing the LED flash lamp of a smartphone, envisioning the implementation of simple customer-level investigation tools ready to be integrated with commercial portable devices. In the end, as a third recognition checkpoint at the nanoscale, we exploit the random distribution of plasmonic nanometric units resulting from sub-percolation threshold sputtering deposition of Ag. Such a process gives rise to a rough surface distribution of Ag nano-islands which can be investigated by means of AFM measurements. The AFM map can then be digitalized and used as a morphological nanoscale barcode acting as a recognition tag. The intrinsic impossibility of the operator to gain control over the morphological disposition of the Ag nano-islands ensures the unclonability of the morphological barcode which can be classified as "manufacturer resistant". We demonstrated, by means of software image recognition techniques, the robustness of the morphology-based recognition process. In particular, we found that the tag recognition expectation value (TREV) in the case of the comparison between two images inherent to AFM analyses of the same area (proper recognition) is about 76%, being 690 times larger than the TREV stemming from a comparison of AFM scanning of two different areas (counterfeit case). In the end, we demonstrate the robustness of the proposed PUF to errors potentially occurring while operating the AFM apparatus and, specifically, in the case in which the tag is non-coaxially rotated with respect to the predefined analysis position. Even in this case, tag recognition percentages of 20% minimum are found, a value that is 180 times larger than the counterfeit TREV. The integrated and multipronged approach we propose here holds great promise in revolutionizing the PUF blockchains turning the randomness intrinsic in nanotechnology processes into the key to reach out incomparable anti-counterfeit labeling strength. The unique values of hue, saturation, and lightness obtainable with the proposed MIN multilayers foster their application in new-generation trademarks which could intrinsically be endowed with robust anti-counterfeit characteristics. Our plasmonic PUF can be easily deposited over whatever metallic back-reflector, opening to a cheap tagging of a plethora of devices and goods. The straightforward deposition technique by which our plasmonic PUFs are produced is scalable from wafer-to-chip size envisioning the integration of our tags in consumer-level electronics and food safety applications. In the end, the exceptional and irreproducible randomness of the nano-islands' formation process can be potentially employed as a morphological cryptography platform, with potential disruptive impact in novel distributed ledger technologies where high cyber-security standards are paramount.

References

1. Y. Gao, S. F. Al-Sarawi, and D. Abbott, Physical unclonable functions, *Nat. Electron.* **3**, 81–91 (2020).
2. B. Halak, *Physically Unclonable Functions*, Springer International Publishing (2018). doi:10.1007/978-3-319-76804-5.

3. A. Riahi Sfar, E. Natalizio, Y. Challal, and Z. Chtourou, A roadmap for security challenges in the Internet of Things, *Digi. Commun. Netw.* **4**, 118–137 (2018).

4. B. Gassend, D. Clarke, M. van Dijk, and S. Devadas, Silicon physical random functions, In (pp. 148–160). Association for Computing Machinery (ACM) (2002). doi:10.1145/586110.586132.

5. T. McGrath, I. E. Bagci, Z. M. Wang, U. Roedig, and R. J. Young, A PUF taxonomy, *Appl. Phys. Rev.* **6**, 11303–11303 (2019).

6. C. Herder, M. D. Yu, F. Koushanfar, and S. Devadas, Physical unclonable functions and applications: A tutorial, *Proc. IEEE* **102**, 1126–1141 (2014).

7. R. Maes and I. Verbauwhede, Physically unclonable functions: A study on the state of the art and future research directions, In *Information Security and Cryptography* (pp. 3–37). Springer International Publishing (2010). doi:10.1007/978-3-642-14452-3_1.

8. B. C. Grubel *et al.*, Secure authentication using the ultrafast response of chaotic silicon photonic microcavities, In *Conference on Lasers and Electro-Optics (2016), Paper SF1F.2.* SF1F.2 Optica Publishing Group (2016). doi:10.1364/CLEO_SI.2016.SF1F.2.

9. B. C. Grubel *et al.*, Secure communications using nonlinear silicon photonic keys, *Opt. Express* **26**, 4710 (2018).

10. B. Shao *et al.*, Crypto primitive of MOCVD MoS2 transistors for highly secured physical unclonable functions, *Nano. Res.* **14**, 1784–1788 (2021).

11. A. Scholz *et al.*, Hybrid low-voltage physical unclonable function based on inkjet-printed metal-oxide transistors, *Nat. Commun.* **11**, 5543 (2020).

12. C.-E. Yin and G. Qu, Temperature-aware cooperative ring oscillator PUF, In *2009 IEEE International Workshop on Hardware-Oriented Security and Trust,* pp. 36–42 (2009). doi:10.1109/HST.2009.5225055.

13. A. Maiti and P. Schaumont, Improving the quality of a physical unclonable function using configurable ring oscillators, In *2009 International Conference on Field Programmable Logic and Applications,* pp. 703–707 (2009). doi:10.1109/FPL.2009.5272361.

14. A. Maiti, J. Casarona, L. McHale, and P. Schaumont, A large scale characterization of RO-PUF. In *2010 IEEE International Symposium on Hardware-Oriented Security and Trust (HOST),* pp. 94–99 (2010). doi:10.1109/HST.2010.5513108.

15. A. T. Erozan *et al.*, Inkjet-printed EGFET-based physical unclonable function—design, evaluation, and fabrication, *IEEE Trans. Very Large Scale Integr. (VLSI) Syst.* **26**, 2935–2946 (2018).

16. J. Rajendran *et al.*, Nano meets security: Exploring nanoelectronic devices for security applications, *Proc. IEEE* **103**, 829–849 (2015).

17. B. Yoon *et al.*, Recent functional material based approaches to prevent and detect counterfeiting, *J. Mater. Chem. C* **1**, 2388–2403 (2013).

18. R. Arppe-Tabbara, M. Tabbara, and Sørensen, T. J. Versatile and validated optical authentication system based on physical unclonable functions, *ACS Appl. Mater. Interf.* **11**, 6475–6482 (2019).

19. J. Kim *et al.*, Anti-counterfeit nanoscale fingerprints based on randomly distributed nanowires, *Nanotechnology* **25**, 155303 (2014).

20. B. R. Anderson, R. Gunawidjaja, and H. Eilers, Initial tamper tests of novel tamper-indicating optical physical unclonable functions, *Appl. Opt., AO* **56**, 2863–2872 (2017).

21. J. Feng *et al.*, Random organic nanolaser arrays for cryptographic primitives, *Adv. Mater.* **31**, 1807880 (2019).

22. K. Nakayama, Optical security device providing fingerprint and designed pattern indicator using fingerprint texture in liquid crystal, *Opt. Eng.* **51**, 040506 (2012).

23. J. Fei and R. Liu, Drug-laden 3D biodegradable label using QR code for anti-counterfeiting of drugs, *Mater. Sci. Eng.: C* **63**, 657–662 (2016).

24. X. Li and Y. Hu, Luminescent films functionalized with cellulose nanofibrils/CdTe quantum dots for anti-counterfeiting applications, *Carbohydr. Polym.* **203**, 167–175 (2019).

25. R. Bao *et al.*, Green and facile synthesis of nitrogen and phosphorus co-doped carbon quantum dots towards fluorescent ink and sensing applications, *Nanomaterials* **8**, 386 (2018).

26. Y. Jiang *et al.*, A neutral dinuclear Ir(III) complex for anti-counterfeiting and data encryption, *Chem. Commun.* **53**, 3022–3025 (2017).

27. S. Kalytchuk, Y. Wang, K. Poláková, and R. Zbořil, Carbon dot fluorescence-lifetime-encoded anti-counterfeiting, *ACS Appl. Mater. Interf.* **10**, 29902–29908 (2018).

28. F. Chen *et al.*, Unclonable fluorescence behaviors of perovskite quantum dots/chaotic metasurfaces hybrid nanostructures for versatile security primitive, *Chem. Eng. J.* **411**, 128350 (2021).

29. X. Zheng *et al.*, Inkjet-printed quantum dot fluorescent security labels with triple-level optical encryption, *ACS Appl. Mater. Interf.* **13**, 15701–15708 (2021).

30. A. Wali *et al.*, Biological physically unclonable function, *Commun. Phys.* **2**, 39 (2019).

31. L. D. McCarthy, R. A. Lee, and G. F. Swiegers, Modulated digital images for biometric and other security applications, In *Optical Security and Counterfeit Deterrence Techniques V*, Vol. 5310, pp. 103–116. SPIE (2004).

32. M. Song *et al.*, Colors with plasmonic nanostructures: A full-spectrum review, *App. Phys. Rev.* **6**, 041308 (2019).

33. U. Cataldi, *et al.* Growing gold nanoparticles on a flexible substrate to enable simple mechanical control of their plasmonic coupling, *J. Mater. Chem. C* **2**, 7927–7933 (2014).

34. J. Olson *et al.*, High chromaticity aluminum plasmonic pixels for active liquid crystal displays, *ACS Nano* **10**, 1108–1117 (2016).

35. M. L. Tseng *et al.*, Two-dimensional active tuning of an aluminum plasmonic array for full-spectrum response, *Nano Lett.* **17**, 6034–6039 (2017).

36. X. Zhu, W. Yan, U. Levy, N. A. Mortensen, and A. Kristensen, Resonant laser printing of structural colors on high-index dielectric metasurfaces, *Sci. Adv.* **3**, e1602487 (2017).

37. N. S. King *et al.*, Fano resonant aluminum nanoclusters for plasmonic colorimetric sensing, *ACS Nano* **9**, 10628–10636 (2015).

38. A. S. Roberts, A. Pors, O. Albrektsen, and S. I. Bozhevolnyi, Subwavelength plasmonic color printing protected for ambient use, *Nano Lett.* **14**, 783–787 (2014).

39. J. S. Clausen *et al.*, Plasmonic metasurfaces for coloration of plastic consumer products, *Nano Lett.* **14**, 4499–4504 (2014).

40. M. K. Hedayati *et al.*, Design of a perfect black absorber at visible frequencies using plasmonic metamaterials, *Adv. Mater.* **23**, 5410–5414 (2011).

41. A. Kristensen, *et al.* Plasmonic colour generation, *Nat Rev Mater* **2**, 1–14 (2016).

42. S. Yokogawa, S. P. Burgos, and H. A. Atwater, Plasmonic Color Filters for CMOS Image Sensor Applications, *Nano Lett.* **12**, 4349–4354 (2012).

43. K. J. Savage *et al.*, Revealing the quantum regime in tunnelling plasmonics, *Nature* **491**, 574–577 (2012).

44. A. Kuzyk, *et al.* DNA-based self-assembly of chiral plasmonic nanostructures with tailored optical response, *Nature* **483**, 311–314 (2012).

45. V. Caligiuri *et al.*, Biodegradable and insoluble cellulose photonic crystals and metasurfaces, *ACS Nano* **14**, 9502–9511 (2020).

46. A. Bojesomo *et al.*, Toward physically unclonable functions from plasmonics-enhanced silicon disc resonators, *J. Lightwave Technol.* **37**, 3805–3814 (2019).

47. A. F. Smith, P. Patton, and S. E. Skrabalak, Anti-counterfeit labels: Plasmonic nanoparticles as a physically unclonable function for responsive anti-counterfeit nanofingerprints (Adv. Funct. Mater. 9/2016), *Adv. Funct. Mater.* **26**, 1305 (2016).

48. S. A. Maier, *Plasmonics: Fundamentals and Applications*, New York: Springer Science+Business Media LLC (2007).

49. D. J. Barber and I. C. Freestone, An Investigation of the origin of the colour of the Lycurgus cup by analytical transmission electron microscopy, *Archaeometry* **32**, 33–45 (1990).

50. T. Chung *et al.*, Nanoislands as plasmonic materials, *Nanoscale* **11**, 8651–8664 (2019).

51. M. Šubr, M. Petr, O. Kylián, J. Kratochvíl, and M. Procházka, Large-scale Ag nanoislands stabilized by a magnetron-sputtered polytetrafluoroethylene film as substrates for highly sensitive and reproducible surface-enhanced Raman scattering (SERS), *J. Mater. Chem. C* **3**, 11478–11485 (2015).

52. M. Kang, J.-J. Kim, Y.-J. Oh, S.-G., Park, and K.-H. Jeong, Nanoplasmonics: A deformable nanoplasmonic membrane reveals universal correlations between plasmon resonance and surface enhanced Raman scattering (Adv. Mater. 26/2014), *Adv. Mater.* **26**, 4509 (2014).

53. J. N. Anker *et al.*, Biosensing with plasmonic nanosensors, *Nat. Mater.* **7**, 442–453 (2008).

54. P. A. Thiel, M. Shen, D. J. Liu, and J. W. Evans, Coarsening of two-dimensional nanoclusters on metal surfaces, *J. Phys. Chem. C* **113**, 5047–5067 (2009).

55. K.-C. Lee, S.-J. Lin, C.-H. Lin, C.-S. Tsai, and Y.-J. Lu, Size effect of Ag nanoparticles on surface plasmon resonance, *Surf. Coat. Technol.* **202**, 5339–5342 (2008).

56. V. Caligiuri, M. Palei, G. Biffi, and R. Krahne, Hybridization of epsilon-near-zero modes via resonant tunneling in layered metal-insulator double nanocavities, *Nanophotonics* **8**, 1505–1512 (2019).

57. V. Caligiuri, M. Palei, G. Biffi, S. Artyukhin, and R. Krahne, A semi-classical view on epsilon-near-zero resonant tunneling modes in metal/insulator/metal nanocavities, *Nano Lett.* **19**, 3151–3160 (2019).

58. V. Caligiuri *et al.*, One-dimensional epsilon-near-zero crystals, *Adv. Photonics Res.* **2**, 2100053 (2021).

59. V. Caligiuri *et al.*, Angle and polarization selective spontaneous emission in dye-doped metal/insulator/metal nanocavities, *Adv. Opt. Mater.* 1901215 (2019) doi:10.1002/adom.201901215.

60. Z. Li, S. Butun, and K. Aydin, Large-area, lithography-free super absorbers and color filters at visible frequencies using ultrathin metallic films, *ACS Photonics* **2**, 183–188 (2015).

61. K. Halterman and M. Alidoust, Waveguide modes in Weyl semimetals with tilted dirac cones, *Opt. Express* **27**, 36164–36164 (2019).

62. K. Halterman, M. Alidoust, and A. Zyuzin, Epsilon-near-zero response and tunable perfect absorption in Weyl semimetals, *Phys. Rev. B* **98**, 085109 (2018).

63. S. Feng and K. Halterman, Coherent perfect absorption in epsilon-near-zero metamaterials, *Phys. Rev. B — Condens. Matter Mater. Phys.* **86**, 165103 (2012).

64. N. Maccaferri *et al.*, Enhanced nonlinear emission from single multilayered metal-dielectric nanocavities resonating in the near-infrared, *ACS Photonics* **8**, 512–520 (2021).

65. G. E. Lio, A. Ferraro, M. Giocondo, R. Caputo, and A. De Luca, Color gamut behavior in epsilon near-zero nanocavities during propagation of gap surface plasmons, *Adv. Opt. Mater.* **8**, 2000487 (2020).

66. G. E. Lio *et al.*, Leveraging on ENZ metamaterials to achieve 2D and 3D hyper-resolution in two-photon direct laser writing, *Adv. Mater.* **33**, 2008644 (2021).

67. D. G. Lowe, Object recognition from local scale-invariant features, In *Proceedings of the Seventh IEEE International Conference on Computer Vision*, 20–27 September 1999, Kerkyra, Greece, Vol. 2, pp. 1150–1157 (1999).

68. D. G. Lowe, Distinctive image features from scale-invariant keypoints, *Int. J. Comput. Vis.* **60**, 91–110 (2004).

Chapter 8

Laser-Assisted Micromachining and Applications

L. Criante[*,‡], R. Ramos-García[†,§], and S. Bonfadini[*]

*Center for Nano Science and Technology, Istituto Italiano di Tecnologia,
via Rubattino 81, 20134 Milano, Italy*
†*Coordinación de óptica, Instituto Nacional de Astrofísica,
Óptica y Electrónica, Tonantzintla, Puebla 72840, Mexico*
‡*luigino.criante@iit.it*
§*rgarcia@inaoep.mx*

Ultrashort pulse laser micromachining is a clear technological breakthrough with exciting potential for many applications and has led to impressive progress in the study of light–matter interactions. In this context, the laser-assisted wet etching fabrication technique has opened new frontiers in the fabrication of 3D, monolithic mini-microfluidic structures that can be completely buried in the substrate (typically fused silica). In addition to the benefit of an inert substrate (strategic for bio applications), the absence of the sealing step to join two halves together to achieve a working platform and the high mechanical strength offer numerous advantages, including the ability to achieve maximum fluid injection pressure. Here is a challenging application of a large-volume (over $9 \, \text{mm}^3$) monolithic glass chamber, manufactured to match needle-free injectors (NFI). The unique optical transparency of the device allowed the full dynamics of thermocavitation to be studied in detail, from bubble formation to ejection mechanism, providing valuable data for future optimizations. With appropriate chamber design and fabrication, high-speed jets (average value up to ∼70 m/s) of thermocavitation bubbles were demonstrated using a Continuous-Wave (CW) laser. High-speed camera analysis has shown a maximum bubble wall velocity of ∼10–25 m/s for almost any combination of incident laser parameters.

1. Introduction

Fabrication involves producing structures, equipment, and components by casting, machining, forming, welding, and assembling. Among them, machining, considered as the removal and/or addition of material from the substrate, may be categorized as two types depending on the dimensions of the feature produced: macromachining and micromachining. Let us not be fooled by the simplest correlation which would like to associate the former processing with elements that can be clearly measured with the naked eye, approximately up to 1 mm, while the latter with elements that have a final size below 1 mm and up to the range of 0.5–1 μm. Because micromachining literally means removal of material at the micro/nano level with high precision and no restrictions on the size of the workpiece.

Based on the intrinsic mechanism and basic requirements, the type of energy required and the means of energy transfer, modern micromachining processes can be classified into four major areas: Mechanical — where mechanical forces are used to propel abrasive particles; Thermal — where material is removed by generating intense and high localized heating followed by melting and vaporization; Chemical — where material is removed by chemical and electrochemical reactions; and Hybrid. Laser micromachining falls in the latter category.

Arguably, one of humankinds' greatest inventions of the last century, lasers have become part of many aspects of our lives, from medicine and advanced industry to the most innovative areas of research such as nuclear fusion. Their power lies in their ability to generate large electromagnetic fields and extract large amounts of energy in the form of coherent light. Moreover, their use in micromachining has proven to be extremely versatile, as the principle behind their operation is driven by the physics of light–matter interaction: photon energy and wavelength, material band, absorption and energy transfer.[1] The process of laser micromachining is truly adaptable. Beyond the type of operating principle used, key elements are driven by the concept of energy: the material bands, the energy required, and the tools of transferring this energy. The main parameters are therefore the size and depth of the focal point and the energy density used both as a function of time and space. Modification of the chemical and physical properties of a material by a laser usually requires a very high intensity. Therefore, the most suitable sources for this type of processing are pulsed ones.

Among all the available regimes (ns, ps, fs pulse width) femtosecond (fs) lasers represent a clear technological breakthrough with exciting potential for many applications and have led to impressive advances in the study of light–matter interactions.[2–4] Understanding the different timescales involved in converting the laser pulse energy into a structural change provides insight into why ultrashort laser pulses are well suited to micromachining applications. Although laser-induced damage has been

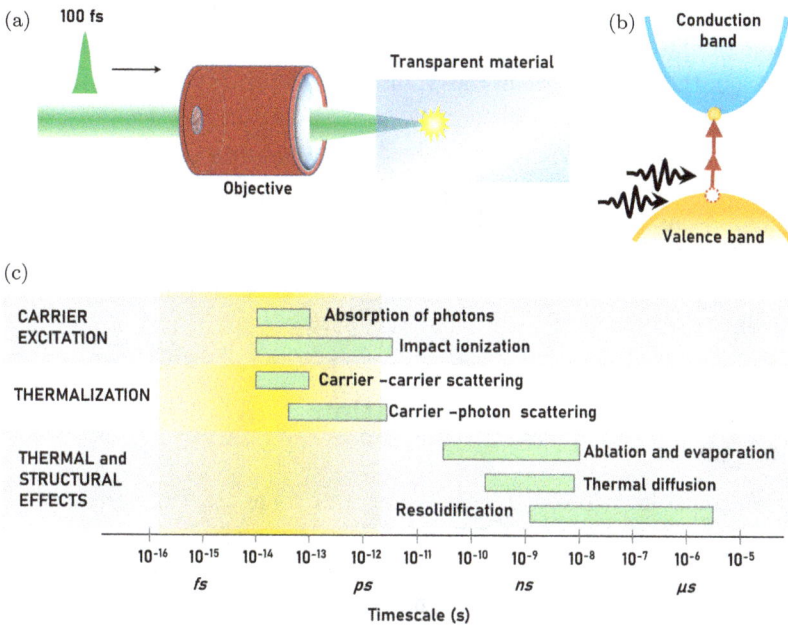

Figure 1. Femtosecond micromachine process. (a) Schematic representation of the laser incident on a transparent material. (b) Diagram of the excitation of electrons on the conduction band. (c) Timescales of the physical phenomena associated with the interaction of a femtosecond laser pulse with transparent materials. The green bar represents a typical timescale for the relevant process. Although the absorption of light occurs on the femtosecond timescale (yellow region), the material can still continue to undergo changes microseconds later.

studied since the early days of the laser,[5] the changes in material properties caused by femtosecond laser pulses are fundamentally different from the damage caused by laser pulses longer than one picosecond (ps). For sub-picosecond pulses, the timescale over which the electrons are excited is smaller than the electron–phonon scattering time (about 1 ps). Thus, a femtosecond laser pulse ends before the electrons have thermally excited any ions (Fig. 1(c)). As a direct result, heat diffusion outside the focal area is minimized, reducing the thermal effect on material and increasing the precision of the modification.[6,7]

In addition, femtosecond laser processing is a deterministic process even in transparent materials (i.e., transparent to the wavelength used for the modification) because no defect electrons are required to seed the absorption process, indeed enough seed electrons are generated by nonlinear ionization from the first tens of femtoseconds of the pulse itself.[8,9]

With a sufficiently high energy input into the target material, significant nonlinear absorption takes place and localized energy deposition occurs on the surface or in the volume of (semi)transparent materials. The nonlinear nature of absorption is responsible for the "threshold" material modification process: combined with tight focusing, it is possible to confine the absorption to the focal volume inside the bulk of the material involved, resulting in minimally sized micromachined volumes (down to sub-μm^3).[10] This spatial confinement, combined with laser-beam scanning or sample translation, makes it possible to micromachine geometrically complex structures in 3D.

2. Femtosecond Laser Irradiation: Outline of Theory

The extreme high peak intensities at the focus of femtosecond laser pulses have opened the possibility for a wide range of new applications, from precision scalpels for delicate life science[11] to drive sources for tabletop particle accelerators.[12] One of the most promising applications of femtosecond lasers is micromachining in transparent materials, such as glasses,[13] crystals,[14] and polymers.[15] In dielectric transparent materials, the energy bandgap is much larger than the energy carried by the fs-photon in the ultra-violet (UV)-near infra red (IR) spectrum. In the absence of impurities, carriers are initially generated by multiphoton absorption, which promotes electrons from the valence band to the conduction band (Fig. 1(b)).

In more detail, when a femtosecond laser pulse with a sufficiently high peak intensity is focused on a material (Fig. 1(a)), optical breakdown is observed.

This is possible when the electric field of the incident beam is close to the valence electron binding electric field of the atoms (10^9 Vm^{-2}), which corresponds to a laser intensity[4] around 5×10^{20} Wm^{-2}. Part of the laser pulse energy is transferred to the electrons in the short-pulse duration. After a few picoseconds, the laser-excited electrons transfer their energy to the lattice by thermalizing with ions. As a result of the irradiation, the material can undergo a phase or structural change, leaving a permanent change.

Following the energy transfer concept, three main processes may occur:

(1) *Multiphoton absorption ionization*: this phenomenon is caused by the absorption of more than one photon by an electron in the valence band,

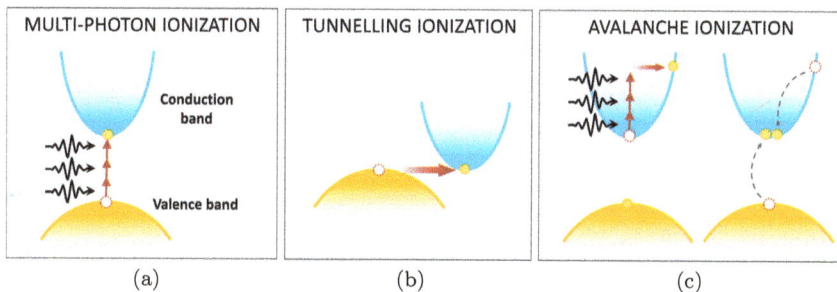

Figure 2.   Schematic representation of the nonlinear processes: (a) multiphoton absorption. (b) Tunneling ionization. (c) Avalanche ionization.

thereby promoting it to the conduction band (Fig. 2(a)). For instance, in the case of fused silica, the energy gap E_{gap} is about 9 eV and the energy of a photon at 515 nm (for example) is 2.4 eV. Therefore, optical breakdown can only be achieved when 4–5 photons are absorbed at the same time. The rate of occurrence of this phenomenon is strongly related to the beam intensity (I), as follows:

$$R_n(I) = \sigma_n I^n, \tag{1}$$

where σ_n is the absorption cross-section for n photons.

(2) *Tunneling ionization*: the high intensity electric field of the incident laser beam reduces the Coulomb potential energy barrier and levels out the two energy bands (Fig. 2(b)). Therefore, the probability that electrons can pass from the valence band to the conduction band for tunneling effect is different from zero. Tunneling ionization and multiphoton absorption are two competing processes. To predict the physical phenomenon, the Mstislav Keldysh parameter can be used as follows:

$$\gamma = \frac{\omega_f}{e}\sqrt{\frac{m_e c n_0 \epsilon_0 E_g}{I_f}} \tag{2}$$

here I_f and ω_f are the laser intensity and frequency in the focal spot, m_e is the effective electron mass, e is its fundamental charge, c is the speed of light, n_0 is the linear refractive index, and ϵ_0 is the permittivity of free space. For $\gamma \ll 1.5$, tunneling ionization prevails, while for $\gamma \gg 1.5$ multiphoton absorption is the dominant effect.[16–18]

(3) *Avalanche ionization*: this phenomenon can only occur if free electrons are already present at the bottom of the conduction band. As shown

schematically in Fig. 2(c), the incident photons can be absorbed by the free electrons, increasing their kinetic energy. As the energy value rises above E_{gap}, the electrons can collide with the valence band electrons, pushing them into the conduction band. This increases the population of free electrons in the conduction band and the probability of the process repeating. The time (t) evolution of the free charge density can be described as

$$\eta_{AV}(t) = \eta_0 2^{\rho t}, \tag{3}$$

where η_0 is the initial free electron density and ρ is the impact ionization probability.

When the density of excited electrons reaches a value above 10^{29} m^{-3} the electrons behave as a plasma with a natural frequency that is resonant with the laser — leading to reflection and absorption of the remaining pulse energy.[19,20] Under these conditions, the material becomes strongly absorbing. Once the threshold is exceeded, material damage occurs only in the 3D focal spot of the laser beam.[21] In this way, complex 3D geometries can be inscribed within the substrate with high spatial resolution. The uniqueness and versatility of this fabrication process lies in the ability to precisely control the effect of laser irradiation on the material through the writing parameters. In fact, the change in the material depends on both exposure parameters — e.g., energy, pulses' time length, repetition rate, writing speed, polarization, wavelength — and the material properties — i.e., the energy bandgap.

For glass materials, (such as fused silica substrates), the change in laser fluence is sufficient to achieve three different morphological effects,[22,23] summarized in Fig. 3:

(1) *REGIME 1*: Fluences slightly above the optical breakdown threshold can induce small isotropic refractive index variations in the material $\Delta n \sim 10^{-3}$, due to the relaxation and densification of the material. This weak modification can be used to write optical waveguides buried in the substrate with arbitrary 3D geometry (Fig. 3(c)), featuring good performances.[24]

(2) *REGIME 2*: Increasing the fluence to reach an intermediate level, the chemical bonds between the silica and oxygen atoms are broken and unique nanostructures — known as nanogratings or nanocracks — appear. These self-assembled periodic structures show an alternation

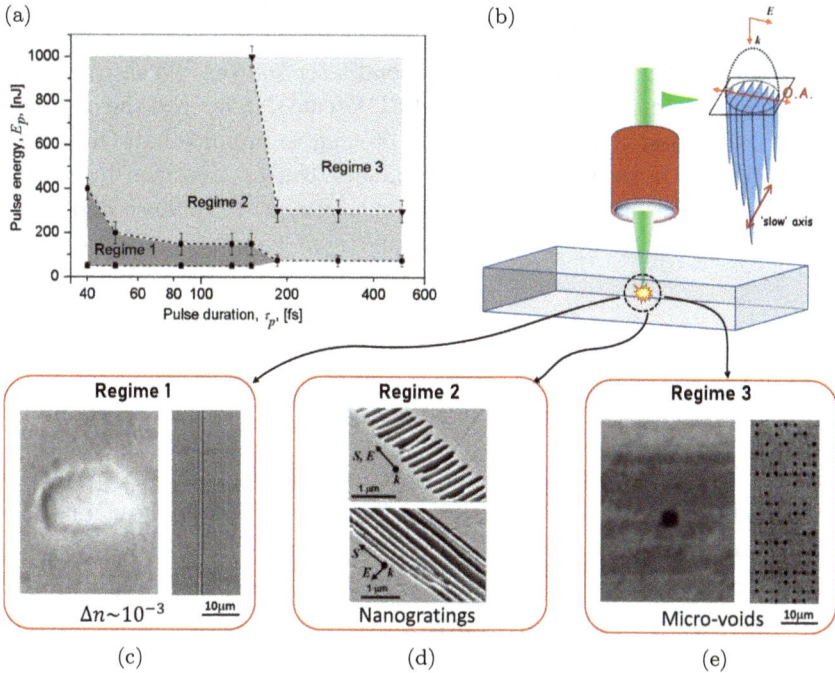

Figure 3. Structural modification induced in fused silica by the fs-laser. (a) Regimes of the three morphological effects as a function of pulse duration vs pulse energy of the fs-laser. (b) Light intensity distribution in the laser spot due to the interaction between incident photon and electronic structure of the lattice. The latter begins to behave like plasma, which resonates at the frequency of the laser. (c) Weak anisotropic index variations ($\Delta n \sim 10^{-3}$): this modification of the material structure is not sufficient to provide the chemical etching step a favorable substrate for material subtraction. (d) Nanogratings' generation: structural modifications can be inducted at any height within the sample, allowing volume irradiation (e) Material ablation: high fluence generates micro-explosions, resulting in micro-holes surrounded by higher density material (Refs. 22, 23, and 25).

(period $\sim\lambda/2n$) of high- and low-density material — i.e., light and dark in Fig. 3(d) — oriented perpendicular to the writing polarization of the laser beam.

(3) *REGIME 3*: At high fluence values (peak intensities $>10^{14}$ Wcm^{-2}) pressures exceeding Young's modulus of the material occur in the focal spot of the beam, creating shock waves and micro-explosions. In this case, the irradiated material presents central micro-holes surrounded by higher density material. In this regime, the fs laser micromachining is typically used for surface ablation.

3. Glass Micro-chip Fabrication

Technological development of femtosecond laser sources, which can easily achieve peak intensities greater than $10\,\mathrm{TW/cm^2}$, has opened the opportunity of using laser writing-assisted wet etching to produce Lab-On-a-Chip devices buried in fused silica. The technique can be identified by several names, although one of its most familiar acronyms is Femtosecond Laser Irradiation followed by Chemical Etching (FLICE) and it consists of two steps:

(1) The permanent modification of the morphological, physical, and chemical properties of the substrate, by femtosecond laser irradiation.
(2) Selective removal of the modified material through chemical etching.

This can be described as a subtractive 3D printing process which places no constraints on the three-dimensional complexity of the fabricated mini and microstructures.

3.1. *Selective chemical etching*

To enable selective removal of fused silica material by laser-assisted chemical bath, we take advantage of the intermediate one fluence working point (Regime 2, Fig. 3(d)): the material undergoes a refractive index change, shifting from isotropic to birefringent, due to the creation of 3D nanogratings directly buried in glass substrate. The properties of such structures can be precisely controlled by acting on the laser writing parameters. Indeed, it has been observed that nanogratings are oriented orthogonal to the electric field ($\vec{E}$) — i.e., to its polarization — and when arranged parallel to the writing direction ($\vec{S}$), the etching rate can be improved by orders of magnitude[25] (Fig. 4(b)). On the other hand, when the nanostructures are oriented orthogonal to the writing direction, the etching rate is significatively hindered[25] (Fig. 4(c)). Thus, the control over the beam polarization greatly affects the outcome of the fabrication, requiring the use of several different polarization angles within the same geometry. Therefore, when approaching the fabrication of glass Lab-On-a-Chip by exploiting direct writing laser techniques, the generation of nanogratings plays a key role in promoting the access of the etchant solution to the material to be removed, thus improving the diffusion of the reaction products (Fig. 4(a–c)). In addition to this physical modification, the chemical properties of the material also change. Indeed, the high pressure and stress generated in the

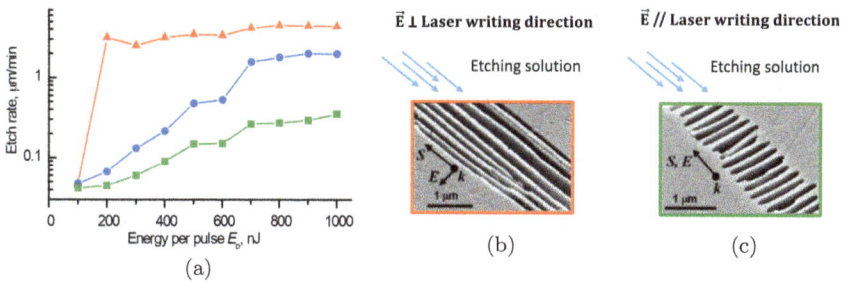

Figure 4.   (a) Microfluidic channel HF etching rate as a function of the pulse energy for different ($\theta = 90°, 45°, 0°$) writing beam polarization relative to writing direction ($\vec{S}$). (b) Electric field ($\vec{E}$) orthogonal to writing direction ($\vec{S}$), in this case the nanogratings' arrangement promotes the etching procedure. (c) $\vec{E}$ is parallel to writing direction ($\vec{S}$). The nanogratings' disposition hinders the etching solution, thus drastically reducing etching performances (Ref. 25).

focal spot volume causes a reduction in the Si-(O)-Si bridging angle.[26] This modification leads to an increased susceptibility of the material to etchant solutions, resulting in a large difference in etch rate between modified and unmodified materials. As a result of a wet etching, the creation of hollow 3D structures monolithically embedded in a fused silica substrate has been easily realized.

To compare the action of different etchant solutions, the selectivity index has been introduced as follows:

$$S = \frac{l/h + r_0}{r_0}, \tag{4}$$

where l is the channel length in μm, h is the etching time in hours, and r_0 is the etching rate of unmodified fused silica in μm/h.[27] This quantity describes the efficiency of the etching solution: the higher the selectivity, the greater the ability of the solution to remove the modified material, while leaving the unmodified areas almost intact.

In this type of application, hydrofluoric (HF) acid in aqueous solution has found widespread use as an etching agent. HF is known to have a high etching rate ($\sim$500–1000 μm/h) but with a moderate selectivity[28] since it also reacts with pure silica with the reaction

$$SiO_2 + 6HF \leftrightarrow H_2SiF_6 + 2H_2O \tag{5}$$

showing an unmodified etching rate of around 20–30 μm/h.

So, when designing the microfluidic geometries (channel network, reaction chambers, accesses, etc.), attention should be paid to the relationship between the etching rate of the modified material compared to the unmodified material as a function of the etching agent.

It should also be stressed that the difference between the two rates can be controlled by writing[29] and etching parameters,[28,30,31] such as reagent concentration, temperature, and others. For example, the terminations of inlet accesses or straight microchannels from which the etching process starts are exposed to the etching solution for longer times, resulting in an unwanted conical shape effect (Fig. 5(a)). To overcome this effect, compensation must be included in the design of the geometry (Fig. 5(b)). This property can prevent the fabrication of accurate self-standing microstructures — such as pillars — or small channels — e.g., bottlenecks and filtering channels — but other solutions have been considered to overcome this problem. Potassium hydroxide (KOH) aqueous solution is a best alternative/combination to HF. KOH almost doubles the selectivity value, but has a limited rate.[28] With the aim of maximizing the yield of the fabrication technique, a useful operational strategy may be as follows: whenever both high volumes of modified material to be removed and fabrication accuracy to be maintained were required, the combination of HF and KOH effects was exploited by performing multiple etching steps.[27] HF bathing has allowed high volumes to be removed in relatively short times,

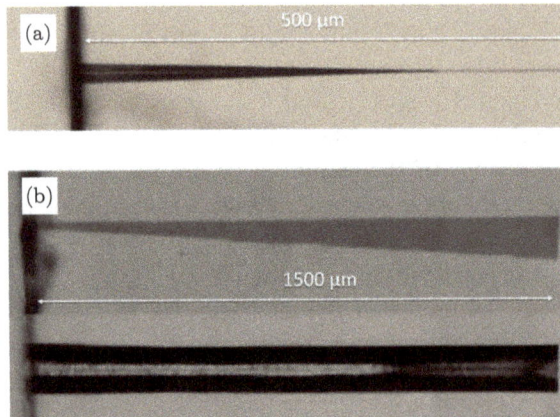

Figure 5.   (a) Conical shape effect due to the isotropic etching behavior of the HF water-based solution. (b) Compensation design to overcome the conical shape effect in the pre-etching (Up) and post-etching case (Down).

allowing areas where a greater accuracy is required to be reached quickly. The etching process was then completed in a KOH solution to open the remaining unetched areas with better precision.

3.2. *Femtosecond micromachining setup*

Figure 6 schematically shows a typical femtosecond laser micromachining setup. Its pulsating heart consists of an amplified Yb:KGW femtosecond laser system (Pharos, Light Conversion) with a fundamental emission wavelength of 1030 nm. Several parameters, including pulse duration (240 fs-10 ps), repetition rate (1 kHz–1 Mhz), pulse energy (up to ~0.2 mJ), and average power (up to ~10 W) are user controlled. The generation of ultrashort and high-power pulses is implemented by a standard chirped pulse amplification mechanism, which consists of the following:

(1) Short-pulse oscillator,
(2) Pulse stretcher\compressor module,
(3) A regenerative amplifier.

The active material for both the laser oscillator and the regenerative amplifier is a potassium-gadolinium tungstate (KGW) crystal doped with 5% ytterbium (Yb) populated with diodes and emission wavelength in the

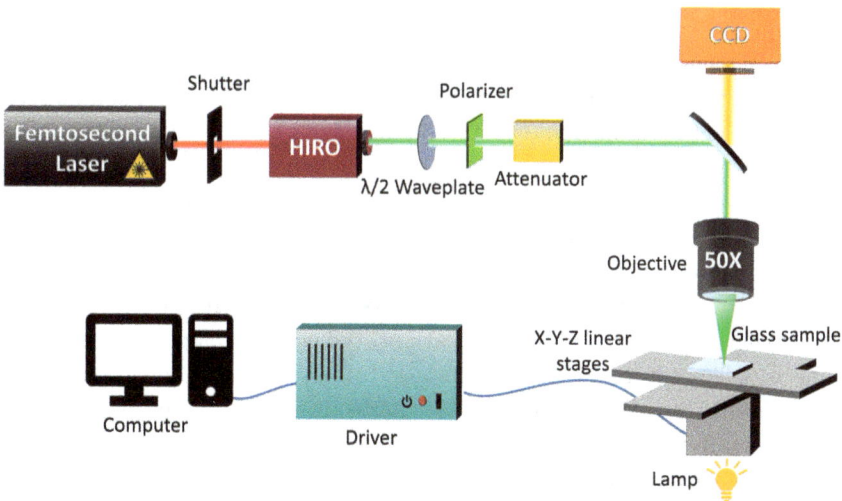

Figure 6.   Femtosecond micromachine system scheme.

near-IR ($\lambda = 1030$ nm). Stable pulses at 80 fs and 67 MHz repetition-rate are generated with the oscillator operating in Kerr-lens mode-locking, but only a fraction of these is selected by the Pockels BBO cell to enter in the amplifier. To reduce its bandwidth and possible distortion, the pulse must be stretched in time before the amplification and then compressed again by a pair of gratings. Finally, an electro-optical shutter allows the repetition rate at which the laser pulses leave the system to be selected without changing the characteristics of the laser cavity. From the laser output to the focusing system, there are several optical elements that control the laser beam characteristics. The first important element is the harmonics generator (HIRO) which allows, in a very user-friendly way, the generation and selection of one of the additional harmonics beside the fundamental (2^{a}H = 515 nm, 3^{a}H = 342 nm, 4^{a}H = 257 nm).

This is achieved inside the HIRO by means of three different BBO crystals in a collinear geometry, taking into account the phase-matching conditions. The second harmonic (515 nm) is typically used in the manufacture of fused silica chips using FLICE technique. Other important beam control elements are power attenuators and polarization rotators, one for each wavelength line, controlled by software. Both tools consist of a half-waveplate that can rotate the laser incident polarization and then the attenuator has a pair of Brewster mirrors to let only the Transverse Electric (TE) component pass thorough, while the Transverse Magnetic (TM) one is blocked. By rotating the incident light polarization, it is possible to fine-tune the percentage of intensity that is transmitted. Finally, the laser light is statically focused through an objective lens inside the substrate. Depending on the substrate and working conditions, different objectives can be selected according to the resolution (typical spot size from $3\,\mu$m $\times$ $3\,\mu$m $\times$ $10\,\mu$ m to sub-micron $0.8\,\mu$m $\times$ $0.8\,\mu$m $\times$ $4\,\mu$m) and the depth of processing required, (from a few hundred microns for oil immersion lenses to centimetres for a long-distance objective lens). The 3-axis computer-controlled motion stages were linked by CAD-based software to an integrated acoustic-optic modulator, which is typically used to move the sample relative to the laser beam with high accuracy. In applications, typically industrial, where write time is a key parameter to be minimized, a second write line based on a galvanic mirror can be implemented: the beam moves vibration-free while the substrate is stationary.

In summary, the interaction of femtosecond laser and transparent materials is a flexible, versatile, and relatively inexpensive way to efficiently fabricate multi-dimensional (3D) index-modified structures without the need for complex photolithographic processes. Chemical etching assisted by femtosecond laser micromachining (widely known as the FLICE technique) is an innovative, simple, and maskless fabrication technique that enables rapid prototyping of a 3D chip thanks to its inherent ability to create buried microstructures.[32,33] Moreover, writing the structure directly into an inert substrate (such as fused silica) has several advantages over the conventional microfluidic device platform (via soft lithography) or microchannel molding (by standard photolithography) in terms of robustness and the ability to develop innovative 3D geometries with corresponding performance. In fact, despite the advantages of mass-scale production (once the mold is defined) and cost-effectiveness, the necessary use of composite polymer materials (PDMS, PMMA SU-8) as substrate in lithography present some critical limitations: (i) the device cannot be manufactured monolithically and therefore requires a sealing step to glue two halves together to obtain a working platform; (ii) the critical mechanical weakness of the geometry and low stiffness of composite polymer materials limits the maximum injection pressure of the fluid; (iii) the gas permeability and low molecular weight chains of composite polymer materials may affect some biological studies. In addition, the ability to rapidly prototype opto-microfluidic chips using fs-laser-assisted micromachining may prove to be a crucial element in the seamless execution of design cycles.

Furthermore, the fused silica material is well known for its optical properties, such as transparency, low background fluorescence, complete inertness — and therefore compatibility with biological samples — presenting relatively low non-specific adsorption and non-permeability to gases.[34] Moreover, its stiffness (shear modulus $\sim$30 GPa) allows the creation of self-standing devices capable of withstanding high input pressures without deformation or mechanical weaknesses. All these characteristics make this substrate an ideal material for microfluidic applications involving the processing of chemical and biological samples or the integration of optical analysis[35] to study the dynamics of the physical concepts that drive some applications, as shown in the following.

4. Needle-free Injectors in a Microfluidic Platform: A Challenging Application for the FLICE Fabrication Technique

As described in the previous section, FLICE has proven to be an excellent tool for selective material removal by enabling the creation of 3D hollow structures that can be monolithically buried in the substrate (typically fused silica). However, for increasing volumes of material to be removed (beyond a few mm^3) in a monolithic device, this direct fabrication technique may suffer from the limiting factor of the input channel/reaction chamber aspect ratio (length/volume), unless an appropriate writing strategy is employed. As evidence of this, a challenging application of large-volume glass monolithic chamber fabrication tuned to needle-free injectors (NFIs) is presented here. To better understand the need and the concept behind these devices, it is necessary to provide some background on how an NFI works.

NFIs typically consist of three main components: (1) a chamber containing the drug to be delivered, (2) an energy source for propelling the drug, and (3) a nozzle from which the fluid is ejected. Commercially available injectors use either compressed air, loaded springs, piezo actuators, or electrical discharge as the propulsion mechanism, producing a jet that is thin and powerful enough to penetrate the skin.[36-38] However, a recent innovation in drug delivery, which does not rely on electromechanical means, has been demonstrated based on the generation of laser-induced cavitation bubbles.[39] The expansion of the bubble drives the liquid out of the chamber through the nozzle, resulting in a more stable jet than those produced by electromechanical means.[40]

The first demonstrations of laser-based NFI were performed on capillary tubes using nanosecond laser pulses. One end of the tube was closed and the laser focused near the open aperture; bubble expansion produced thin (<100 μm) but fast (~600–700 m/s) jets due to kinematic focusing at the exit, which occurs when a liquid surface concave toward the gas is impulsively accelerated by the liquid converging toward the center of the curvature.[39] However, these injectors are not practical for real-world applications because they are partially emptied with each shot, requiring continuous refilling. More interesting, but less common due to their design and manufacturing complexity, are injectors consisting of a chamber containing the liquid and a nozzle from which the liquid is ejected. Mi-ae Park et al.[41] fabricated a chamber divided in two by a flexible membrane to separate the cavitating liquid from the drug to be injected. 3D printing technologies have also been explored to fabricate similar chamber designs, but their limited spatial resolution resulted in excessive roughness in the outlet channel, which severely affected the quality and stability of the jets.[42] The fabrication of chamber using computer numerical controlled lathes also does not achieve the resolution required for NFIs.

To move forward in the optimization of an NFI, it could be crucial to fabricate the chamber in transparent materials: this would allow a detailed study of the bubble dynamics and its morphology, as they are related to the velocity and the ejection volume of the jets. One of the first efforts to fabricate an NFI microfluidic device with a transparent chamber was a hybrid approach. Glass substrates were employed for the spherical-like chamber. In contrast, silicon was chosen for the straight outlet channel, primarily for its compatibility with a well-established plasma dry-etching technique. This combination allowed for confinement within the device and generation of high-speed jets. Although this type of etching has produced nozzles with smooth surfaces, resulting in stable jets, the cavities created in the glass were too small and emptied each time a jet was fired, limiting the practical development of the injector.[43] Moreover, the uncontrolled isotropic nature of this wet etching in glass made it difficult to produce complex 3D structures.

Progress in the development of the chamber injector, demands a significant improvement in the manufacturing process to ensure robustness, rapid prototyping, and design freedom, in terms of nozzle shapes, geometries, and sizes, which should be monolithically buried in the same transparent substrate (fused silica). In this way, the overall geometry of the final device can be easily optimized according to various design parameters, taking advantage of numerical simulations (e.g., COMSOL Multyphysics).

Exploiting FLICE technique, a large-volume glass chamber was fabricated that allows up to 20 shots before refilling (see Fig. 7(a)). In addition, the unique microfluidic configuration of the device allows continuous and automatic refilling of the cavity via an inlet port connected to an infusion pump. The nozzle (here with an internal diameter of 200 μm) is designed to increase the velocity of the jet with a conical shape ending in a 100 μm long cylindrical channel from which the jet is ejected. The entire volume of the syringe chamber was irradiated with 600 nJ/pulse (λ = 515 nm, P_{avg} = 300 mW, repetition rate = 500 kHz), moving the substrate at a speed of 1 mm/s. As the volume of material to be removed posed a challenge for the production of a monolithic glass-buried version of the chip (about 9 mm^3), considerable attention was paid to the development of writing trajectories and light polarization to minimize wet etching times. The chemical treatment was carried out by immersing the sample in an ultrasonic bath of 20% aqueous hydrofluoric acid solution for 10 hours.

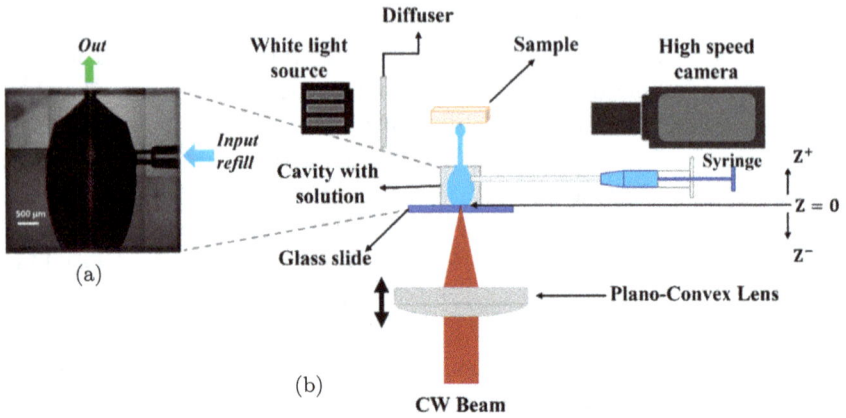

Figure 7. (a) Image of the buried chamber of the needle-free injector in continuous flow. The volume of internal material removed is about $9\,\text{mm}^3$; (b) Experimental setup for the generation of bubbles by thermocavitation and high-speed jets. The laser beam is focused on the glass–liquid interface of the chamber. Bubble and jet dynamics are captured using a high-speed camera.

4.1. *Thermocavitation bubble generation*

Most laser-based NFIs use nanosecond pulsed lasers, as optical breakdown of the liquid is required to form bubbles, but this mechanism usually requires very high intensities[44] ($>10^9$ W/cm^2). Although technological advances have reduced the size of pulsed laser heads, their cost has not been reduced to the same extent. Typically, the cost of nanosecond pulsed lasers exceeds tens of thousands of US dollars, making NFI prohibitively expensive. Therefore, to reduce the cost of NFIs, cheaper laser sources such as CW lasers must be used, but these lasers do not reach the intensity levels to achieve optical breakdown. The question might turn to: is it possible to use CW lasers to create cavitation bubbles in liquids? We will provide a positive answer.

CW lasers typically rely on thermal effects for bubble formation. Thermal effects are a single-photon absorption phenomenon compared to the multiphoton absorption required for optical breakdown. Light absorption can occur in the bulk (either intrinsic or extrinsic by the addition of dopants such as dyes or nanoparticles) or on a substrate (dielectric or metallic). We shall focus solely on the bulk absorption mechanism of heating. The absorption of light in highly absorbent solutions rapidly raises their temperature beyond the boiling point without doing so. Around the spinodal limit ($\sim$300°C for pure water), the liquid reaches a metastable state (superheated water) where any perturbation to the liquid density causes an explosive liquid-to-vapor phase transition producing a rapidly expanding vapor bubble.[45,46] In this case, the bubble is essentially formed at the absorbing substrate–liquid interface, and it evolves attached to the glass substrate, assuming a hemispherical shape. The radius of the bubble depends on the intensity of the laser at the glass–solution interface. At high intensities, the bubbles are the smallest because the heating rate is so high that the spinodal limit is reached in a time scale smaller than or comparable to the heat diffusion time. On the contrary, at low intensities, the heating rate is less than the heat diffusion time, producing larger bubbles. Thermocavitation is a self-organizing phenomenon, which means that when the laser is continuously on, bubble generation (and collapse) occurs in a quasi-periodic manner, allowing the cavitation frequency (defined as the number of bubbles generated per second) to be measured. For single-shot operation of NFI, either an electronic shutter is placed on the beam path or the laser current is time-modulated to control the exposure time to ensure a single cavitation bubble. Thermocavitation is attractive for needleless applications because bubble size and periodicity can be controlled with light intensity. So, this means that delivered volume and number of shots per second are light controlled. Further information on thermocavitation can be found in Refs. 45 and 46.

To generate the thermocavitation bubble, a schematic of the experimental setup is shown in Fig. 7(b). A collimated beam from a CW fiber-coupled laser (IPG Photonics model YLR-5-1064-LP operating at 1064 nm) is focused onto the lower glass–liquid interface of the chamber using a 5 cm focal length lens. The fully transparent chamber is laterally illuminated by a high-power halogen lamp to visualize and record the bubble dynamics captured by a high-speed camera (Phantom v311). The chamber was filled with a saturated solution of copper nitrate (13.78 g copper nitrate dissolved in 10 ml deionized water). The absorption coefficient of the solution at the operating wavelength is $\sim$130 cm^{-1}, which means that light is essentially absorbed near the face entrance. The beam waist radius at the glass–liquid interface (focal point of the lens, $z = 0$) was $\omega_0 \sim 22\,\mu$m, and the corresponding Rayleigh distance ($z_R \sim 1.5$ mm) was calculated using the equations for the transmission of Gaussian beams through a thin lens.[45] Vertical displacement of the lens holder changes the focal position inside ($z > 0$) or outside ($z < 0$) the chamber. This parameter changes the beam spot and therefore the light intensity at the glass-liquid interface. Placing the focal position at different distances, z changes the beam waist according to the following equation:

$$\omega(z) = \omega_0(1 + (z/z_R)^2)^{\frac{1}{2}}.$$

In our experiments, the focal point position was placed at different z-positions and varied from 10 mm ($\omega \sim 152.2\,\mu$m) to 34 mm ($\omega \sim 510.5\,\mu$m) at $z = 4$ mm intervals. For $z < 10$ mm, the cavitation frequency is highest, but the radii of the bubbles were so small that only a small disturbance of the liquid near the nozzle was generated, i.e., no jets are generated for $z < 10$ mm.

Figure 8 shows the importance of fabricating the chamber in a transparent material, as the dynamics of the thermocavitation bubble can be followed. At the time $t = 0\,\mu$s, the laser is switched on, but the cavitation bubble is not created immediately, it takes some time (cavitation time) $\sim$50 μs for the experimental conditions (laser power of 590 mW, $z = 23\,\mu$m and spot size of $= 346.9\,\mu$m). The cavitation time is also dependent on the intensity: the higher the intensity, the shorter the cavitation time. The bubble reaches its maximum radius at $t = 814\,\mu$s and then starts to collapse, reaching its minimum radius at $t = 1073\,\mu$s. At this moment, a high-pressure wave is emitted, easily audible as a loud explosion, which mechanically dissipates most of its energy.[46] During the expansion phase of the bubble, the jet is generated (not shown) and when it begins to collapse, air enters the chamber (the dark region near the exit channel), as shown at time $t = 1073\,\mu$s. The missing liquid is obviously equal to the ejected liquid.

From the video analysis, the bubble dynamics can be obtained for several intensities at the glass–liquid interface (see Fig. 9). Note that the lowest intensity (largest spot) produces the largest bubbles $\sim$1 mm radius, while the highest intensity (smallest spot) produces the smallest bubbles $\sim$0.25 mm radius. This is a unique feature of thermocavitation bubbles. From the data in Fig. 9, the speed of the bubble wall was obtained by taking

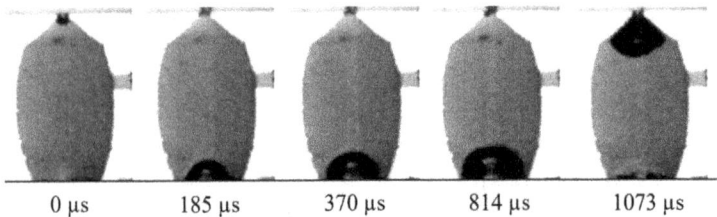

| 0 μs | 185 μs | 370 μs | 814 μs | 1073 μs |

Figure 8. Bubble formation and collapse inside the transparent chamber for a laser power of 590 mW, $z = 23\,\mu$m, and a spot size of $\omega = 346.9\,\mu$m. The bubble achieves its maximum size at 814 μs and collapses at 1073 μs emitting a high-pressure wave.

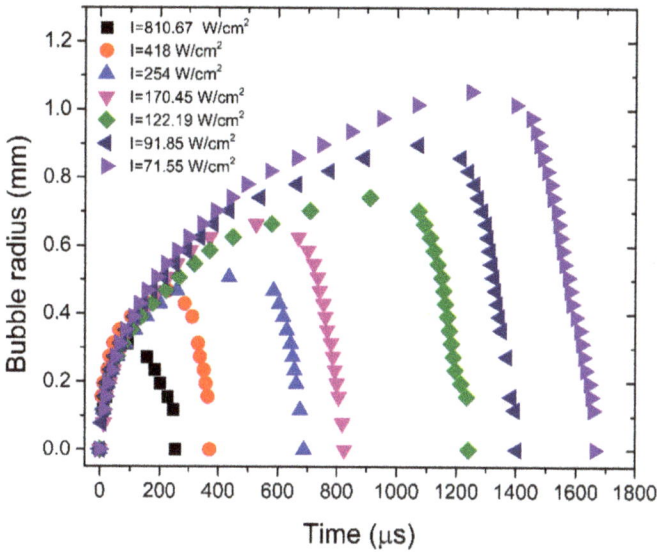

Figure 9. Bubble radius dynamics for different intensities calculated at the glass–liquid interface (or different beam sizes).

the time derivative and the resultant values are in the range 10–25 m/s, i.e., the speed does not vary significantly while the bubble radius does in the same intensity range. Bubble growth and collapse are asymmetric in time, with collapse being faster than growth. The larger the bubbles, the greater the acoustic signal, i.e., the mechanical energy released.[46,47]

An interesting fact is that the temporal dynamics and the maximum bubble radius obtained by thermocavitation (CW lasers) are not so different from those obtained by optical breakdown (pulsed lasers).[44,48,49] This fact further favors the use of CW lasers over pulsed lasers for NFIs.

Bubble dynamics and bubble size determine not only the jet dynamics but also the ejected volume. Figure 10 shows a liquid jet exiting the nozzle (200 μm diameter) which has a unique morphology: a thinner jet followed by a thicker body. The thinner jet is typically faster than the thicker body and could, in principle, help to facilitate skin penetration and drug delivery. To achieve this type of jet, the formation of a meniscus at the channel or even inside the chamber is critical. The pressure pulse that reaches the meniscus causes the fluid to focus.[39] Typical average jet velocities achieved with thermocavitation are in the range of 50–70 m/s, 4–5 times higher than the bubble wall velocity, due to the clever design of the chamber and nozzle geometry. Although kinematic focusing was first reported for

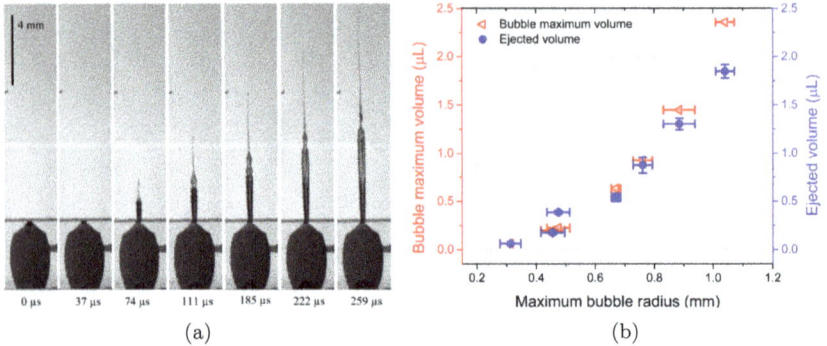

Figure 10.   Water jet ejected from the chamber as a result of bubble expansion. The jet morphology consists of thin and faster jet and a slower but thicker jet. The overall jet speed is ∼70 m/s.

short-pulse laser-assisted jet generation, a comparable effect has also been demonstrated with a CW laser.

A critical parameter for NFI performance is the ability to deliver the desired volume per shot. Typical values for laser-based injectors are in the range of 0.1–1 μl per shot, with the lower limit dominated by short-pulse lasers because kinematic focusing significantly reduces the jet diameter.

This is well below the dose required for medical applications, which is typically 2–3 ml. Despite the low dose delivered by laser-based injectors, there is a niche for low dose delivery.[50] Figure 10(b) shows the maximum bubble radius and the corresponding volume of liquid ejected: the larger the bubble, the larger the volume ejected, ranging from ∼0.1 μl (for the smallest bubble) to ∼2 μl for the largest. For the size of the chamber presented here, up to 2 μl is ejected per shot and can deliver up to 20 shots before refilling. As thermocavitation is a self-organizing phenomenon, the cavitation frequency can be extended up to several kHz by increasing the intensity. In principle, a working medical dose (ml) can easily be achieved in less than one second.

Finally, injection into fresh *ex vivo* porcine skin is demonstrated. The skin was cut into cubes of 1.5 cm side length and placed at a stand-off distance of 5 mm. Immediately after injection, the residual solution at the injection site was removed to prevent diffusion into the skin. Figures 11(a) and 11(b) show the *ex vivo* porcine skin before and after injection. A transverse section of the skin around the injected fluid was made to analyze the fluid penetration, which was ∼4 mm and laterally diffused to almost ∼7 mm.

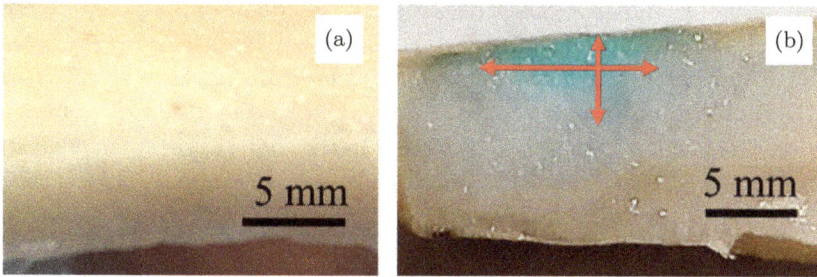

Figure 11. (a) Freshly cut porcine skin before the injection and (b) Transversal cut to the skin after injection to show the liquid penetration into skin. The blue coloration is produced by the copper nitrate solution.

Apparently, this type of pattern is common on skin compared to skin agar phantoms where the injected liquid is more confined.

The use of copper nitrate solution as a light-absorbing material is a perfect candidate to demonstrate the working principle and capability of thermocavitation-based injectors. However, copper nitrate is a toxic and corrosive solution, so a non-toxic solution must be found to determine the depth of penetration and the extension of drug diffusion more accurately in the skin. The most viable option for thermocavitation-based injectors is to divide the chamber into two compartments separated by an elastic and impermeable membrane.[41] One chamber contains the solution in which thermocavitation takes place, while the second one contains the drug to be injected. This prevents thermal damage to the drug. Copper nitrate could be replaced by pure water, but a laser emitting at ~3 μm must be used as the water absorption coefficient is the highest.[51] Alternatively, if more common CW lasers operating in the visible range are used and no external dyes or dopants are added, then a thin layer of metal (e.g., titanium) might be deposited at the bottom of the chamber substrate to create the thermocavitation bubbles.

5. Conclusions

Femtosecond laser micromachining offers unique opportunities for three-dimensional, material-independent, subwavelength processing. It enables the fabrication of three-dimensional photonic and microfluidic devices, even on the same platform, with far greater ease than lithography, and the field is maturing at an extraordinary pace. Thanks to one of its spin-off fabrication techniques — FLICE — it has been possible to produce a large-volume monolithic transparent chamber, avoiding joints and bonding, with obvious advantages in terms of robustness, lack of leakage, and resistance to high pressure. This allowed us to study in detail the dynamics of bubble generation inside the chamber in relation to the shape of the liquid ejection jet. High-speed jets were obtained from thermocavitation bubbles using a CW laser. High-speed camera analysis has shown that the maximum bubble wall velocity is ~10–25 m/s for almost any combination of laser parameters, but with proper chamber design and fabrication, jets with an average speed of ~70 m/s can be produced. The ejected volume can be controlled from ~0.1 μl to ~2 μl simply by changing the focal position within the chamber. The volume delivered by our injector is well below that of electromechanical injectors, which can deliver up to 1 ml per shot. However, the delivery of small volume doses can have certain advantages in terms of the administration of some types of drugs, faster injection rate, greater depth of drug dispersion, and no visible damage to the skin. The resulting jets are characterized by a finer jet (tip) followed by a thicker jet (body). These jets penetrate up to 4 mm into *ex-vivo* porcine skin.

As laser micromachining holds great promise beyond microfluidics and photonics, the transfer of ultrashort (fs) laser pulse technology to the industrial world could enable innovative applications in 3D micromachining and drive the laser industry market.

References

1. G. Steinmeyer, D. H. Sutter, L. Gallmann, N. Matuschek, and U. Keller, Frontiers in ultrashort pulse generation: Pushing the limits in linear and nonlinear optics, *Science* **286**(5444), 1507–1512 (1979).
2. T. Südmeyer, S. V. Marchese, S. Hashimoto, C. R. E. Baer, G. Gingras, B. Witzel *et al.*, Femtosecond laser oscillators for high-field science, *Nat. Photonics* **2**(10), 599–604 (2008).
3. T. Ditmire, J. Zweiback, V. P. Yanovsky, T. E. Cowan, G.Hays, and K. B. Wharton, Nuclear fusion from explosions of femtosecond laser-heated deuterium clusters, *Nature* **398**(6727), 489–92 (1999).
4. R. R. Gattass and E. Mazur, Femtosecond laser micromachining in transparent materials, *Nat. Photonics* **2**(4), 219–25.
5. N. Bloembergen, A brief history of light breakdown, *J. Nonlinear Opt. Phys. Mater.* **25**(6/4):377–85.
6. B. N. Chicbkov, C. Momma, S. Nolte, F. Yon, Alvensleben, and A. Tiinnermann, Femtosecond, picosecond and nanosecond laser ablation of solids, *Appl. Phys. A.* (1996).

7. X. Liu, D. Du, and G. Mourou, Laser ablation and micromachining with ultrashort laser pulses, *IEEE J. Quant. Electron.* (1997).

8. D. Du, X. Liu, G. Korn, J. Squier, and G. Mourou, Laser-induced breakdown by impact ionization in SiO_2 with pulse widths from 7 ns to 150 fs, *Appl. Phys. Lett.* **64**(23), 3071–3073 (1994).

9. B. Stuart, M. Feit, S. Herman, A. Rubenchik, B. Shore, and M. Perry, Nanosecond-to-femtosecond laser-induced breakdown in dielectrics, *Phys. Rev. B Condens. Matter. Mater. Phys.* **53**(4), 1749–61 (1996).

10. E. N. Glezer, M. Milosavljevic, L. Huang, R. J. Finlay, T. H. Her, J. P. Callan *et al.*, Three-dimensional optical storage inside transparent materials: Errata, *Opt. Lett.* **22**(6), 422 (1997).

11. I. Ratkay-Traub, T. Juhasz, C. Horvath, C. Suarez, K. Kiss, I. Ferincz *et al.*, Ultra-short pulse (femtosecond) laser surgery: Initial use in LASIK flap creation, *Ophthalmol. Clin. North Am.* **14**(2), 347–55 (2001).

12. S. P. D. Mangles, C. D. Murphy, Z. Najmudin, A. G. R. Thomas, J. L. Collier, A. E. Dangor *et al.*, Monoenergetic beams of relativistic electrons from intense laser–plasma interactions, *Nature* **431**(7008), 535–538 (2004).

13. J. Qiu, K. Miura, K. Hirao, Femtosecond laser-induced microfeatures in glasses and their applications, *J. Non Cryst. Solids* **354**(12–13), 1100–1111 (2008).

14. J. Thomas, M. Heinrich, J. Burghoff, S. Nolte, A. Ancona, and A. Tünnermann, Femtosecond laser-written quasi-phase-matched waveguides in lithium niobate, *Appl. Phys. Lett.* **91**(15) (2007).

15. S. Kawata, H. B. Sun, T. Tanaka, and K. Takada, Finer features for functional microdevices, *Nature* **412**(6848), 697–698 (2001).

16. R. Osellame, G. Cerullo, and R. Ramponi, Femtosecond laser micromachining: Photonic and microfluidic devices in transparent materials, *Top. Appl. Phys.* **123**, 3–18 (2012).

17. L. V. Keldysh, Ionization in the field of a strong electromagnetic wave, *J. Exptl. Theoret. Phys. (USSR).* **20**, (1965).

18. C. B. Schaffer, A. Brodeur, and E. Mazur, Laser-induced breakdown and damage in bulk transparent materials induced by tightly focused femtosecond laser pulses [Internet], *Meas. Sci. Technol.* **12**, 1784–1794 (2001).

19. B. C. Stuart, M. D. Feit, A. M. Rubenchik, B. W. Shore, and M. D. Perry, Laser-induced damage in dielectrics with nanosecond to subpicosecond pulses, *Phys. Rev. Lett.* **74**(12), 2248–2251 (1995).

20. N. Bloembergen, Laser-induced electric breakdown in solids, *IEEE J. Quant. Electron.* **10**(3), 375–386 (1974).

21. S. Bonfadini, T. Nicolini, F. Storti, N. Stingelin, G. Lanzani, and L. Criante, Femtosecond laser-induced refractive index patterning in inorganic/organic hybrid films, *Adv Photonics Res.* **3**(8), 2100257 (2022).

22. R. Taylor, C. Hnatovsky, and E. Simova, Applications of femtosecond laser induced self-organized planar nanocracks inside fused silica glass, *Laser Photonics Rev.* **2**, 26–46 (2008).

23. K. Itoh, W. Watanabe, S. Nolte, and C. B. Schaffer, Ultrafast processes for bulk modification of transparent materials, *MRS Bull.* **31**(8), 620–625 (2006).

24. R. Osellame, H. J. W. M. Hoekstra, G. Cerullo, and M. Pollnau, Femtosecond laser microstructuring: An enabling tool for optofluidic lab-on-chips, *Laser Photonics Rev.* **5**, 442–463 (2011).

25. C. Hnatovsky, R. S. Taylor, E. Simova, V. R. Bhardwaj, D. M. Rayner, and P. B. Corkum, Polarization-selective etching in femtosecond laser-assisted microfluidic channel fabrication in fused silica, *Opt. Lett.* **30**(14), 1867 (2005).

26. A. Marcinkevičius, S. Juodkazis, M. Watanabe, M. Miwa, S. Matsuo, and H. Misawa *et al.*, Femtosecond laser-assisted three-dimensional microfabrication in silica, *Opt. Lett.* **26**(5), 277 (2001).

27. V. Stankevič, Formation of rectangular channels in fused silica by laser-induced chemical etching, *Lithuanian J. Phys.* **54**, 136–141 (2014).

28. S. Kiyama, S. Matsuo, S. Hashimoto, and Y. Morihira, Examination of etching agent and etching mechanism on femotosecond laser microfabrication of channels inside vitreous silica substrates, *J. Phys. Chem C.* **113**(27), 11560–11566 (2009).

29. D. A. Yashunin, Y. A. Malkov, and A. N. Stepanov, Fabrication of micro-capillaries in fused silica using axicon focusing of femtosecond laser radiation and chemical etchingion/ms, *Quantum Electron (Woodbury).* **43**(4), 300–303 (2013).

30. V. Maselli, R. Osellame, G. Cerullo, R. Ramponi, P. Laporta, and L. Magagnin *et al.*, Fabrication of long microchannels with circular cross section using astigmatically shaped femtosecond laser pulses and chemical etching, *Appl. Phys. Lett.* **88**(19) (2006).

31. C. Hnatovsky, R. S. Taylor, E. Simova, P. P. Rajeev, D. M. Rayner, V. R. Bhardwaj *et al.*, Fabrication of microchannels in glass using focused femtosecond laser radiation and selective chemical etching, *Appl. Phys A.* **84**(1–2), 47–61 (2006).

32. F. Simoni, S. Bonfadini, P. Spegni, S. Turco, D. E. Lo. Lucchetta, and L. Criante, Low threshold Fabry-Perot optofluidic resonator fabricated by femtosecond laser micromachining, *Opt. Express* **24**(15), 17416–17423 (2016).

33. M. Massetti, S. Bonfadini, D. Nava, M. Butti, L. Criante, G. Lanzani *et al.*, Fully direct written organic micro-thermoelectric generators embedded in a plastic foil, *Nano Energy* **75**, 104983 (2020).

34. P. N. Nge, C. I. Rogers, and A. T. Woolley, Advances in microfluidic materials, functions, integration, and applications, *Chem. Rev.* **113**(4), 2550–2583 (2013).

35. F. Storti, S. Bonfadini, A. Donato, and L. Di Criante, 3D in-plane integrated micro reflectors enhancing signal capture in lab on a chip applications, *Opt. Express* **30**(15), 26440 (2022).

36. A. Ravi, D. Sadhna, D. Nagpaal, and L. Chawla, Needle free injection technology: A complete insight, *Int. J. Pharm. Investig.* **5**(4), 192 (2015).

37. J. Baxter and S. Mitragotri, Needle-free liquid jet injections: Mechanisms and applications, *Exp. Rev. Med. Dev.* **3**, 565–74 (2006).

38. S. Mitragotri, Current status and future prospects of needle-free liquid jet injectors, *Nat. Rev. Drug Discov.* **5**(7), 543–548 (2006).

39. Y. Tagawa, N. Oudalov, C. W. Visser, I. R. Peters, D. van der Meer, C. Sun *et al.*, Highly focused supersonic microjets, *Phys. Rev X* **2**(3) (2012).
40. M. Moradiafrapoli and J. O. Marston, High-speed video investigation of jet dynamics from narrow orifices for needle-free injection, *Chem. Eng. Res. Des.* **117**, 110–121 (2017).
41. M. A. Park, H. J. Jang, F. V. Sirotkin, and J. J. Yoh, Er:YAG laser pulse for small-dose splashback-free microjet transdermal drug delivery, *Opt. Lett.* **37**, 3894–3896 (2012).
42. R. Zaca-Morán, J. Castillo-Mixcóatl, N. E. Sierra-González, J. M. Pérez-Corte, P. Zaca-Morán, J. C. Ramírez-San-Juan *et al.*, Theoretical and experimental study of acoustic waves generated by thermocavitation and its application in the generation of liquid jets, *Opt. Express* **28**(4), 4928 (2020).
43. C. Berrospe-Rodriguez, C. W. Visser, S. Schlautmann, R. Ramos-Garcia, and D. F. Rivas, Continuous-wave laser generated jets for needle free applications, *Biomicrofluidics* **10**(1), (2016).
44. A. Vogel, K. Nahen, D. Theisen, and J. Noack, Plasma formation in water by picosecond and nanosecond Nd:YAG laser pulses-part I: Optical breakdown at threshold and superthreshold irradiance, *IEEE J. Sel. Top. Quant. Electron.* **2**, 847–860 (1996).
45. B. E. A. Saleh, *Teich MCarl. Fundamentals of Photonics.* Wiley (1991), p. 966.
46. J. P. Padilla-Martinez, C. Berrospe-Rodriguez, G. Aguilar, J. C. Ramirez-San-Juan, and R. Ramos-Garcia, Optic cavitation with CW lasers: A review, *Phys. Fluids* **26**(12) (2014).
47. J. C. Ramirez-San-Juan, E. Rodriguez-Aboytes, A. E. Martinez-Canton, O. Baldovino-Pantaleon, A. Robledo-Martinez, N. Korneev *et al.*, Time-resolved analysis of cavitation induced by CW lasers in absorbing liquids, *Opt. Express* **18**(9), 8735 (2010).
48. P. K. Kennedy, D. X. Hammer, and B. A. Rockwell, Laser-induced breakdown in aqueous media, *Prog. Quant. Electr.* **21**, 1997.
49. A. Vogeland and W. Lauterborn, Acoustic transient generation by laser-produced cavitation bubbles near solid boundaries, *J. Acoust. Soc. Am.* **84**(2), 719–31 (1988).
50. J. Schoppink and D. Fernandez Rivas, Jet injectors: Perspectives for small volume delivery with lasers, *Adv. Drug Delivery Rev.* **182** (2022).
51. K. J. Linden, C. P. Pfeffer, J. G. Sousa, N. D'Alleva, A. Aslani, G. Gorski *et al.*, 3-μm CW lasers for myringotomy and microsurgery, In *Photonic Therapeutics and Diagnostics*, Vol. IX. San Francisco: SPIE (2013), p. 85652P.

Chapter 9

Novel Photo-sensitive Materials for Microengineering and Energy Harvesting

D. Sagnelli[*,¶], A. Vestri[*], A. D'Avino[*], M. Rippa[*], V. Marchesano[*],
F. Ratto[*], A. De Girolamo Del Mauro[†], F. Loffredo[†],
F. Villani[†], G. Nenna[†], G. Ardila[‡], P. Meneroud[§], J. Gauthier[§],
S. Duc[§], M. Thomachot[§,‖], F. Claeyssen[§], and L. Petti[*]

[*]*Institute of Applied Sciences and Intelligent Systems of CNR,
Pozzuoli, Italy*
[†]*ENEA, Portici Research Centre, Piazzale E. Fermi 1,
Napoli, Portici, Italy*
[‡]*University Grenoble Alpes, Univ. Savoie Mont Blanc,
CNRS, Grenoble INP, IMEP-LaHC, Grenoble, France*
[§]*Cedrat Technologies, 59 Chemin du Vieux Chêne,
38240 Meylan, France*
[¶]*domenico.sagnelli@isasi.cnr.it*
[‖]*mathieu.thomachot@cedrat-tec.com*

According to the European parliament, Research Europe's journey toward technological sovereignty should cover Key Enabling Technologies (KETs) which include Advanced Materials. The new active materials, that are presented hereafter and that have been supported by the UE R&D funded PULSE-COM project, are photomobile polymers. These active materials are characterized by their ability to generate significant strokes under light excitation, which allows the development of new actuation devices. The recent progress made in the manufacturing of new photomobile films offers innovative solutions for light-induced motion actuators and devices. Indeed, such films can be assimilated as transducers thanks to their ability to convert light into displacement. By adjusting the incident light parameters (wavelength, exposure time ...) the photomobile films' actuation can be controlled to answer many applications requesting high displacements and low forces. In these regards, the behavior of the photomobile films was characterized prior to their integration in more complex devices. Then, several proof-of-concepts of these devices were manufactured to try to bring new functionalities to the market such as light-driven optical switch, optical micro-valve, and deflector. Applications in the harvesting field have also been preliminarily explored and discussed.

1. Introduction

Smart materials that respond to external stimuli such as temperature, light, pressure, and electric or magnetic fields have gained significant interest in recent times. These materials possess the unique ability to adjust their size, rigidity, shape, or other physical characteristics in response to their environment, allowing for the development of innovative and dynamic products.[1-9]

In the medical field, for example, smart materials are being used to develop implanted devices such as pacemakers and prosthetic joints that can respond to the body's impulses and adapt accordingly.[10,11] Another area of interest for medical applications is the creation of hydrogels, which are materials that can change their structure or volume in response to stimuli such as temperature or electric fields, enabling the targeted release of drugs.[12,13]

Among these smart materials, photomobile materials have emerged as a novel class that has attracted considerable attention due to their capability to respond to light stimuli and produce precise movements. They represent a rapidly evolving group of stimuli-responsive materials that can convert light energy into mechanical work[14-19] (Fig. 1) and exhibit properties that allow them to undergo controlled and reversible deformations upon exposure to specific wavelengths of light. These materials have diverse potential applications ranging from optoelectronics, microengineering, biomedical to environmental science. Over the past few decades, their scientific and technological interest has grown exponentially, driven by the need to address challenges in fields such as soft robotics,[14,17] artificial muscles,[20] and adaptive optics.[21,22] (Fig. 2). It is important to underline that such material revolution will have to further develop toward a more sustainable approach integrating smart materials with, for example, biomass-extracted monomers[23] that are promising for high-end applications or cheaper biopolymers like GMO starches.[24]

The development of photomobile materials was spurred by the observation of light-induced deformations in polymers containing azobenzene.[25,26] Azobenzene, a photochromic molecule that can undergo reversible trans–cis isomerization upon exposure to light, was first synthesized in the 19th century,[27] and has since garnered significant interest from researchers. By incorporating azobenzene moieties into various polymer backbones, scientists have created photomobile materials with impressive actuation capabilities. The successful development of these systems has motivated

(a)   (b)   (c)

(d)

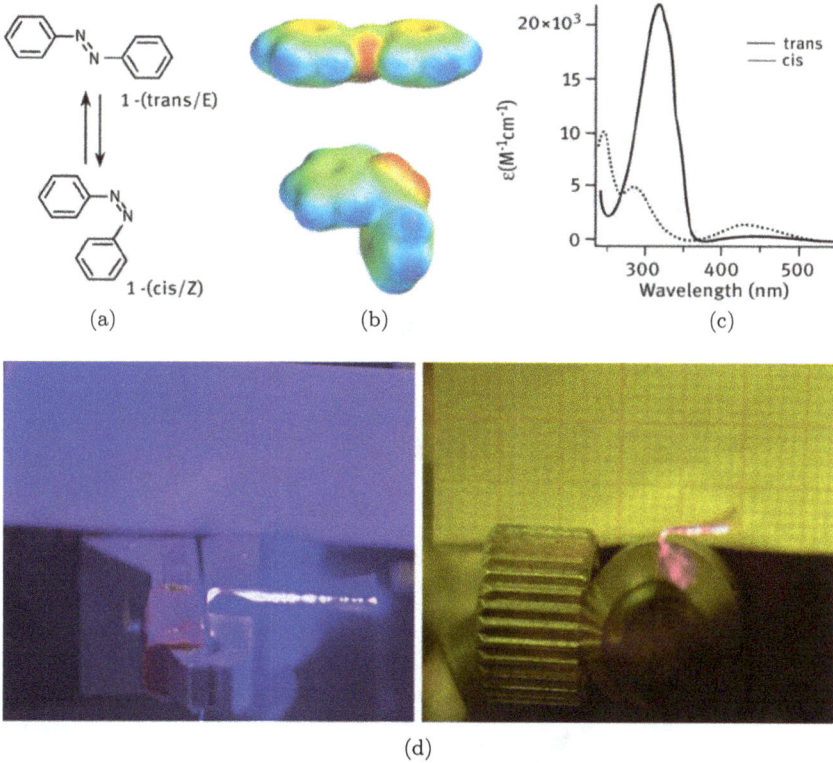

Figure 1. (a) Structures of trans and cis isomers of azobenzene.[19] (b) Space filling models that are colored by electrostatic potential (red — negative to blue — positive).[19] (c) Electronic absorption spectra of the isomers trans and cis in solvent.[19] (d) Photomobile material using azobenzene as engine.

scientists to continue exploring ways to improve their properties even further. Azobenzene is a prime example of a molecule that can serve as the driving force for these photo-responsive materials.[25,28] Its unique ability to change its shape and structure in response to UV-light makes it an attractive candidate for the development of smart materials that can respond to light.[29] Examples of ongoing research include tailoring the materials' response to specific wavelengths of light, improving their fatigue resistance, and enhancing their mechanical performance.[30] Advancements in the properties of photomobile materials have the potential to unlock a variety of exciting new applications for these materials. Overall, the development of photo-responsive materials based on azobenzene holds great promise for a wide range of technological applications. One promising

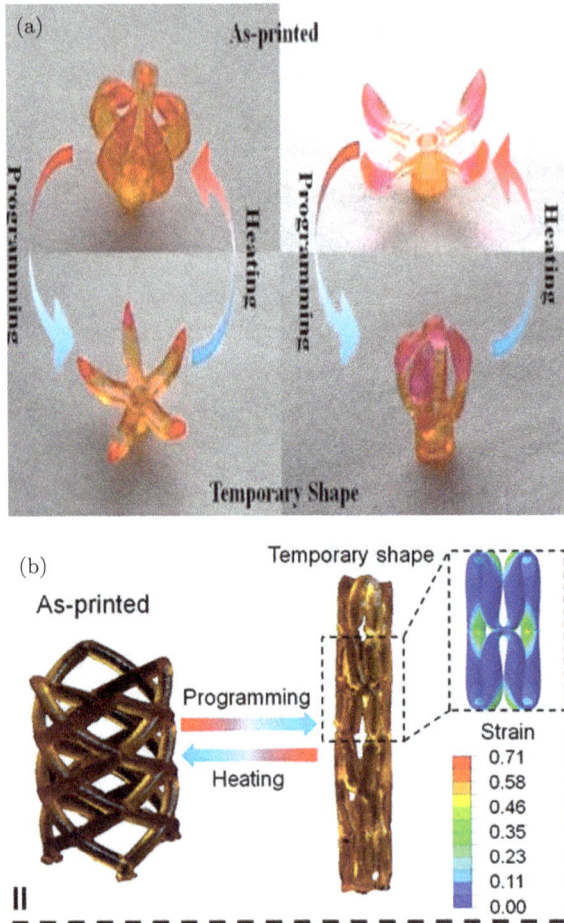

Figure 2. (a) Soft robot example of a grappling arm.[90] (b) Early example of thermosensitive artificial muscle.[90]

possibility is the use of these materials in creating smart windows that can automatically adjust their transparency in response to sunlight.[31] Another exciting application involves the development of responsive coatings that can adapt to their environment.[32] Azobenzene's photo-responsive properties have been used in many fields, including the creation of light-responsive surfaces and textiles.[33–35]

In the field of optoelectronics, azobenzene has the potential to be used in the development of light-responsive switching mechanisms, such as optical switches and thin films that respond to light.[36,37] It is also being studied for its potential use in the creation of intelligent materials that respond to changes in light, such as self-cleaning surfaces or UV-resistant fabric.[33-35] Azobenzene is being researched for its potential medical applications, such as in the development of light-activated delivery devices that can target specific cells or tissues in the body, increasing their efficacy and reducing the likelihood of side effects.[38,39] It is also being investigated as a potential component in materials that can convert light into electrical energy, which could lead to the creation of highly efficient and affordable solar cells.[40] Azobenzene is a highly adaptable and promising photo-sensitive molecule that has the potential of making an impact in various scientific and technological fields. As research and development in this area continue, we can expect to see further developments and exciting new uses soon. Building on the pioneering work with azobenzene-containing polymers, the research community has expanded its focus to encompass other photo-responsive molecules and polymer architectures, including spiropyrans, diarylethenes, and photo-responsive liquid crystalline elastomers.[41,42] The ongoing progress and expansion of photomobile materials have enhanced their flexibility and suitability, enabling them to be utilized in diverse applications within different fields. Further exploration and enhancement of these materials will undoubtedly generate novel and imaginative applications yet to be projected.

2. Photo-responsive Polymers

Photo-responsive polymers, as already explained, exhibit a change in their properties, such as conformation, mechanical behavior, or optical characteristics, in response to light stimuli. The ability to harness light as an external stimulus offers a non-contact, reversible, and precise means of controlling the material's behavior, making photo-responsive polymers particularly appealing for numerous innovative technologies. The underlying mechanism behind the photo-responsive behavior of these polymers typically relies on the incorporation of photochromic moieties into the polymer structure.[2,16] These moieties, when exposed to specific wavelengths of light, undergo reversible photochemical reactions or isomerizations, which in turn induce changes in the polymer's properties. One prominent example of such a photochromic moiety as already discussed is azobenzene. Polymers containing azobenzene moieties can exhibit controlled and reversible deformations, actuations, or changes in their optical properties, depending on the position and orientation of azobenzene within the polymer chain. An alternative

approach to designing photo-responsive polymers involves the creation of
bilayer polymer systems, often referred to as photomobile polymer materials
(PMPs). These materials consist of two layers with distinct thermal expansion coefficients and light-responsive properties, enabling them to bend,
twist, or curl when exposed to light. By carefully selecting and tailoring
the properties of each layer, it is possible to design PMPs with highly
controlled and specific responses to light stimuli. Overall, the vast array
of photo-responsive polymers demonstrates the immense potential of this
class of materials for a myriad of applications. With continued research and
development, these materials are poised to make significant contributions
to the advancement of various fields, ultimately shaping the future of
light-responsive technologies.

2.1. *Mechanisms of photo-responsiveness*

Photo-responsive materials can be found to rely on various mechanisms
that enable them to change their properties upon exposure to light.
These mechanisms play a critical role in the design and function of smart
materials, allowing for tunable characteristics that facilitate their use
across a wide range of applications. Some of the most prominent mechanisms of photo-responsiveness include photo-isomerization, bilayer systems,
Spiropyran-merocyanine photo-isomerization, photodimerization, and photothermal effects discussed in the following sections.

2.1.1. *Photo-isomerization*

As previously mentioned, azobenzene moieties can undergo reversible trans–
cis isomerization upon exposure to specific wavelengths of light. This
light-induced isomerization leads to changes in the molecular geometry,
dipole moment, and optical properties of the material. These changes can
manifest in various ways, such as deformations, actuations, or alterations
in optical behavior, depending on the nature of the polymer matrix and the
positioning of the azobenzene units within the material.

The azobenzene chromophore is normally used in the construction
of photo-isomerization-based PMP. Cyanostilbene and its derivatives
have interesting features to make intelligent optoelectronic materials.[44]
P4VP(Z-TCS)x is a series of hydrogen-bonded supramolecular polymers
made of poly(4-vinyl pyridine) (P4VP) as the main chain and a — cyanostilbene derivative (Z-TCS) as side groups. Under UV irradiation, the polymers

exhibit good photo-induced deformation ability in terms of both bending speed and bending angle.[43] Furthermore, due to the uneven distribution of Z and E-cyanostilbene isomers along the radial direction of the fibers, the shape of these fibers formed by UV irradiation can be eliminated via external stretching force and recovered under heating. To the best of our knowledge, this is the first report of photo-induced deformation showed by a polymer based on — cyanostilbene, which broadens the applications of the — cyanostilbene derivative as a photodeformable material.[43]

These polymers demonstrated remarkable photo-induced deformation capabilities when subjected to UV irradiation. Other examples are the electrocyclic ring formation and cleavage reaction of diarylethenes, electrocyclic ring formation and cleavage isomerization of furylfulgide, intramolecular hydrogen transfer reaction of salicylideneanilines, intramolecular bond linkage isomerization of a nitropentaamminecobalt(III) complex, $2\pi + 2\pi$ cycloaddition reaction, and $4\pi + 4\pi$ cycloaddition reaction.

2.1.2. *Bilayer systems*

Bilayer photomobile polymer materials (PMPs) are innovative materials consisting of two distinct layers,[45] each exhibiting unique thermal expansion coefficients and photo-responsive properties (Fig. 3). These materials have gained significant attention in recent years for their simple synthesis procedure.[46] The two layers in bilayer PMPs are carefully chosen to have different thermal expansion coefficients, which means that they expand or

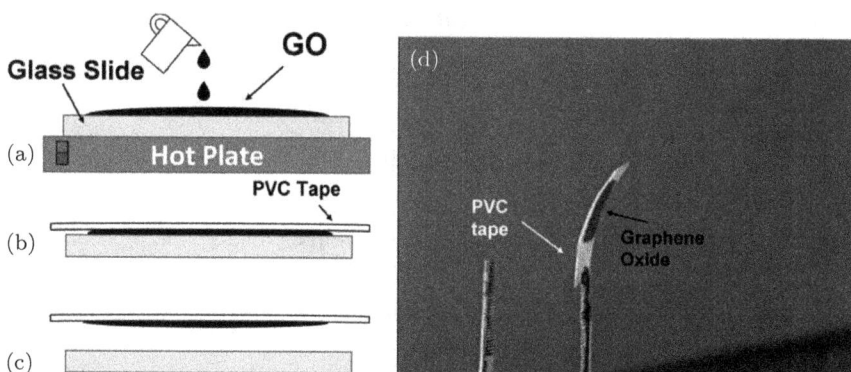

Figure 3. (Left) Sketch of the workflow. Graphene Oxide (GO) is deposited on a glass substrate (60°C). The PVC tape is placed in contact with the dried GO film, and the bilayered structure is peeled from the substrate. (Right) Final result.[48]

contract at different rates when subjected to changes in temperature. Additionally, these layers are engineered to exhibit photo-responsive properties, meaning that they can change their shape or dimensions in response to exposure to light. An interesting example is shown by Zhang *et al.* (2020) with soft bilayer actuators with a straightforward fabrication procedure, variable molecular alignment, and multi-stimulus response. Smart bilayer membranes were tailored into a variety of delicate biomimetic devices, including a bionic butterfly, a bionic leaf, and a foot robot, offering a wide range of applications in biomimetic and intelligent soft robotics disciplines.[46] When bilayer PMPs are exposed to light, the differential response of the two layers results in a variety of mechanical deformations, such as bending, twisting, or curling of the material. This is because the layers expand or contract at different rates in response to the light stimulus, leading to stress buildup at the interface between the layers. The stress, in turn, causes the material to change its shape in a controlled manner. The precise control of the material's shape change is achieved by carefully tuning the properties of the individual layers, such as their thickness, composition, and photo-responsive characteristics. By adjusting these parameters, researchers can develop bilayer PMPs with specific deformations in response to different wavelengths, intensities, or durations of light exposure. This kind of actuator is used for high-tech applications such as nanopositioners. This nanopositioner is very internally dynamic due to its capacity to dynamically modify IR intensity/temperature to vary the mechanical stiffness of the polymer chains in specific composite locations.[47]

2.1.3. *Spiropyran-merocyanine photo-isomerization*

Spiropyran-merocyanine photo-isomerization is a fascinating photochemical process in which spiropyran molecules undergo a reversible structural transformation to merocyanine upon exposure to light[49,50] as shown in Fig. 4. The phenomenon involves significant changes in molecular structure, color, and polarity, and has garnered considerable attention for its potential applications in diverse fields such as materials science, chemistry, and biotechnology. During the photo-isomerization process, spiropyran molecules absorb light energy, which triggers the transformation of their closed, colorless form to the open, colored merocyanine form. This change in structure is accompanied by a shift in color, often from colorless to a vibrant hue, as well as an increase in polarity. Importantly, this transformation is reversible, meaning that when the light stimulus is removed or altered,

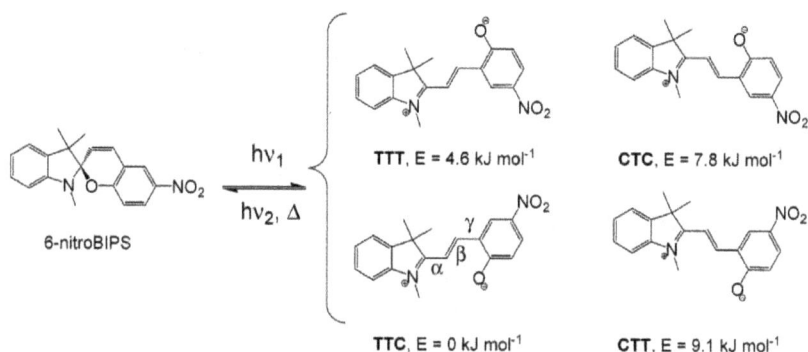

Figure 4. After 6-nitroBIPS spiropyran is exposed to UV light, several merocyanine isomers, including the TTC and TTT forms, are produced.[50]

the merocyanine molecules can revert to their original spiropyran form. The unique photo-responsive properties of spiropyran-merocyanine systems can be harnessed for a wide range of applications, including: Hydrogels containing spiropyran moieties, which can undergo reversible changes in swelling behavior in response to light exposure,[51,52] photo-switchable adhesives,[53] as a molecular optical switch,[54] and more.

2.1.4. *Photodimerization*

In certain photo-responsive materials, exposure to light initiates a process known as reversible photodimerization.[55] This phenomenon involves the formation and cleavage of covalent bonds between molecules upon exposure to specific wavelengths of light. This reversible reaction has attracted considerable interest for its potential applications in the development of smart materials, soft robotics, sensors, and photopatterning techniques. When photo-responsive materials undergo photodimerization, their properties can be significantly altered. Some of these changes include variations in mechanical properties, solubility, or swelling behavior. As a result, photodimerization can be harnessed to create photo-responsive crosslinked networks that exhibit dynamic and adaptive characteristics. Moreover, the photodeformation of P4VP using photodimerization has been explored. In one instance, coumarin was incorporated as a photoreactive component, leading to photomechanical effects driven by hydrogen bonding in polymer composites that used P4VP as a host polymer. Consequently, the composite film curved toward the light source due to the photodimerization of coumarin pendant groups.[56,57]

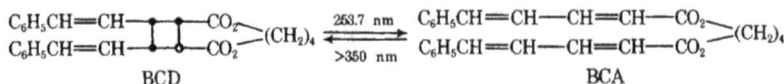

Figure 5. Example of photodimerization of 1,4 butanediol dicinnamylideneacetate (BCA) and its photo-cycled compound BCD.

2.1.5. Photothermal effects

Some materials exhibit photo-responsive behavior due to the photothermal effect, in which the absorption of light energy results in localized heating. This heating can induce changes in the material's mechanical properties, cause phase transitions, or drive photothermal actuation. Photothermal-responsive materials can find applications in soft robotics, controlled drug delivery systems, and adaptive surfaces. These photo-responsive mechanisms represent just a few examples of the diverse range of strategies employed by photo-responsive materials to convert light energy into a tangible response. With continued research and innovation, the development of new photo-responsive mechanisms and materials is poised to drive the advancement of smart materials and their applications across various fields.

2.2. Azobenzene in photomobile polymers

Azobenzene, a highly versatile photochromic molecule, has garnered significant attention in the field of smart materials due to its unique structure and properties. Composed of two phenyl rings connected by a nitrogen double bond (N = N), azobenzene exists in two primary isomeric forms: trans and cis.[58] In its stable trans configuration, the phenyl rings are positioned opposite one another, whereas in the cis configuration, they are adjacent. The key to azobenzene's remarkable photo-responsive behavior lies in its ability to undergo reversible trans–cis isomerization upon exposure to UV-light.

The energy levels of azobenzene play a crucial role in its photo-isomerization process. The highest occupied molecular orbital (HOMO) and the lowest unoccupied molecular orbital (LUMO) are responsible for the molecule's electronic transitions.[59,60] When azobenzene absorbs light, typically in the UV region, around 320–360 nm, the π–π^* transition occurs, promoting an electron from the HOMO to the LUMO. This transition results in a structural transformation from the trans to the cis isomer, which usually has an energy difference of approx. 50 kJ/mol higher than the trans isomer.[61]

This process is accompanied by a change in the molecule's molecular geometry, dipole moment, and optical properties. Subsequently, when exposed to visible light or heat, the molecule reverts to its original trans configuration through the less energetic $n-\pi^*$ transition (usually around 420–500 nm).[62] This reversible isomerization allows for the controlled and repeatable modulation of the material's properties, making azobenzene an attractive candidate for a wide range of applications, including photo-responsive polymers, molecular switches, and optical data storage devices.

Incorporating azobenzene moieties into the polymer or as dopants in various materials enables the transfer of the molecule's photo-responsive characteristics to the resulting composite material. Consequently, these materials can exhibit light-induced deformations, actuations, or changes in optical properties. The continued study and understanding of azobenzene's structure, energy levels, and properties have opened the door to the development of innovative, light-responsive technologies with the potential to revolutionize numerous industries. And the incorporation of azobenzene into diverse polymer matrices has enabled the development of photo-responsive materials with a wide range of attributes and applications.[20,21] A significant factor that influences the properties of these materials is the position of the azobenzene moieties within the polymer chain and how the elastomer is prepared. In fact, the most performing photomobile polymers containing azobenzene are liquid-crystal-based elastomers (LCEs). LCEs are a unique class of materials that exhibit properties of both elastomer and liquid crystals.[63] Because of their unique characteristics, they can undergo large and reversible deformations in response to various stimuli, such as changes in temperature, light, or mechanical stress. There are several reasons why LCEs can outperform other materials:

- *Large deformations*: LCEs can undergo large and reversible deformations, which makes them ideal for use in applications that require stretchable or bendable materials.
- *Anisotropic behavior*: LCEs have anisotropic behavior, meaning they exhibit different physical properties in different directions. This makes them useful in applications that require materials with directional properties.
- *Responsiveness to stimuli*: LCEs are highly responsive to external stimuli, such as temperature or light. This makes them useful in applications

that require materials to change their physical properties in response to changing conditions.

- *Low density*: LCEs typically have a lower density than other plastics, which can make them useful in applications where weight is a concern.

Overall, the unique combination of properties exhibited by LCEs makes them highly versatile and useful in a wide range of applications, including soft robotics, biomedical devices, and smart textiles.

By harnessing the photo-induced isomerization process and carefully tailoring the position of azobenzene within the polymer structure, these materials can exhibit controlled and reversible deformations, actuations, or changes in optical properties. The versatility and tunability of azobenzene-based photo-responsive LCEs render them ideal candidates for a myriad of applications, including soft robotics, adaptive optics, molecular switches, and optical data storage devices. For example, in the frame of the PULSE-COM project (H2020-FETOPEN-2018-2020, 863227) various liquid-crystal mixtures with 100% azobenzene moieties were tested to understand the thermo/mechanical phenomena that regulate their photo-actuation.[2,64] Particularly, three different 100% azo-based LC mixtures were developed, using an azo linear monomer (L) and an azo crosslinker (C) in different ratios (L:C molar ratio of 8:1, 1:1, or 1:8). The nematic temperature of the LC monomer mixtures (in absence of the initiator) was characterized using a dedicated transmittance measurement experiment (scheme reported in Fig. 6), making DSC measurements inconclusive.

In particular, the state of order (S) of the LC monomer mixtures was estimated to evaluate the orientational order of the LCs as a function of temperature, using a heater chamber with an optical window to perform the linear dichroism measurements (Fig. 6(a)). Birefringence measurements (Fig. 6(b)) of LC monomer mixtures were also performed to narrow down the range of nematic temperatures. Both kinds of measurements were performed for all LC formulations to investigate the effect of the ratio between linear and cross-linker monomers on the LCs' organization. Based on the reported results, the range of nematic temperatures for the mix with the highest content in linear monomer (LC81) was 81.5–83°C; while 87–88°C was the range for the mix with the lowest content in linear monomer (LC18). At an equal molar ratio between the two LC monomers (LC11), the nematic temperature was estimated only on the observation of the direct transmittance, characterized by a slight peak at 84°C. In fact, for this formulation, the state-of-order calculation and the birefringence

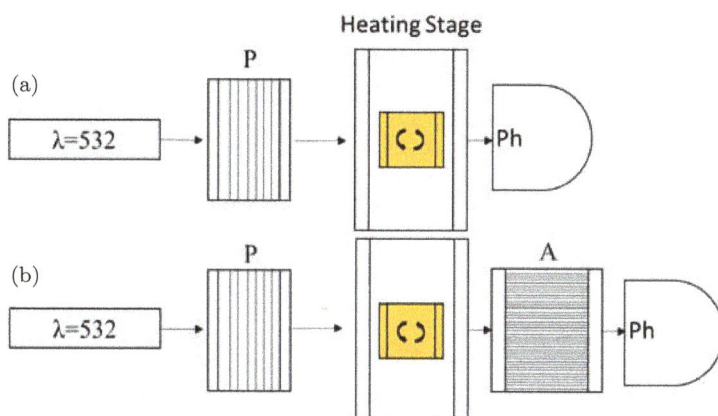

Figure 6.   Schematic representation of the transmittance measurement setup used for LCEs' characterization. (a) Configuration used to measure the state of order of the mixtures (the reactor cell is oriented parallel or orthogonal to the table). (b) Crossed polarized configuration (the reactor cell is oriented 45° as compared to the table). P and A are polarizer and analyzer, respectively. Ph is a photodiode used to measure transmittance.

measurement were less informative, probably because the presence in equal amounts of the two monomers hinders the fast and stable achievement of an equilibrium state.

The polymerization of the 100% azo elastomers was realized using an initiator not typically reported for the synthesis of azo-based LCEs. In particular, the use of dicumyl peroxide allowed an efficient and easy polymerization of the 100% azo-LCEs, preventing the formation of gas bubbles during the process and avoiding the resulting need for a vacuum oven. The birefringence properties of 100% azo-LCEs were tested after synthesis. LC11 and LC18 films were semi-transparent and able to rotate the polarized light, as expected (LC81 was too opaque to allow birefringence measurements).

The optical properties of the PMPs were also investigated by spectrophotometric analysis. In particular, the percentage absorbance estimated for LC81 was higher than 90% in the wavelength range 200–500 nm, decreasing down to a few percentage values in the near-infrared region, proving that a 457 nm laser can be used as light source to investigate the PMP bending behavior.

The three PMPs were then characterized for their mechanical response to polarized light at 457 nm. LC81 was the best one at converting light energy into mechanical movement, most likely because of its low degree of cross-linking. In fact, LC11 and LC18 have a cross-linker content 4- and 8-times higher than LC81, respectively, resulting in more brittle and rigid materials. The bending speed of LC81 was also evaluated, showing a fast and dose-dependent response to the 457 nm polarized light (up to 0.02 m/s using a power density of 6 W/cm^2). It is also interesting to report that the LC81 material returned to its starting position after laser switching off as fast as it bent upon laser illumination. Moreover, this material was also proved to be sharply polarization-dependent, changing its bending behavior when the impinging laser was polarized parallel or perpendicular to the LC molecular long axis.

The results collected during the characterization of 100% azo PMP suggested that the cross-linker linear monomer ratio greatly influences the nematic temperature of the monomer mixture and plays a key role in determining the bending behavior of the final PMP. Furthermore, thanks to the peculiar polarization-dependent responsiveness of this kind of PMPs, they could be exploited for the development of polarization-selective switches.

2.3. *Doping of photomobile polymers*

Doped and laminated photochromic polymers have emerged as an intriguing area of research, with promising potential for a wide range of applications in the realm of smart materials.[30,65–67] The incorporation of nanoparticles into LCEs has resulted in the creation of photo-responsive materials with enhanced properties and functionalities, which is a particularly intriguing development in this field.[68–71] Researchers were able to fine-tune the optical and mechanical properties of the resulting nanocomposite materials by doping LCEs with various types of nanoparticles, such as gold or silver, allowing for precise control over their response to light stimuli.[72,73] The benefits of nanoparticle or dye-doped LCEs are numerous, and go beyond their ability to manipulate light transmission, reflection, and absorption characteristics selectively. When compared to their undoped counterparts, these materials have improved mechanical strength, thermal stability, and response times. For example, the effect of ZnO nanoparticles on the thermal and photomechanical characteristics of azobenzene-based LCEs was examined by Sagnelli *et al.*[2] in the frame of the PULSE-COM project (H2020-FETOPEN-2018-2020, 863227). The findings showed that the inclusion of ZnO nanoparticles increased polymer's thermal and mechanical stability (Fig. 7).

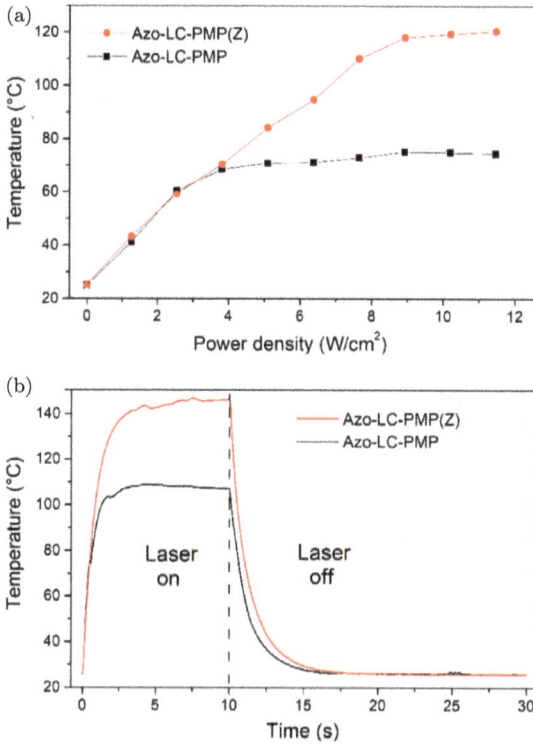

Figure 7. Thermometric measurement of the bare and doped azo-based PMPs. (a) Temperature at the max bending versus the power density of the laser irradiation, (b) time-based trend of temperature during and at the end of laser irradiation with a power density of 11.5 W/cm^2.[2]

In comparison to Azo-LC-PMP, ZnO-doped materials were shown to have better traction capability and light-to-mechanical-energy conversion (Fig. 8).[2] The traction capability was measured using an Instron universal testing machine with a 0.1 kN cell at a 0 mm/sec speed. Upon being placed between the clamps in traction mode, the sample was irradiated with a 457 nm laser having a surface area of 4.99 mm^2. The power density on the surface was 12.02 mW/mm^2. Furthermore, the maximum bending angle and bending speed in photomechanical studies were improved for the ZnO-doped PMP. Then the ZnO nanoparticles increased the amount of visible radiation that was absorbed, which may have had a role in the improvement in photomechanical response that was reported.[26] Soft robotic actuators that can be precisely controlled using light are real-world examples of potential applications for nanoparticle-doped LCEs, as are smart windows

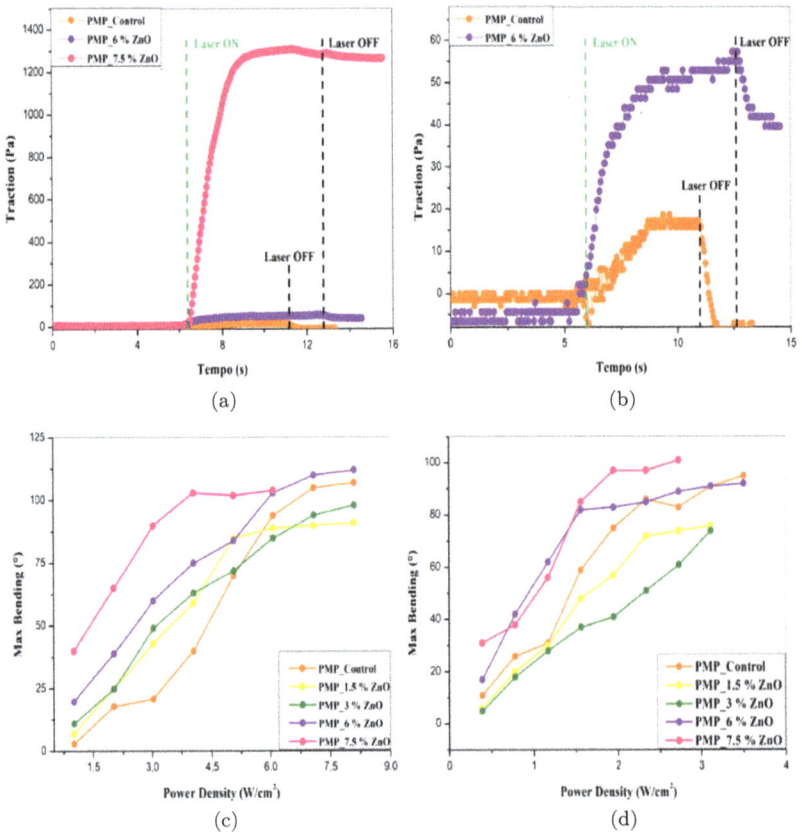

Figure 8. Traction capability measurement of: (a) Samples doped with 0%, 6%, and 7.5% ZnO nanoparticles. (b) Close = up of the PMP_Control and PMP_6% ZnO graphs to observe their entire behavior. (c) Bending capability of undoped and doped PMPs under laser 457 nm (d) under laser 405 nm.

that regulate sunlight transmission to optimize energy efficiency and indoor comfort, and tunable optical devices for telecommunications or sensing applications. Furthermore, the adaptability of these materials enables the development of advanced devices with tailored properties and responses to specific wavelengths of light.

An interesting study focuses on the preparation and characterization of an azobenzene-based photomobile polymer (azo-LC-PMP) and its composites with varying concentrations of carbon black (CB). The composites were characterized through optical, morphological, and UV/Visible analyses to investigate the effects of CB concentration on film uniformity, nematic alignment, surface roughness, and light absorption properties. The photo-responsive behavior of the composites was assessed using lasers at different wavelengths in the visible range. Results showed that low CB concentrations improved the homogeneity of CB dispersion and preserved nematic alignment. The introduction of 0.03 wt% CB was sufficient to prompt photomobile behavior, with the best performance observed at 0.1 wt% CB concentration. These findings suggest that carbon-based materials in PMP films can be used to exploit the entire solar spectrum bandwidth, opening new perspectives for applications in photonic devices.

Researchers can engineer materials with highly desirable properties by using the right combination of nanoparticles and LCEs, expanding the realm of possibilities for light-responsive technologies.[74-76] As we continue to push the boundaries of doped and laminated photochromic polymers, there is enormous potential for innovation in this fast-moving field, which has the potential of changing the way we think about light-responsive technologies and their real-world applications.

3. Microengineering Applications of Photomobile Materials

3.1. *Introduction to microengineering*

The earliest evidence of prehistoric human activities shows that engineering has always been present in human life: a 2.4-million-year-old stone artifact and a bone with man-made cutmarks were found in Algeria, showing the ability of humans (or at least hominins) to intentionally transform existing materials into tools, use these tools to process another material, to answer a need: eat. Thanks to the progress of science, engineering has flourished exponentially since prehistoric ages, paving the way for our insatiable need for development, knowledge, and exploration. According to the École Polytechnique Fédérale de Lausanne (EPFL), "Microengineering is the art and science of creating, designing, integrating and manufacturing miniature components, instruments and products". Back in prehistorical ages, people manufactured needles with bones, to sew clothes and answer another need: protecting their bodies from the environment. Considering that a needle can be used to achieve a millimetric operation that is not directly accessible to human fingers, we can consider this as an early example of microengineering in human history. Since then, many technical improvements were made according to growing human needs.

The need to measure time and position led to many developments in mechanics and instrumentation, when the first sundials, clepsydras, or astrolabes were created, requiring fine and precise parts. With the expanding development of maritime exploration from the 15th century, an important need was to know the position of a ship. The measurement of the longitude needed either complex calculations or a precise time measurement tool that could be transported in a ship and that could stay precise despite the variations of temperature. This challenged watch manufacturers, until the invention of the first portable and reliable chronometer in the 18th century, which involved an incremental step in the miniaturization of the mechanical components.

In 1786, a diffraction grating was made by David Rittenhouse, by stringing hairs between two finely threaded screws. The first developments of interferometry started with the evidence of the wave nature of light by Young's experiment in 1803. In 1858, Léon Foucault developed a method for polishing a telescope mirror in a spherical or parabolical shape with a nanometric scale precision. These important steps were fundamental for the future improvements of micro-engineering, as the progress of miniaturization is intimately linked with the possibility of observing and measuring micro and nanoscale phenomena.

The first transistor was made in 1947. This breakthrough drastically changed the scale of electronic devices that were made with vacuum tubes. This was followed by the invention of the integrated circuit by Jack Kilby in 1958, paving the way for modern computer hardware design and manufacturing. After mechanics, electronics was one of the first disciplines that could benefit from the progress of miniaturization.

In 1959, Richard Feynman gave a lecture titled "There's Plenty of Room at the Bottom", in which he mentioned the possibility of manipulating individual atoms. This lecture later inspired the first developments of nanotechnologies in the 1980s.

After the invention of the first CMOS circuits in the 1960s and the microelectronics industry's boom, the first microelectromechanical systems (MEMS) were developed in the 1970s and first commercialized in the 1980s. Applications in microfluidics, which first started in the 1950s with ink-jet printing, also benefited from micro fabrication processes development, addressing other industries such as microbiology.

In the 1970s, the term "photonics" appeared for the first time, followed by an intensive development of industrial applications starting in the 1980s and fostered by the development of telecommunications. Light, which was initially used to develop instruments and techniques that helped the development of micro-engineering, was in turn micro engineered itself. In early 2000s, the term "plasmonics" appeared to describe a new discipline involving the interaction of photons and free electrons in nanoparticles.

Microengineering has played an important role in technical developments since the first apparition of engineering. Changing the scale of work offers great opportunities for improving devices by making them more reliable, compact, precise, fast, affordable, and by reducing their environmental footprint. Surprisingly, despite all the progress made in science and technology, there are still many disciplines that remain unexplored. The work described in the next paragraphs aims at exploring some of them.

3.2. *Photo-responsive properties in microengineering*

A photo-responsive material, in our case a photomobile polymer, converts light energy into mechanical energy. This rather unexplored path of energy conversion can be used for developing a new kind of innovative sensors or actuators at microscale. To start such a development, we need to understand and measure how the material does the conversion. What is the material's spectral response, what stroke is generated, what force is produced, how fast is the material moving, how repeatable it is, how precise it is, are just some examples of the many questions that an engineer will have in mind before starting a design.

To answer some of these questions, a dedicated test bench was developed by Cedrat Technologies in the frame of the PULSE-COM, H2020-FETOPEN-2018-2020, 863227 project. The bench is based on the following principle: the photopolymer trans form (thermodynamically stable) can be converted into the cis form (metastable) by using a UV wavelength of 300–400 nm, and that visible illumination at >400 nm converts the molecule back to the trans form.

The test bench (Fig. 9) thus includes two ultra-violet (UV) ($\lambda = 385$ nm) and blue ($\lambda = 470$ nm) light sources. The light beam is focalized by a lens on the PMP sample, which is clamped and suspended. The sample is equipped with a reflector and an interferometer laser measures the obtained stroke. The typical measured power density at the PMP sample position is 20 mW/cm^2 for the blue light and 30 mW/cm^2 for the ultra-violet light.

The active material is a PMP-azo with a copper layer, which has been produced conjointly by the Institute of Applied Sciences and Intelligent Systems (CNR/ISASI) and Department of Sustainability

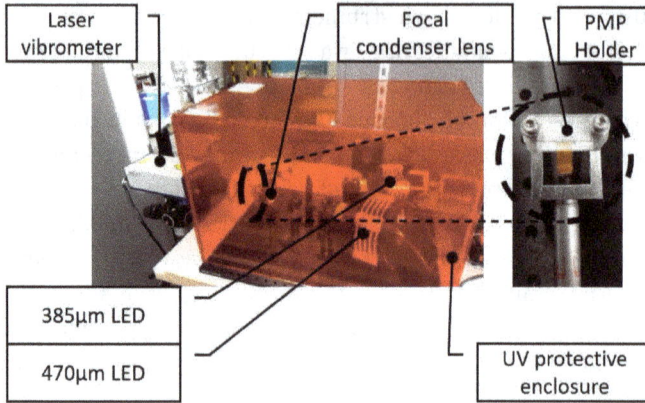

Figure 9.	PMP samples measurement test bench.

(ENEA/SSPT).[2,45,64,77] The ultra-violet light forces the PMP to bend, whereas the blue light straightens the PMP back.

Sequences of successive alternance of blue and ultra-violet light have been set at different frequencies from 1 Hz to 25 Hz.

Alternative stroke amplitudes reach 200–300 μm at low frequencies around 1 Hz and naturally decrease down to a few tens of microns when the frequency is increased up to 25 Hz. It is also noticeable that the average bending of the PMP increases from 500 μm to 800 μm according to the driving pattern (Fig. 10). This phenomenon is attributed to the increase of temperature of the sample while it is exposed to light power.

The control of the PMP displacement can be simplified by limiting the mean power, which will reduce the warming impact and thus the mean displacement shift. For more demanding applications, this position shift can be considered, which will add more complexity in the control.

PMP response is known to be slow compared to other actuation technologies such as piezo ceramics. PMP actuators thus have a low bandwidth, which means the PMP stroke decreases when the oscillation frequency increases (Fig. 11).

Applications compatible with a slow response (typically less than 1 Hz) can benefit from actuation capability of several hundreds of micrometers. For applications requiring larger bandwidth in tens of hertz, much smaller strokes, typically in the range of micrometers, should be accepted.

The controllability of a PMP in closed loop was evaluated by mounting a strain gauge sensor on it (Fig. 12) and comparing the strain gauge signal

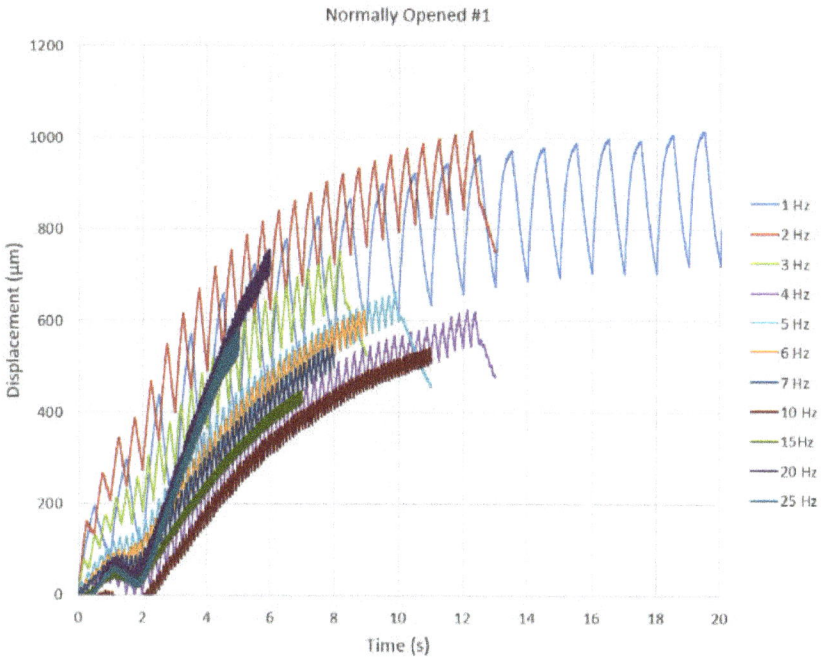

Figure 10.   Displacement of the PMP samples in time, at different frequencies.

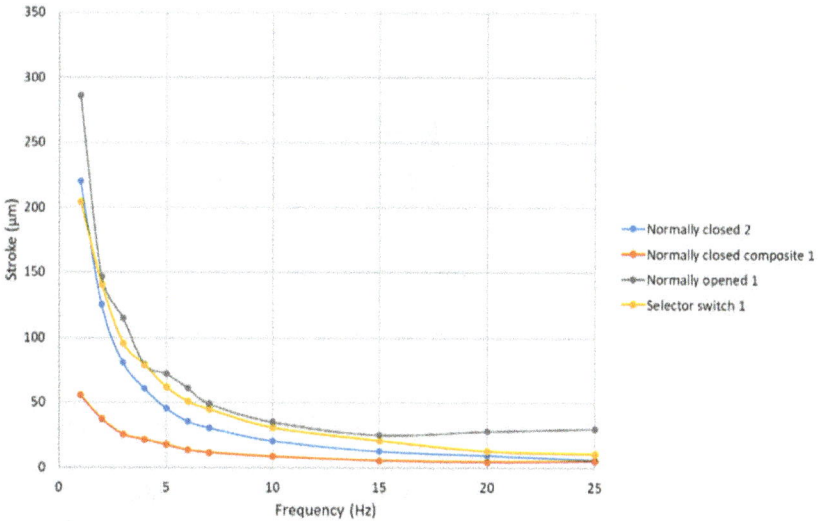

Figure 11.   PMP stroke vs frequency measured on several prototypes.

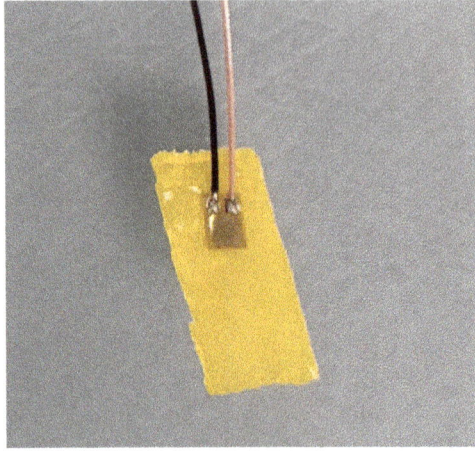

Figure 12.   Strain gauge sensor mounted on a PMP.

Figure 13.   PMP sample with strain gauge and a reflective tape.

with the position of the PMP sample with a laser interferometer hitting a reflective tape (Fig. 13).

The strain gauge was used in a Wheatstone bridge configuration and showed good linearity (Fig. 14). An alternative option for the control sensor is the use of a piezoelectric layer made of nanowires.[78,79] With such a

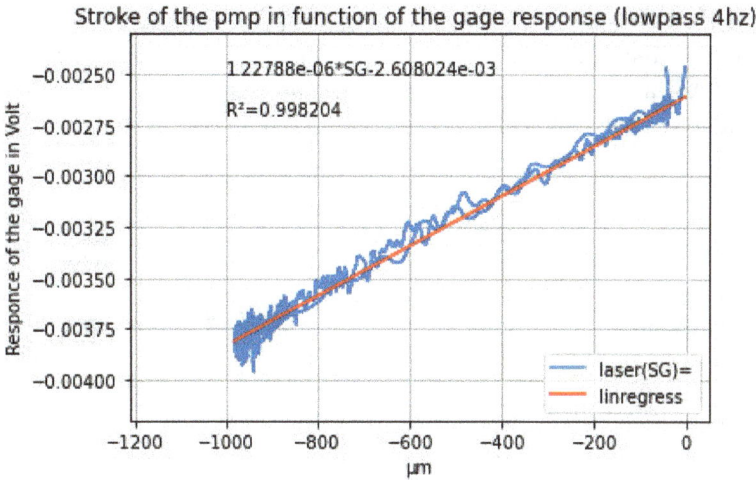

Figure 14.   Strain gauge signal and laser interferometer signal vs PMP stroke.

kind of piezoelectric material, vibration harvesting applications can even be considered.[80]

3.3. *Applications of photomobile materials in micro valves, micro switches, and micro steering mirrors*

In the frame of the PULSE-COM, H2020-FETOPEN-2018-2020, 863227 project, three different applications have been explored at the prototype level: a micro switch, a micro valve, and a closed-loop controlled steering mirror.

3.3.1. *Micro switches*

The principle of a switch is to open and close a conductive path in a power circuit thanks to a specific command. The energy used by PMP to change state is effectively assumed to be low at term. By using light for the switch activation, we suppress any conductor link between the power circuit and the command, which simplifies many electromagnetic compatibility (EMC) and electric insulation issues. The large stroke of PMP is particularly well suited for switch application, as it allows a large air gap in open state, which significantly reduces any risk of spark in case of noisy peak voltage.

Several electric switch configurations were designed and manufactured by Cedrat Technologies, using azobenzene PMP samples provided by CNR/ISASI and ENEA/SSPT: a normally open switch (Figs. 15 and 16), a normally closed switch (Figs. 17 and 18), and a selector switch (Figs. 19 and 20).

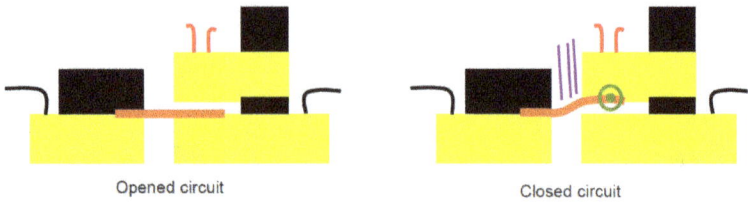

Figure 15.   Normally open switch principle.

Figure 16.   Normally open switch prototype.

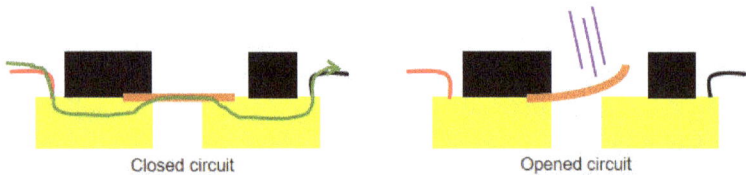

Figure 17.   Normally closed switch principle.

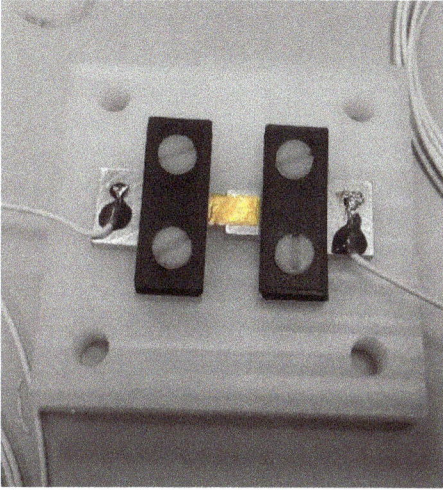

Figure 18.   Normally open switch prototype.

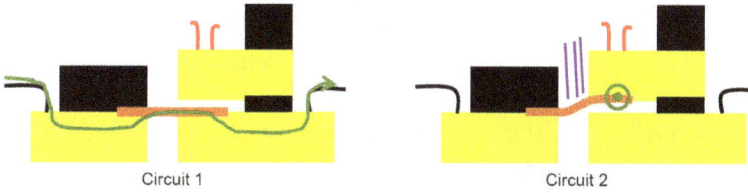

Circuit 1

Circuit 2

Figure 19.   Selector switch principle.

Figure 20.   Selector switch prototype.

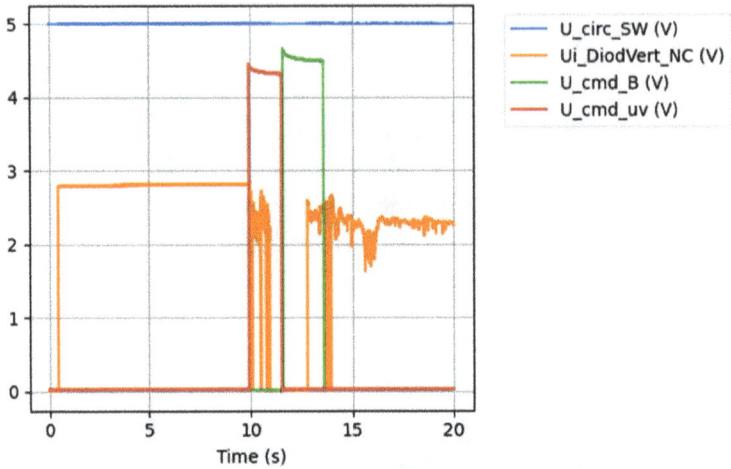

Figure 21.   Evolution of switch current (orange) in time (arbitrary units).

The main difficulty with PMP equipped with copper layer for switch application is that the sample remains partly bent, and its state depends on its history. It makes it non reproducible, with a hysteresis. A second issue is that a contact requires enough pressure to be effective and the force generated by PMP stays low. Despite difficulties in getting repeatable and robust behavior on the prototypes, we were able to observe a successful switch functionality on the normally closed switch prototype by monitoring the current passing through the switch at different states (Fig. 21). In Fig. 21, the red line shows the switch opening order, while the green shows the switch closing order.

There is still a high step to get reproductible switch actuation, which will probably show a response time in the order of the second.

3.3.2. *Micro valves*

Light-controlled valves offer new possibilities for demanding applications where the command must be physically separated from the actuator. Such requirements can be found in instrumentation, biology, medical devices, and the nuclear industry. As no local source of power is needed, the integration, maintenance, and durability of such devices could be a key advantage compared to other solutions. By using an optical command, the device does not need electricity, which makes it easily compliant with ATEX, ESD, or magnetic environments.

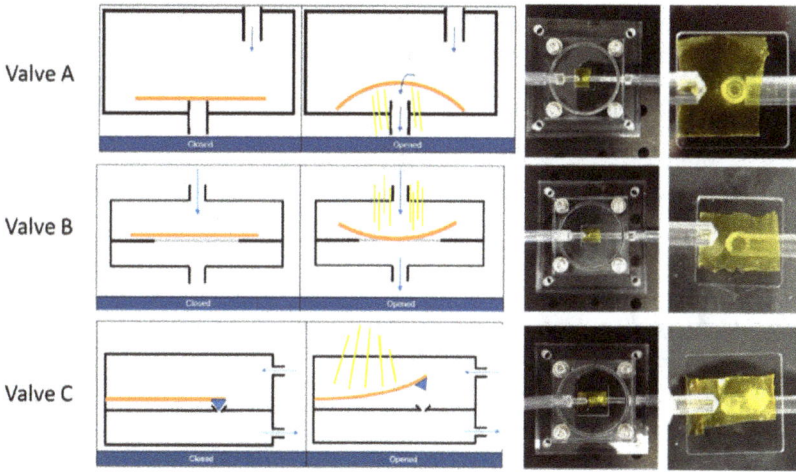

Figure 22.   Tested microvalve concepts.

Figure 23.   Light-driven microvalve test bench.

Three microvalve concepts (Fig. 22) were designed and prototyped by Cedrat Technologies, using PMP-azo with PDMS-protective layer produced by the laboratories CNR/ISASI and ENEA/SSPT. The prototypes were tested on a dedicated work bench (Fig. 23).

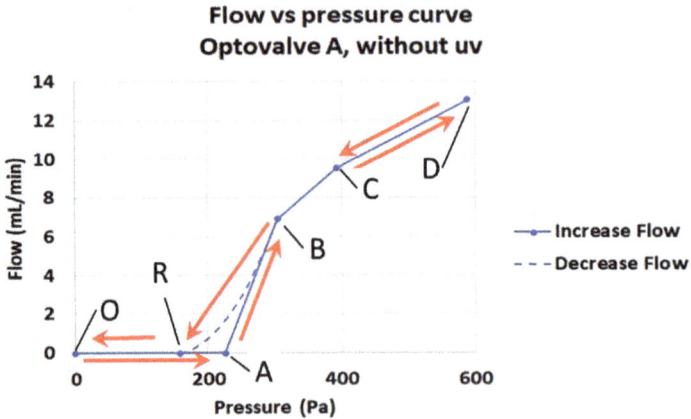

Figure 24.   Typical flow vs pressure on an optovalve.

It has been found that the fluid circuit is nonlinear at low pressure, as shown in Fig. 24. A minimum of pressure is necessary to start the fluid flow, which is attributed to the fluid tension at the fluid circuit output. For higher pressure than the point A, the fluid starts to flow and the measurement at low pressure gives nonlinear response while going through points A, B, C, and D. At decreasing pressure, the fluid stops flowing at point R, whose value is about 50–100 Pa lower than point A. Physically, this is compatible with the inertial force necessary to stop the fluid. The optovalves have been tested at pressure around R and A, to evaluate whether the valve can pass from a blocked state to flow state.

The best results were obtained with prototype "C" (Figs. 22, 25, 26 and 27), showing a 10 times ratio between the open state flow and the closed state leakage flow (Fig. 24).

Prototype "C" has been tested under fluid pressure in the range of 7900 Pa. While all other valves let an important continuous flow of water pass through, prototype "C" stayed closed. This is a good result, as it means it can effectively stop the fluid flow. Probably the pressure presses the valve cone against the valve seat and friction keeps it blocked. At this stage, a small leakage flow is still observed and measured at $72\,\mu l/min$. When applying the UV lamp at maximal 2.45 W light power, the flow gets significantly higher at about $750\,\mu l/min$. The operation has been repeated several times. This clearly shows a valve effect made with the optovalve including PMP material. However, it is not clear what impact the self-heating of the device has on the leakage value.

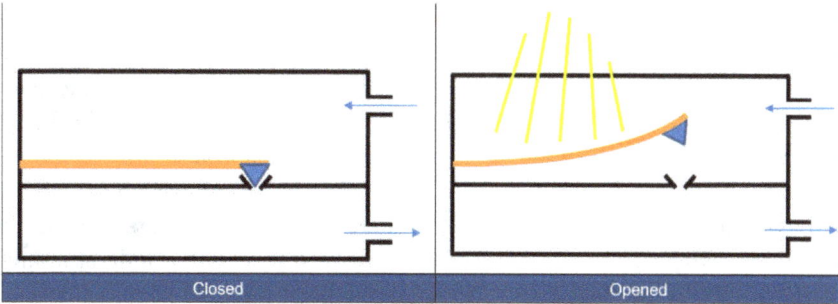

Figure 25.   Principle of light-controlled valve.

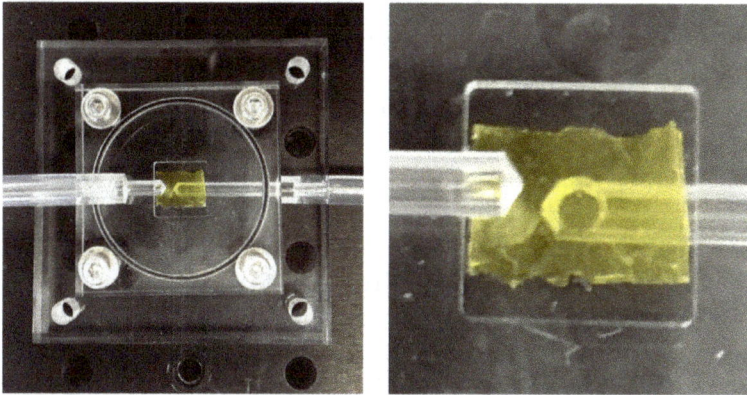

Figure 26.   Light-controlled valve prototype.

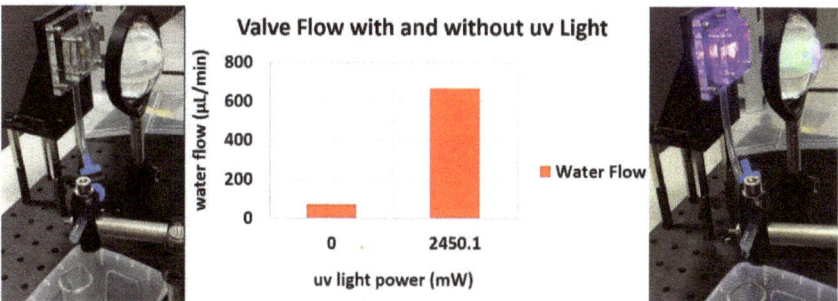

Figure 27.   Measurement of the valve flow in closed and open state.

The design of an optovalve made with photomobile polymer is still a challenge. As the valve includes a housing and liquid, the light power reaching the PMP material is damped by absorption, reducing the efficiency of the device. Even though enough light power would be used, the PMP generates low forces, which should supersede fluid pressure or frictions in the valve seat. It is likely that PMP thermal and thermomechanical effects also contribute to the flow variation.[9] There are still many tests to be done for the research of a valve effect. One issue that limits the understanding of the results is the fact that it has not yet been possible to visualize the behavior of the PMP exposed to light beam inside the optovalve.

3.3.3. *Closed-loop-controlled steering mirror*

Steering mirror mechanisms are widely used in many applications such as pointing, image stabilization, or tracking. They are currently built with technologies like conventional servomotors, voice coils, and magnetic- or piezo-actuators. The angular stroke varies from a few mrad up to 360°, and the control bandwidth varies from a few Hertz to several hundred Hertz, depending on the application. Controlling a steering mirror optically offers a unique opportunity to have a system that does not rely on a local power supply and control electronics, reducing its complexity and its volume drastically. This possibility of offering such an innovative solution on the market was explored by Cedrat Technologies, by designing and testing a prototype, in the frame of the PULSE-COM, H2020-FETOPEN-2018-2020, 863227 project.

A PMP sample was equipped with a mirror to perform a closed-loop-controlled deviation of a laser beam. The deviation of the light was measured by an optical device, using a reflective tape, and injected into a digital PID controller (Fig. 28).

The PMP used is the normally closed composite sample, which is a PMP of type azobenzene covered by copper on one side. The control of PMP bending is performed by the command of either the UV LED or the blue LED depending on the side of the bending. The bending amplitude is driven by the light power through a PWM signal at 5 kHz (Fig. 29). The measured position is compared to the order and processed by an in-house digital controller board UC55, which generates the PWM command.

The closed loop is made with a basic *PI* (proportional integral) control with $P = 2$ and $I = 500$, refreshed at 100 Hz.

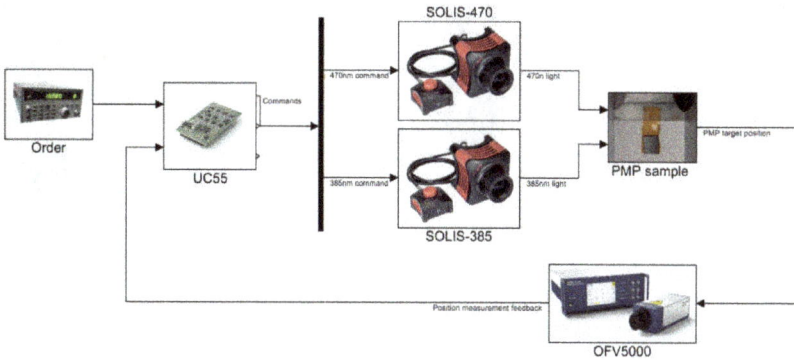

Figure 28. Principle of the PMP closed-loop control for laser beam deviation.

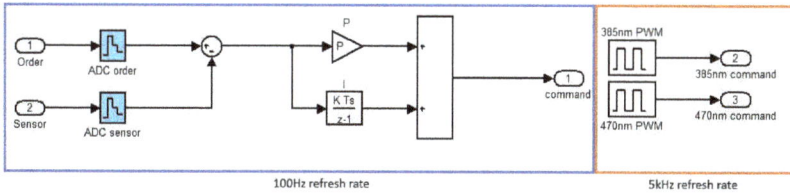

Figure 29. PMP steering mirror control strategy.

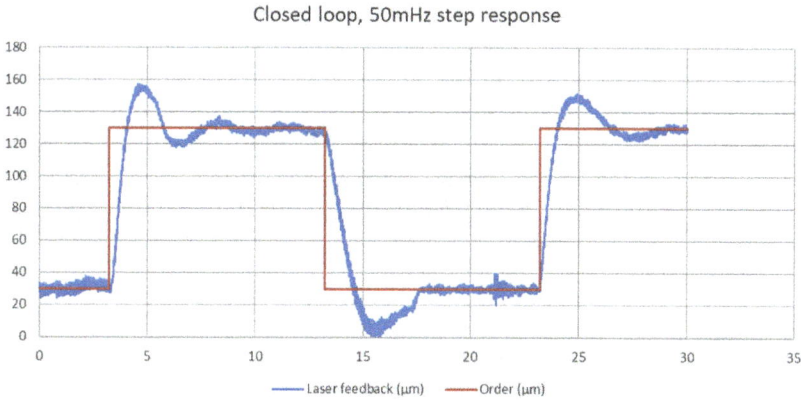

Figure 30. PMP sample closed-loop response.

The system's response was measured in time, and compared to the order, showing successful controllability of the system (Fig. 30).

As the PMP response is slow, typically a few seconds, the repetitive step signal is set at 50 mHz frequency. The stroke observed is 100 μm for a

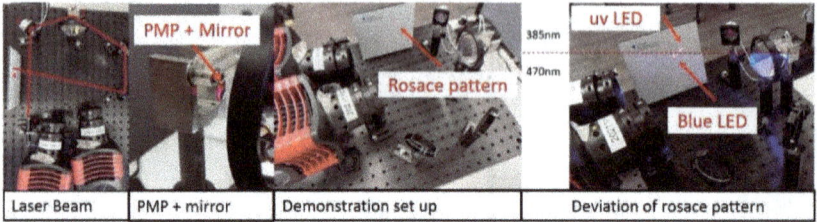

Figure 31.   Controlled deviation of a rosace pattern with a PMP sample.

1 cm long sample. Considering the position of the reflective tape, this means an average bending angle of 12.5 mrad (0.7°). There are nonlinearities on PMP actuators, limiting the efficiency of such control strategy. Activating (385 nm) and Relaxing (470 nm) LED effects are not symmetrical, they have different dynamics, and the LED power is not linear with the effect on the PMP. The power of both LED generates a parasitic displacement (due to the thermal effect of the material). Basically, a full power relaxing (470 nm) LED does not provide an extremum position. Shutting of this LED will cool down the PMP and allow it to move further in the relaxing direction.

To demonstrate a beam deviation in a real case, the PMP sample was used to deviate a laser beam describing a rosace pattern thanks to a fast-steering mirror from Cedrat Technologies (Fig. 31). The PMP sample was able to deviate the mean position of the laser beam, showing its complementarity with mechanisms performing at lower strokes and higher speeds.

As already seen before, the stroke of PMP is large compared to the component dimensions, but generated forces are low and response time is in the order of a few seconds. Therefore, optical deflector applications are a good target for PMP actuators, as the light beam deviation does not oppose significant resistance force on the PMP. The load of the PMP is limited to the mass of the mirror, which can be minimized. The main limitation seems to be the nonlinearities, which are explained by thermal effects and the di-symmetries of actuation between the different light excitations. Nevertheless, one very positive point is that a servocontrol has been successfully implemented with a PI linear control on an adapted range of actuation. As the time response is significant, the bandwidth of the servocontrol stays low. The corresponding advantage is that the PMP actuation would be a very soft actuation without noisy vibration. This is very interesting in optics applications.

An additional advantage of the PMP used in deflector application is that the stroke obtained is large versus the mass (or volume) of the actuator. A 12.5 mrad angular actuation was recorded, for a typical PMP actuator volume of 5 mm × 10 mm × 60 μm (3 mm^3). This volume should be increased to consider the necessary housing, but then the expected volume may be 10 mm × 15 mm × 5 mm, which is about 1 cm^3. The TT60SM is a Cedrat Technologies piezo FSM product that drives a mirror with a 20 mrad angular stroke, so has the same range of values. Its dimensions are Φ 55 mm × 25 mm, which is a volume of about 60 times higher. Even Micro Electro Mechanical Systems (MEMS), which are known to be small devices, require a typical 20 mm × 20 mm × 10 mm size.[10] which is still 4 times more than the PMP deflector.

With large angular strokes, small volumes, and no need for local power source nor electrical connections, PMP steering mirrors offer a unique combination of advantages that will certainly find new applications in a context where data transmission is migrating from electric and radiofrequencies toward optical wavelengths.

4. Toward Piezoelectric Applications with Photomobile Materials

4.1. *Introduction to the combination of piezoelectric materials and photomobile polymers*

In the literature, there is no evidence of photo-induced piezo-actuators based on photomobile polymeric films. Thus, combining these two concepts to obtain highly efficient photo-piezoelectric devices stimulated by various light sources or even sunlight could lead to groundbreaking applications. To our knowledge, only one group has explored the coupling of light-induced strain and piezoelectric materials to generate an electric signal,[81] where a light/heat-sensitive layer (~1.5 μm thick) of carbon nanofibers (CNF) was integrated onto a PZT (piezo-transducer) layer (~200 nm thick) on silicon, forming a cantilever (350 μm long, 30 μm wide). The CNF layer bends approximately 5 μm in the presence of light/heat, producing ~0.2 mV of AC superimposed on a ~0.2 mV of DC voltage (sunlight exposure), which is not suitable for real applications. In essence, transforming photonic energy into mechanical work (using a polymeric film) and then, using the same device, converting mechanical work into electricity (through a piezo-composite) could provide further application possibilities, including converting sunlight into energy. In the literature, the main sources of vibratory energy investigated for piezoelectric harvesting over the past decade include fluid sources, human body motion, animal activity, infrastructure vibration, and vehicle motion. Examples of energy harvesting systems from fluids are airflow using flutter-style harvesters, hydropower piezoelectric harvesters using fluttering flags, wearable devices like shoe insoles, energy harvesting from vehicles, structures using bridge traffic harvesters, and more.[82] Aiming to achieve power outputs ranging from $10\,\mu\mathrm{W}$ to $1\,\mathrm{mW}$ from a single PMP–PZL unit could enable the development of truly applicable IoT systems. Utilizing sunlight energy requires working within the framework of solar concentrators for both photovoltaic and thermodynamic applications. Consequently, it is crucial to develop innovative optics concepts for light concentration to maximize solar energy conversion into mechanical movement of PMP films. To target the new field of photo-activated piezoelectricity, an interdisciplinary approach that transcends traditional boundaries is essential. This vision is necessary to create suitable PMP and piezo-composite layers (PZL) to achieve the desired performance; to identify radically new solutions and strategies to effectively couple the PMP–PZL device, demonstrating innovative features in various fields of active flexible optoelectronics; and finally, to integrate the PMP–PZL device into optoelectronic systems for advanced industrial implementation, as already stated in a pioneering patent.[83] A schematic working principle of the operation of a possible prototype is shown in Fig. 32. Light comes from one side, reaches the device, causing it to bend, transferring the strain to the piezo, and generating an electrical signal. This signal can then be used in several ways, or the PMP can be used alone in specific applications. The PZL integrated into the flexible polymer substrate can also function as a conventional mechanical-electrical transducer in the absence of changing light. To enhance the functionalities of the final device (proper combination of photomobile polymer and piezoelectric composite), it is important to focus on the following aspects of our materials: the absorbed light depends on the compound composition of the PMP film and the appropriate geometries to maximize the signal based on the application. For instance, this type of system, where the PMP is attached to the PZL device, may generate energy due to the heating of the PZL from the light source if proper precautions are not taken, and the electrical signal would be a combination of the thermal and mechanical effects. As we will discuss later, other geometries could offer better solutions to optimize the PMP–PZL system.

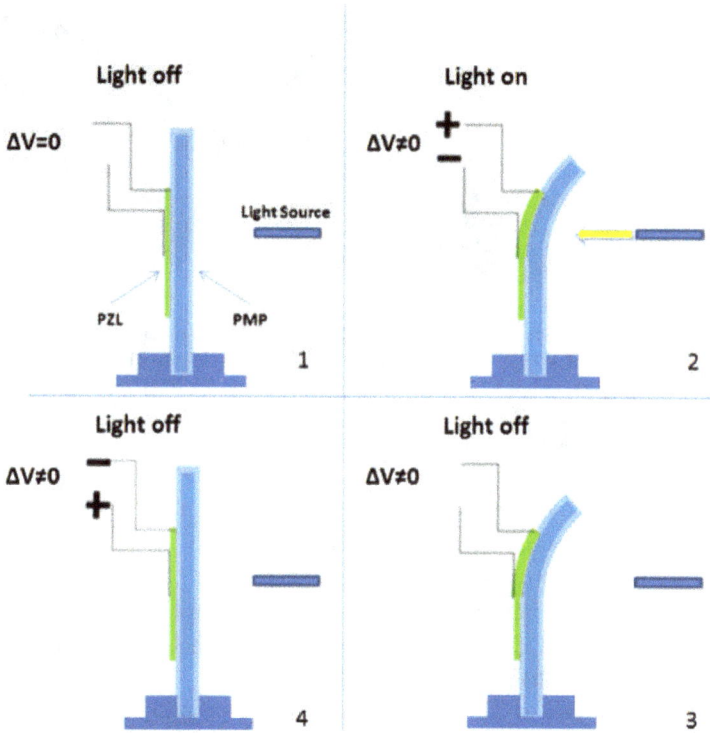

Figure 32.   Schematic working principle of the operation of the prototype.

4.2. *Photo-responsive possible architectures with piezoelectrics*

Smart material actuators can exhibit reversible shape changes in response to external stimuli (such as light, moisture, pH, heat, and electrical current) to accomplish a function or carry out a task, for example, mimicking plant motions[84] or serving as energy transducers.[85] A liquid-crystalline polymer (LCP) actuator, able to self-organize its molecular alignment, represents arguably the most promising candidate for realizing the above-mentioned applications as mentioned earlier in the manuscript. In this section, we will show all the experiments related to the potential strategies that we could adopt to utilize the PMP as an actuator in a PMP–PZL system and how we could increase the efficiency of our system. As already mentioned, as a first design, it is possible to simply attach the PMP to the PZL device, but it might not always be the appropriate architecture. A perpendicular

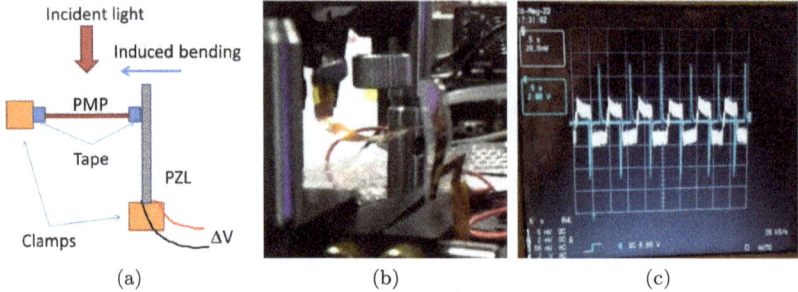

Figure 33. (a) Second strategy to combine PMP/PZL. Schematic of the combination principle. (b) Photo of a PZL (from UGA/CTEC) and PMP-Azo (From CNR/ENEA). (c) (White) Voltage signal of PMP/PZL devices, integrating a PMP/copper layer with a commercial thick (205 μm) PZL device (L-mode configuration). (Green) optical chopper signal with a duty cycle of around 0.9 s.

integration is shown in Fig. 33(a). In this configuration, the PMP is attached to the PZL layer using, for example, Kapton tape. To ensure the PMP movement, we used two laser lights with wavelengths of 405 nm and 457 nm. The incident light, in this configuration, only reaches the PMP and not the PZL, avoiding the PZT heating. The bending of the PMP pulls the PZL device, then a piezoelectric signal is generated.

As the first experiment, to demonstrate the feasibility of the idea, the PMP was attached to the edge of the PZL (205 μm thick), see Fig. 33(a), in a perpendicular configuration that we define here as L-mode. Modulating the laser light with an optical chopper of around 1.8 s, we obtained the voltage signal reported in Fig. 33(b), which, considering the top and the bottom of the related noise signal, ranges from 20 mV to −20 mV.

Clearly, the L-mode structure is not the only possible solution, and other critical points must be considered, such as the thickness of the PZL films and the distance between the anchor point of the PMP and the fulcrum where the PZL is located. For this reason, we have carried out a series of experiments related to moving the PMP–PZL connection point away from the point where the PZL is secured, which functions as a fulcrum for a sort of pendulum. Here, a first experiment is reported, in which we extend the PZL by 10 cm with double-sided adhesive Kapton, allowing us to reach a peak-to-peak voltage difference of about 1.2 V. This experiment, at this stage, is a single-movement-based experiment, and we do not obtain a signal that repeats over time, but it provides us with the opportunity to see a significant improvement compared to the previous configurations which resulted in a maximum peak-to-peak voltage difference of 40 mV.

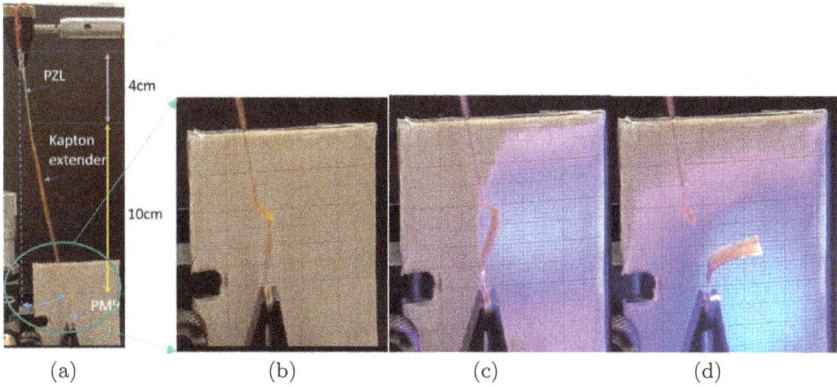

Figure 34. Setup of the pendulum configuration and four frames related to (a) starting configuration, (b) zoom image of the actuating part, (c) intermediate position after the switching on of the laser sources, and (d) the release of the PZL extensor by the PMP film.

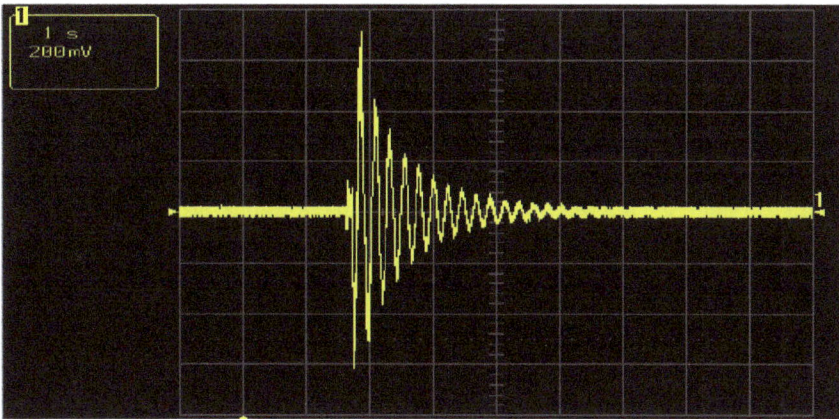

Figure 35. Oscilloscope image of the voltage behavior vs time after the release of the PZL extensor by the PMP film.

In Fig. 34, it is possible to observe the setup of the pendulum configuration and three frames related to the actuation of the PZL and its release. In Fig. 35, the oscilloscope image of the voltage behavior vs time after the release of the PZL extensor by the PMP and the consequent creation of the voltage signal are shown.

It is important to consider that when we extend the PZL, it is crucial to not create a second fulcrum at the junction between the PZL and the

Kapton to avoid dissipating part of the energy due to the increase in the total momentum. Also, it might be appropriate to try to make longer PZLs to maximize the voltage and prevent the creation of the second fulcrum. It is also necessary to maximize the signal by attaching the PMP in the best possible position and optimizing the thickness of the PZL, as already pointed out.

4.3. *Photomobile materials in energy harvesting applications*

In recent years, energy harvesting has experienced rapid growth, primarily because it offers an excellent alternative to wires and batteries, as well as the ability to recover waste energy. The ongoing advancements in the modern semiconductor industry have led to the creation of smaller, highly integrated, and more energy-efficient devices.[86,87] As a result, power consumption has been reduced to such an extent that it has become viable for energy harvesting techniques to be employed.

It is increasingly thought that the semiconductor industry's future progress is mostly dependent on the capacity to deliver devices with energy autonomy. This advantage will produce gadgets that are (i) maintenance-free, (ii) universally functioning, and (iii) waste-free, which are characteristics sought after by the entire electronics industry. There are numerous energy harvesting techniques that can generate enough energy to power current electronics. Figure 36 shows a summary of the captured output power densities obtained by various energy harvesting techniques.[88] Radiant (photovoltaics) and mechanical approaches, among all conventional energy harvesting systems, provide high power densities over a wide output voltage range. Mechanical energy can be converted into electricity using a variety of processes, including electromagnetic, electrostatic/triboelectric, and piezoelectric. The choice between them is highly dependent on the application, but, among these methods, piezoelectricity is the most widely studied. Here, we will focus fundamentally on the possibility of using piezoelectricity in an alternative manner harnessing solar radiation to transform it into mechanical movements and then using piezoelectricity to generate electrical energy.

Our objective is to create piezo-electro-photonic devices by combining two concepts: (i) converting photon energy into mechanical work using a PMP, which is typically a polymer film whose deformations can be induced and controlled by the wavelength, polarization, intensity, density, and angle of incidence of impinging coherent and/or incoherent light, and (ii) converting mechanical energy into electrical energy using a PZL. Specifically, our goal is to generate an electrical signal or mechanical movement and store the resulting energy with suitable storage systems. The state of the art for our approach is not fully developed in the literature; therefore, we must base our work on the existing knowledge of each component we plan to utilize.

The aim is to develop PMP–PZL systems by merging two concepts: transforming photon energy into mechanical work using a PMP and converting mechanical energy into electrical energy using a PZL device. Our primary objective is to create highly efficient photo-piezoelectric actuators, specifically piezo-actuators stimulated by sunlight, to pioneer a new paradigm in the field of energy harvesting.

Various architectures and geometries could be considered to transfer the mechanical energy from PMP to PZL devices. For this reason, different models and experiments need to be evaluated to maximize this transfer. To ensure the use of such systems for energy harvesting applications, continuous movement of the PMP film under solar irradiation is required. To achieve this, appropriate solutions must be developed, such as partially masking the solar light when necessary, implementing an auto-shadow system to allow the PMP to return to its initial position, or engineering new solutions to create an unstable behavior of the system. In particular, the PMP–PZL device might be able to convert light from a spectrum wider than that accessible to classic photovoltaic (PV) modules, enabling power supply under nonstandard lighting conditions. The proposed system allows for the direct use of movement induced by light, without converting solar energy into heat for subsequent energy storage through devices like pyroelectric, thermoelectric, and HEATec® devices.

The following are the proposed strategies and systems:

- *Optical strategy for solar concentrators*: Over the years, to make better use of the highly efficient yet expensive multijunction solar cells, they began to be combined with solar concentrators. Concentrating solar systems that employ mechanical tracking movements could also be utilized in our prototypes, particularly when high powers and high concentrations are needed. In fact, extremely high light powers may not be necessary, and non-imaging optical systems might suffice (with a concentration factor of 3–10X), using, for instance, optical funnels such as CPC (Compound Parabolic Concentrators).[89]

- The field of concentrator photovoltaics has been under development for decades. Various strategies have been employed to achieve solar concentration in the required modalities related to different needs and applications. The ultimate goal is to take the first steps in concentrator photo-piezo-tronics. Coupling the distinct piezoelectric, semiconductor, and photonic properties in the resulting devices will necessitate a unique approach that is currently difficult to envision but could prove particularly promising for the future.

- *Bistable Systems*: In this section, we will explore potential strategies to achieve continuous movement by creating two unstable positions. From a mechanical and topological standpoint, the most challenging aspect to enhance the produced energy by increasing flexion frequency is to use a bilayer PMP, as illustrated in Fig. 36(a). In this configuration, the PZL is sandwiched between two PMP layers (PMP-A and PMP-B, depicted in red) that exhibit opposite bending directions when exposed to light. When the PMP–PZL membrane is placed in the right dead point, PMP-A is in the focal point and pushes the membrane toward the opposite dead point, where PMP-B is exposed to concentrated light and can move back to the starting point. To prevent the system from getting "trapped" in the middle position where opposing forces could cancel each other out, we need to define the appropriate PMP's length to render the intermediate position unstable and move the photopolymer with the correct starting speed (e.g., appropriate light power) to reach the second focal point. This can increase harvested power because, compared to the HEATtec configuration, sequential flexions are always provided by a single energy source (light), allowing continuous movement of the membrane.

- *Solar Mill System*: To generate electrical energy from mechanical movement, a flexible piezoelectric device can be utilized (Fig. 36(b)) or a permanent magnet generator. This way, the mechanical energy produced by the blades can be converted into electrical energy. The latter energy conversion mechanism is based on Faraday's laws of electromagnetic induction, which dynamically induce an electro-motive force (e.m.f.) into the generator coils as they rotate. There are numerous configurations for an electrical generator, but the one used in a wind power system is the Permanent Magnet DC Generator or PMDC Generator. The PMDC generator is ideal for use as a simple wind turbine generator or mill system like the one proposed in this application. Continuous movement caused by the deformation of the PMP and the resulting change in the moment of force enables the rotation of the entire system. Continuous irradiation

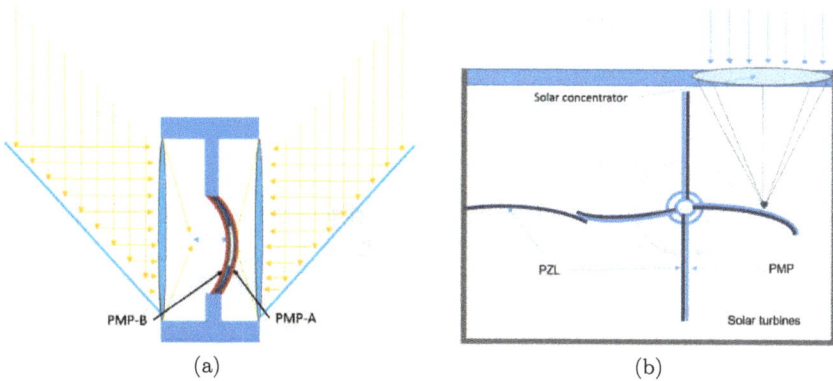

Figure 36. (a) Schematic representation of bi-stable photopolymer-energy harvesting. (b) Schematic representation of photoenergy harvesting mill system.

causes deformation of one or more blades of the system, unbalancing it and inducing a torque force and consequent movement. The energy can be converted both from the deformation of a single blade and from the rotation of the entire system, as depicted in the schematic sketch of Fig. 36. Bending of a blade generates a displacement of its center of mass concerning the rotation fulcrum of the whole system in a counterclockwise direction. This configuration ensures the continuous movement of the entire mill structure due to the blades' ability to return to their initial shape when not irradiated. As illustrated in Fig. 36, the PZL device could also be placed on the border of the structure and stimulated by the blades during the system's rotation to provide an additional energy conversion mechanism. Specifically, the use of multiple photomobile polymer films (the four blades in Fig. 36(b)) and their alternating motion enable energy production even under continuous radiation. Through various doping processes, motion could be induced and controlled by light to work in the entire solar spectrum (UV–vis-NIR) region. The photothermal effect, which is effective and convenient, involves doping the LCP matrix with a light-absorbing and heat-releasing active species such as carbon nanotubes or organic dyes and more. A UV–vis-NIR light-responsive polymer nanocomposite can be achieved by doping polymer-grafted gold nanorods into azobenzene liquid-crystalline dynamic networks (AuNR-ALCNs). The effects of two different photo-responsive mechanisms, the photochemical reaction of azobenzene and the photothermal effect from the surface plasmon resonance of the AuNRs, explain the consequent movement. To dissipate and homogenize the temperature across the entire film, a metal layer could be applied behind the PMP film or between the PMP film and the PZL device.

Acknowledgment

The authors gratefully acknowledge support for this work from the Project PULSE-COM (Grant agreement No. 863227).

References

1. J. I. Lipton, R. MacCurdy, Z. Manchester, L. Chin, D. Cellucci, and D. Rus, Handedness in shearing auxetics creates rigid and compliant structures, *Science (80-).* **360**, 632–635 (2018). doi:10.1126/science.aar4586.
2. D. Sagnelli, M. Calabrese, O. Kaczmarczyk, M. Rippa, A. Vestri, V. Marchesano, K. Kortsen, V. C. Crucitti, F. Villani, F. Loffredo *et al.*, Photo-responsivity improvement of photo-mobile polymers actuators based on a novel Lcs/azobenzene copolymer and zno nanoparticles network, *Nanomaterials* **11**, (2021). doi:10.3390/nano11123320.
3. C. P. Ambulo, J. J. Burroughs, J. M. Boothby, H. Kim, M. R. Shankar, and T. H. Ware, Four-dimensional printing of liquid crystal elastomers, *ACS Appl. Mater. Interfaces* **9**, 37332–37339 (2017). doi:10.1021/acsami.7b11851.
4. M. Tabrizi, T. H. Ware, and M. R. Shankar, Voxelated molecular patterning in three-dimensional freeforms, *ACS Appl. Mater. Interf.* **11**, 28236–28245 (2019). doi:10.1021/acsami.9b04480.
5. H. Koshima (ed.), *Mechanically Responsive Materials for Soft Robotics*, Wiley (2020); ISBN 9783527346202.
6. P. Fu, H. Li, J. Gong, Z. Fan, A. T. Smith, K. Shen, T. O. Khalfalla, H. Huang, X. Qian, J. R. McCutcheon *et al.*, 4D Printing of polymers: Techniques, materials, and prospects, *Prog. Polym. Sci.* **126**, 101506 (2022). doi:10.1016/j.progpolymsci.2022.101506.
7. L. Petti, M. Rippa, A. Fiore, L. Manna, and P. Mormile, Optically induced light modulation in an hybrid nanocomposite system of inorganic CdSe/CdS nanorods and nematic liquid crystals, *Opt. Mater. (Amst).* **32**, 1011–1016 (2010). doi:10.1016/j.optmat.2010.02.022.
8. P. Mormile, L. Petti, M. Abbate, P. Musto, G. Ragosta, and P. Villano, Temperature switch and thermally induced optical bistability in a PDLC, *Opt. Commun.* **147**, 269–273 (1998). doi:10.1016/S0030-4018(97)00551-8.
9. J. Zhou, L. Petti, P. Mormile, and A. Roviello, Comparison of the thermo- and electro-optical properties of doped and un-doped MOM based PDLCs, *Opt. Commun.* **231**, 263–271 (2004). doi:10.1016/j.optcom.2003.12.040.
10. T. A. Saleh, G. Fadillah, and E. Ciptawati, Smart advanced responsive materials, synthesis methods and classifications: From lab to applications, *J. Polym. Res.* **28** (2021). doi:10.1007/s10965-021-02541-x.

11. U. Montanari, D. Cocchi, T. M. Brugo, A. Pollicino, V. Taresco, M. Romero Fernandez, J. C. Moore, D. Sagnelli, F. Paradisi, A. Zucchelli *et al.*, Functionalisable epoxy-rich electrospun fibres based on renewable terpene for multi-purpose applications, *Polymers (Basel)* **13** (2021), doi:10.3390/polym13111804.

12. N. Aktas, D. Alpaslan, and T. E. Dudu, Polymeric organo-hydrogels: Novel biomaterials for medical, pharmaceutical, and drug delivery platforms, *Front. Mater.* **9** (2022). doi:10.3389/fmats.2022.845700.

13. D. Sagnelli, R. Cavanagh, J. Xu, S. M. E. Swainson, A. Blennow, J. Duncan, V. Taresco, and S. Howdle, Starch/poly (glycerol-adipate) nanocomposite film as novel biocompatible materials, *Coatings* **9** (2019). doi:10.3390/coatings9080482.

14. M. Barnes, S. M. Sajadi, S. Parekh, M. M. Rahman, P. M. Ajayan, and R. Verduzco, Reactive 3D printing of shape-programmable liquid crystal elastomer actuators, *ACS Appl. Mater. Interf.* **12**, 28692–28699 (2020). doi:10.1021/acsami.0c07331.

15. L. T. De Haan, A. P. H. J. Schenning, and D. J. Broer, Programmed morphing of liquid crystal networks, *Polymer (Guildf)*. **55**, 5885–5896 (2014). doi:10.1016/j.polymer.2014.08.023.

16. M. Lahikainen, H. Zeng, and A. Priimagi, Reconfigurable photoactuator through synergistic use of photochemical and photothermal effects, *Nat. Commun.* **9**, 1–8 (2018). doi:10.1038/s41467-018-06647-7.

17. Z. Shen, F. Chen, X. Zhu, K. T. Yong, and G. Gu, Stimuli-responsive functional materials for soft robotics, *J. Mater. Chem. B* **8**, 8972–8991 (2020). doi:10.1039/d0tb01585g.

18. R. C. P. Verpaalen, T. Engels, A. P. H. J. Schenning, and M. G. Debije, Stimuli-responsive shape changing commodity polymer composites and bilayers, *ACS Appl. Mater. Interf.* **12**, 38829–38844 (2020). doi:10.1021/acsami.0c10802.

19. V. Marturano, V. Ambrogi, N. A. G. Bandeira, B. Tylkowski, M. Giamberini, and P. Cerruti, Modeling of azobenzene-based compounds, *Phys. Sci. Rev.* **2** (2019). doi:10.1515/psr-2017-0138.

20. A. Kotikian, R. L. Truby, J. W. Boley, T. J. White, and J. A. Lewis, 3D printing of liquid crystal elastomeric actuators with spatially programed nematic order, *Adv. Mater.* **30** (2018). doi:10.1002/adma.201706164.

21. T. J. White, S. V. Serak, N. V. Tabiryan, R. A. Vaia, and T. J. Bunning, Polarization-controlled, photodriven bending in monodomain liquid crystal elastomer cantilevers, *J. Mater. Chem.* **19**, 1080–1085 (2009). doi:10.1039/b818457g.

22. G. Hu, B. Zhang, S. M. Kelly, J. Cui, K. Zhang, W. Hu, D. Min, S. Ding, and W. Huang, Photopolymerisable liquid crystals for additive manufacturing, *Addit. Manuf.* **55**, 102861 (2022). doi:10.1016/j.addma.2022.102861.

23. D. Sagnelli, A. Vestri, S. Curia, V. Taresco, G. Santagata, M. K. Johansson, and S. M. Howdle, Green enzymatic synthesis and processing of poly (cis-9,10-epoxy-18-hydroxyoctadecanoic acid) in supercritical carbon dioxide (scCO2), *European Polymer J.* **161**, 110827 (2021). https://doi.org/10.1016/j.eurpolymj.2021.110827.

24. D. Sagnelli, K. H. Hebelstrup, E. Leroy, A. Rolland-Sabaté, S. Guilois, J. J. K. Kirkensgaard, K. Mortensen, D. Lourdin, and A. Blennow, Plant-crafted starches for bioplastics production, *Carbohydr. Polym.* **152**, 398–408 (2016). doi:10.1016/j.carbpol.2016.07.039.

25. H. Yu, Recent advances in photoresponsive liquid-crystalline polymers containing azobenzene chromophores, *J. Mater. Chem. C* **2**, 3047–3054 (2014). Doi:10.1039/c3tc31991a.

26. U. A. Hrozhyk, V. P. Tondiglia, R. A. Vaia, N. V. Tabiryan, H. Koerner, S. V. Serak, T. J. Bunning, and T. J. White, A high frequency photodriven polymer oscillator, *Soft Matter* (2008).

27. E. Merino and M. Ribagorda, Control over molecular motion using the cis–trans photoisomerization of the azo group, *Beilstein J. Org. Chem.* **8**, 1071–1090 (2012). doi:10.3762/bjoc.8.119.

28. V. Caligiuri, L. De Sio, L. Petti, R. Capasso, M. Rippa, M. G. Maglione, N. Tabiryan, and C. Umeton, Electro-/all-optical light extraction in gold photonic quasi-crystals layered with photosensitive liquid crystals, *Adv. Opt. Mater.* **2**, 950–955 (2014). doi:10.1002/adom.201400203.

29. M. Chen, S. Liang, C. Liu, Y. Liu, and S. Wu, Reconfigurable and recyclable photoactuators based on azobenzene-containing polymers, *Front. Chem.* **8** (2020). doi:10.3389/fchem.2020.00706.

30. G. L. Hallett-Tapley, C. D'alfonso, N. L. Pacioni, C. D. McTiernan, M. González-Béjar, O. Lanzalunga, E. I. Alarcon, and J. C. Scaiano, Gold nanoparticle catalysis of the cis-trans isomerization of azobenzene, *Chem. Commun.* **49**, 10073–10075 (2013). doi:10.1039/c3cc41669k.

31. D. Cao, C. Xu, W. Lu, C. Qin, and S. Cheng, Sunlight-driven photo-thermochromic smart windows, *Sol. RRL* **2**, 1700219 (2018). doi:10.1002/solr.201700219.

32. M. Ji, N. Jiang, J. Chang, and J. Sun, Near-infrared light-driven, highly efficient bilayer actuators based on polydopamine-modified reduced graphene oxide, *Adv. Funct. Mater.* **24**, 5412–5419 (2014). doi:10.1002/adfm.201401011.

33. S. Fan, Y. Lam, J. Yang, X. Bian, and J. H. Xin, Development of photochromic poly(azobenzene)/PVDF fibers by wet spinning for intelligent textile engineering, *Surf. Interf.* **34** (2022). doi:10.1016/j.surfin.2022.102383.

34. S. N. Ramanan, N. Shahkaramipour, T. Tran, L. Zhu, S. R. Venna, C. K. Lim, A. Singh, P. N. Prasad, and H. Lin, Self-cleaning membranes for water purification by co-deposition of photo-mobile 4,4″-azodianiline and bioadhesive polydopamine, *J. Memb. Sci.* **554**, 164–174 (2018). doi:10.1016/j.memsci.2018.02.068.

35. Q. Gao, L. He, Y. Li, X. Ran, and L. Guo, Controllable wettability and adhesion of superhydrophobic self-assembled surfaces based on a novel azobenzene derivative, *RSC Adv.* **7**, 50403–50409 (2017). doi:10.1039/c7ra08465j.

36. T. Ikeda and O. Tsutsumi, Optical switching and image storage by means of azobenzene liquid-crystal films, *Science (80-).* **268**, 1873–1875 (1995). doi:10.1126/science.268.5219.1873.

37. L. Sirleto, L. Petti, P. Mormile, G. C. Righini, and G. Abbate, Fast integrated electro-optical switch and beam deflector based on nematic liquid

crystal waveguides, *Fiber Integr. Opt.* **21**, 435–449 (2002). doi:10.1080/01468030290096903.

38. J. Zhu, T. Guo, Z. Wang, and Y. Zhao, Triggered azobenzene-based prodrugs and drug delivery systems, *J. Control Release* **345**, 475–493 (2022). doi:10.1016/j.jconrel.2022.03.041.

39. J. S. Boruah and D. Chowdhury, Liposome-azobenzene nanocomposite as photo-responsive drug delivery vehicle, *Appl. Nanosci.* **12**, 4005–4017 (2022). doi:10.1007/s13204-022-02666-5.

40. X. Zuo, Y. He, H. Ji, Y. Li, X. Yang, B. Yu, T. Wang, Z. Liu, W. Huang, J. Gou *et al.*, In-situ photoisomerization of azobenzene to inhibit ion-migration for stable high-efficiency perovskite solar cells, *J. Energy Chem.* **73**, 556–564 (2022). doi:10.1016/j.jechem.2022.06.013.

41. J. Keyvan Rad, Z. Balzade, and A. R. Mahdavian, Spiropyran-based advanced photoswitchable materials: A fascinating pathway to the future stimuli-responsive devices, *J. Photochem. Photobiol. C Photochem. Rev.* **51** (2022). doi:10.1016/j.jphotochemrev.2022.100487.

42. M. Irie, Diarylethenes for memories and switches, *Chem. Rev.* **100**, 1685–1716 (2000). doi:10.1021/cr980069d.

43. J. Liao, M. Yang, Z. Liu, and H. Zhang, Fast photoinduced deformation of hydrogen-bonded supramolecular polymers containing α-cyanostilbene derivative, *J. Mater. Chem. A* **7**, 2002–2008 (2019). doi:10.1039/c8ta12030g.

44. L. Zhu and Y. Zhao, Cyanostilbene-based intelligent organic optoelectronic materials, *J. Mater. Chem. C* **1**, 1059–1065 (2013). doi:10.1039/C2TC00593J.

45. R. Castagna, L. Nucara, F. Simoni, L. Greci, M. Rippa, L. Petti, and D. E. Lucchetta, An unconventional approach to photomobile composite polymer films, *Adv. Mater.* **29** (2017). doi:10.1002/adma.201604800.

46. L. Zhang, J. Pan, Y. Liu, Y. Xu, and A. Zhang, NIR-UV responsive actuator with graphene oxide/microchannel-induced liquid crystal bilayer structure for biomimetic devices, *ACS Appl. Mater. Interf.* **12**, 6727–6735 (2020). doi:10.1021/acsami.9b20672.

47. J. Loomis, X. Fan, F. Khosravi, P. Xu, M. Fletcher, R. W. Cohn, and B. Panchapakesan, Graphene/elastomer composite-based photo-thermal nanopositioners, *Sci. Rep.* **3** (2013). doi:10.1038/srep01900.

48. R. Castagna, A. Di Donato, R. Castaldo, R. Avolio, O. Francescangeli, and D. E. Lucchetta, Scotch-tape and graphene-oxide photomobile polymer film, *Photonics* **9** (2022). doi:10.3390/photonics9090659.

49. Y. Kalisky, T. E. Orlowski, and D. J. Williams, Dynamics of solution, *J. Phys. Chem.* **87**, 5333–5338 (1983). doi:10.1021/j150644a006.

50. C. J. Wohl and D. Kuciauskas, Excited-state dynamics of spiropyran-derived merocyanine isomers, *J. Phys. Chem. B* **109**, 22186–22191 (2005). doi:10.1021/jp053782x.

51. C. Li, A. Iscen, L. C. Palmer, G. C. Schatz, and S. I. Stupp, Light-driven expansion of spiropyran hydrogels, *J. Am. Chem. Soc.* **142**, 8447–8453 (2020). doi:10.1021/jacs.0c02201.

52. C. Li, Y. Xue, M. Han, L. C. Palmer, J. A. Rogers, Y. Huang, and S. I. Stupp, Synergistic photoactuation of bilayered spiropyran hydrogels

for predictable origami-like shape change, *Matter* **4**, 1377–1390 (2021). doi:10.1016/j.matt.2021.01.016.

53. K. Imato, K. Momota, N. Kaneda, I. Imae, and Y. Ooyama, Photoswitchable adhesives of spiropyran polymers, *Chem. Mater.* **34**, 8289–8296 (2022). doi:10.1021/acs.chemmater.2c01809.

54. B. Seefeldt, R. Kasper, M. Beining, J. Mattay, J. Arden-Jacob, N. Kemnitzer, K. H. Drexhage, M. Heilemann, and M. Sauer, Spiropyrans as molecular optical switches, *Photochem. Photobiol. Sci.* **9**, 213–220 (2010). doi:10.1039/b9pp00118b.

55. H. Tanaka and K. Honda, Photoreversible reactions of polymers containing cinnamylideneacetate derivatives and the model compounds, *J. Polym. Sci. Polym. Chem. Ed.* **15**, 2685–2689 (1977). doi:10.1002/pol.1977.170151113.

56. M. Nagata and Y. Yamamoto, Synthesis and characterization of photocrosslinked poly(ε-Caprolactone)s showing shape-memory properties, *J. Polym. Sci. Part A Polym. Chem.* **47**, 2422–2433 (2009). doi:10.1002/pola. 23333.

57. A. Lendlein, H. Jiang, O. Jünger, and R. Langer, Light-induced shape-memory polymers, *Nature* **434**, 879–882 (2005). doi:10.1038/nature03496.

58. S. Poyer, C. M. Choi, C. Deo, N. Bogliotti, J. Xie, P. Dugourd, F. Chirot, and J. Y. Salpin, Kinetic study of azobenzene: E/Z isomerization using ion mobility-mass spectrometry and liquid chromatography-UV detection, *Analyst* **145**, 4012–4020 (2020). doi:10.1039/d0an00048e.

59. J. Isokuortti, K. Kuntze, M. Virkki, Z. Ahmed, E. Vuorimaa-Laukkanen, M. A. Filatov, A. Turshatov, T. Laaksonen, A. Priimagi, and N. A. Durandin, Expanding excitation wavelengths for azobenzene photoswitching into the near-infrared rangeviaendothermic triplet energy transfer, *Chem. Sci.* **12**, 7504–7509 (2021). doi:10.1039/d1sc01717a.

60. H. M. D. Bandara and S. C. Burdette, Photoisomerization in different classes of azobenzene, *Chem. Soc. Rev.* **41**, 1809–1825 (2012). doi: 10.1039/c1cs15179g.

61. A. R. Dias, M. E. Minas Da Piedade, J. A. Martinho Simões, J. A. Simoni, C. Teixeira, H. P. Diogo, Y. Meng-Yan, and G. Pilcher, Enthalpies of formation of cis-azobenzene and trans-azobenzene, *J. Chem. Thermodyn.* **24**, 439–447 (1992). doi:10.1016/S0021-9614(05)80161-2.

62. Z. Mahimwalla, K. G. Yager, J. I. Mamiya, A. Shishido, A. Priimagi, and C. J. Barrett, Azobenzene photomechanics: Prospects and potential applications, *Polym. Bull.* **69**, 967–1006 (2012). doi:10.1007/s00289-012-0792-0.

63. Z. Guan, L. Wang, and J. Bae, Advances in 4D printing of liquid crystalline elastomers: Materials, techniques, and applications, *Mater. Horizons* **9**, 1825–1849 (2022) doi:10.1039/d2mh00232a.

64. D. Sagnelli, M. Rippa, A. D. 'Avino, A. Vestri, V. Marchesano, and L. Petti, Development of LCEs with 100% azobenzene moieties: Thermo-mechanical phenomena and behaviors, *Micromachines* **13**, 1665 (2022). doi:10.3390/ MI13101665.

65. L. Lysyakova, N. Lomadze, D. Neher, K. Maximova, A. V. Kabashin, and S. Santer, Light-tunable plasmonic nanoarchitectures using gold

nanoparticle-azobenzene-containing cationic surfactant complexes, *J. Phys. Chem. C* **119**, 3762–3770 (2015). doi:10.1021/jp511232g.

66. Z. Chu, Y. Han, T. Bian, S. De, P. Král, and R. Klajn, Supramolecular control of azobenzene switching on nanoparticles, *J. Am. Chem. Soc.* **141**, 1949–1960 (2019). doi:10.1021/jacs.8b09638.

67. N. Feng, G. Han, J. Dong, H. Wu, Y. Zheng, and G. Wang, Nanoparticle assembly of a photo- and PH-responsive random azobenzene copolymer, *J. Colloid Interf. Sci.* **421**, 15–21 (2014). doi:10.1016/j.jcis.2014.01.036.

68. R. Montazami, C. M. Spillmann, J. Naciri, and B. R. Ratna, Enhanced Thermomechanical properties of a nematic liquid crystal elastomer doped with gold nanoparticles, *Sens. Actuators A Phys.* **178**, 175–178 (2012). doi:10.1016/j.sna.2012.01.026.

69. J. E. Marshall and E. M. Terentjev, Photo-sensitivity of dye-doped liquid crystal elastomers, *Soft Matter* **9**, 8547–8551 (2013). doi:10.1039/c3sm51091c.

70. Z. Li, Y. Yang, Z. Wang, X. Zhang, Q. Chen, X. Qian, N. Liu, Y. Wei, and Y. Ji, Polydopamine nanoparticles doped in liquid crystal elastomers for producing dynamic 3D structures, *J. Mater. Chem. A* **5**, 6740–6746 (2017). doi:10.1039/C7TA00458C.

71. J. Xu, N. Zhao, B. Qin, M. Qu, X. Wang, B. Ridi, C. Li, and Y. Gao, Optical wavelength selective photoactuation of nanometal-doped liquid crystalline elastomers by using surface plasmon resonance, *ACS Appl. Mater. Interf.* **13**, 44833–44843 (2021). doi:10.1021/acsami.1c08464.

72. Y. Sun, J. S. Evans, T. Lee, B. Senyuk, P. Keller, S. He, and I. I. Smalyukh, Optical manipulation of shape-morphing elastomeric liquid crystal microparticles doped with gold nanocrystals, *Appl. Phys. Lett.* **100** (2012). doi:10.1063/1.4729143.

73. B. Qin, W. Yang, J. Xu, X. Wang, X. Li, C. Li, Y. Gao, and Q. E. Wang, Photo-actuation of liquid crystalline elastomer materials doped with visible absorber dyes under quasi-daylight, *Polymers (Basel)* **12**, (2020). doi:10.3390/polym12010054.

74. J. Zhang, J. Wang, L. Zhao, W. Yang, M. Bi, Y. Wang, H. Niu, Y. Li, B. Wang, Y. Gao *et al.*, Photo responsive silver nanoparticles incorporated liquid crystalline elastomer nanocomposites based on surface plasmon resonance, *Chem. Res. Chin. Univ.* **33**, 839–846 (2017). doi:10.1007/s40242-017-7067-0.

75. T. H. Ware, J. S. Biggins, A. F. Shick, M. Warner, and T. J. White, Localized soft elasticity in liquid crystal elastomers, *Nat. Commun.* **7** (2016). doi:10.1038/ncomms10781.

76. M. Bi, Y. Shao, Y. Wang, J. Zhang, H. Niu, Y. Gao, B. Wang, and C. Li, Liquid crystalline elastomer doped with silver nanoparticles: Fabrication and nonlinear absorption properties, *Mol. Cryst. Liq. Cryst.* **652**, 41–50 (2017). doi:10.1080/15421406.2017.1357420.

77. R. Castagna, M. Rippa, F. Simoni, F. Villani, G. Nenna, and L. Petti, Plasmonic photomobile polymer films, *Crystals* **10**, 1–12 (2020). doi:10.3390/cryst10080660.

78. A. J. Lopez Garcia, M. Mouis, A. Cresti, R. Tao, and G. Ardila, Influence of slow or fast surface traps on the amplitude and symmetry of the piezoelectric response of semiconducting-nanowire-based transducers, *J. Phys. D Appl. Phys.* **55** (2022). doi:10.1088/1361-6463/ac8251.

79. T. Jalabert, M. Pusty, M. Mouis, and G. Ardila, Investigation of the diameter-dependent piezoelectric response of semiconducting ZnO nanowires by piezoresponse force microscopy and FEM simulations, *Nanotechnology* **34** (2023). doi:10.1088/1361-6528/acac35.

80. M. Wlazło, M. Haras, G. Kołodziej, O. Szawcow, J. Ostapko, W. Andrysiewicz, D. S. Kharytonau, and T. Skotnicki, Piezoelectric response and substrate effect of ZnO nanowires for mechanical energy harvesting in internet-of-things applications, *Materials (Basel)* **15** (2022). doi:10.3390/ma15196767.

81. Z. Gong, Y. He, Y. H. Tseng, C. Oneal, and L. Que, A Micromachined carbon nanotube film cantilever-based energy cell, *Nanotechnology* **23** (2012). doi:10.1088/0957-4484/23/33/335401.

82. S. Mohsen, H. A. Sodano, and S. R. Anton, A review of energy harvesting using piezoelectric materials: State-of-the-art a decade later, *Smart Mater. Struct.* **28** (2019).

83. R. Castagna, M. Rippa, L. Petti, A. De Girolamo Del Mauro, G. Nenna, and C. G. A. R. Diletto, Energy conversion device and production method. 2018.

84. K. Kumar, C. Knie, D. Bléger, M. A. Peletier, H. Friedrich, S. Hecht, D. J. Broer, M. G. Debije, and A. P. H. J. Schenning, A Chaotic self-oscillating sunlight-driven polymer actuator, *Nat. Commun.* **7** (2016). doi:10.1038/ncomms11975.

85. T. J. White and D. J. Broer, Programmable and adaptive mechanics with liquid crystal polymer networks and elastomers, *Nat. Mater.* **14**, 1087–1098 (2015). doi:10.1038/nmat4433.

86. F. Schwierz, Graphene transistors, *Nat. Nanotechnol.* **5**, 487–496 (2010). doi:10.1038/nnano.2010.89.

87. L. B. Kong, T. Li, H. H. Hng, F. Boey, T. Zhang, and S. Li, *Waste Energy Harvesting* **24**, Springer (2014).

88. O. Puscasu, S. Monfray, G. Savelli, C. Maitre, J. P. Pemeant, P. Coronel, K. Domanski, P. Grabiec, P. Ancey, P. J. Cottinet *et al.*, An innovative heat harvesting technology (HEATec) for above-seebeck performance, *Tech. Dig. — Int. Electron Devices Meet. IEDM* (2012). doi:10.1109/IEDM.2012.6479031.

89. S. Madala and R. F. Boehm, A review of nonimaging solar concentrators for stationary and passive tracking applications, *Renew. Sustain. Energy Rev.* **71**, 309–322 (2017).

90. Q. Ge, A. Sakhaei, H. Lee *et al.*, Multimaterial 4D printing with tailorable shape memory polymers, *Sci. Rep.* **6**, 31110 (2016).

Index